AF905224

MODERN ELECTROCHEMICAL METHODS IN NANO, SURFACE AND CORROSION SCIENCE

Edited by **Mahmood Aliofkhazraei**

Modern Electrochemical Methods in Nano, Surface and Corrosion Science
http://dx.doi.org/10.5772/57202
Edited by Mahmood Aliofkhazraei

Contributors

Florica Manea, Jadwiga Sołoducho, Joanna Cabaj, Margarita Stoytcheva, Sundeep Dhawan, Ahmet Gurses, Metin Açıkyıldız, Mehtap Ejder Korucu, Kübra Güneş, Murat Eyvaz, Ercan Gürbulak, Ebubekir Yüksel, Serdar Kara, Jeong Woo Choi, Ajay Kumar Yagati, Michal Zurek, Rene Kizek, David Hynek, Pavel Kopel, Vojtech Adam, José García-Antón, Rafael Leiva-García, Rita Sanchez Tovar, Clara Escriva Cerdan, Citlalli Gaona, Facundo Almeraya-Calderon, Patricia Zambrano-Robledo, Jorge Uruchurtu, Maria Valnice Boldrin Zanoni, Thais Tasso Guaraldo, Guilherme Garcia Bessegato

Published by InTech

Janeza Trdine 9, 51000 Rijeka, Croatia

Copyright © 2014 InTech

Individual chapters are under their authors' copyright and distributed under the Creative Commons Attribution 3.0 license, which allows users to download, copy and build upon published chapters provided that the authors and publisher are properly credited, which ensures maximum dissemination and a wider impact of the authors' work. Any republication, referencing or personal use of the individual chapters or any of their contents must explicitly identify the original source.

Permission for commercial use of the book as a whole, such as (but not limited to) reprint rights, republication, distribution, sales, translation, and reproduction in any and all forms of media, must always be obtained from InTech.

Notice

Statements and opinions expressed in the chapters are these of the individual contributors and not necessarily those of the editors or publisher. No responsibility is accepted for the accuracy of information contained in the published chapters. The publisher assumes no responsibility for any damage or injury to persons or property arising out of the use of any materials, instructions, methods or ideas contained in the book.

Publishing Process Manager Iva Lipovic

Technical Editor InTech DTP team

Cover InTech Design team

First published June, 2014

Printed in Croatia

Additional hard copies can be obtained from orders@intechopen.com

Modern Electrochemical Methods in Nano, Surface and Corrosion Science, Edited by Mahmood Aliofkhazraei

p. cm.

ISBN 978-953-51-1586-1

We are IntechOpen, the world's leading publisher of Open Access books

Built by scientists, for scientists

4,200+
Open access books available

116,000+
International authors and editors

125M+
Downloads

151
Countries delivered to

Our authors are among the
Top 1%
most cited scientists

12.2%
Contributors from top 500 universities

CLARIVATE ANALYTICS BOOK CITATION INDEX INDEXED

WEB OF SCIENCE™

Selection of our books indexed in the Book Citation Index
in Web of Science™ Core Collection (BKCI)

Interested in publishing with us?
Contact book.department@intechopen.com

Numbers displayed above are based on latest data collected.
For more information visit www.intechopen.com

For EU product safety concerns:
IN TECH d.o.o., Prolaz Marije Krucifikse Kozulić 3, 51000 Rijeka, Croatia, info@intechopen.com or visit our website at intechopen.com.

Meet the editor

Mahmood Aliofkhazraei is a researcher in the corrosion and surface engineering group at the Tarbiat Modares University. Dr. Aliofkhazraei has received several honors, including the Khwarizmi award and the best young nanotechnologist award of Iran. He is a member of the National Association of Surface Sciences, Iranian Corrosion Association and National Elite Foundation of Iran. His research focuses on nanotechnology and its use in surface and corrosion science.

Contents

Preface

Electrochemistry plays an important role in nanotechnology. Besides, it is the fundamental science for many of processes that occur in surface and corrosion science. Its usage in nanotechnology leads to new findings for the mechanism of reactions as well as fabrication of new nanomaterials. As an example, green energy research is dependent upon our ability to store energy for use as it is needed. The performance of storage media like Li batteries is heavily dependent upon their structure at the nanoscale. Electrochemical atomic force microscopy (EC-AFM) and electrochemical scanning tunneling microscopy (EC-STM) provides a turnkey solution, designed for wide chemical solvent compatibility. The EC-AFM can handle Li battery research investigations. These rigs are compatible with even 1 ppm environmental control and provide fast, easy engaging onto black samples, such as Li cathode materials.

This book collects new developments in electrochemical methods and their usage in nano, surface and corrosion science. I would like to show appreciation to all of contributors to this book and thank them for their high quality manuscripts. I wish Open Access publishing of this book helps all researchers benefit from this collection.

Dr. Mahmood Aliofkhazraei
Corrosion and Protection of Materials Group,
Materials Engineering Department,
Tarbiat Modares University,
Iran

Chapter 1

Modern Electrochemistry in Nanobiology and Sensorics

J. Sołoducho and J. Cabaj

Additional information is available at the end of the chapter

http://dx.doi.org/10.5772/58335

1. Introduction

1.1. Overview

Nanotechnology is an area which has promising perspectives for directing fundamental research into successful innovations. Not only to increase the competitiveness of industry but also to create new products that will make positive changes in the lives of citizens, be it in medicine, environment, electronics or any other field [1-4]. Nanoscience and nanotechnology open up new avenues of research and lead to new, useful, and sometimes unexpected applications. Novel materials and new-engineered surfaces allow making products that perform better [5]. New medical treatments are emerging for many diseases [6]. Computers are built with nanoscale components [7] and improving their performance depends upon shrinking these dimensions yet further.

The requirements of nanoscience and nanotechnology have led to modern developments in i.e. electrochemistry. The biological phenomena such as the cellular redox environment, release of neurotransmitters and other signaling substances based on exocytosis, and cellular adhesion are vital elements in modern nanotechnology [8]. Next important problems are the capabilities of electrochemical amperometric and impedance spectroscopic techniques in monitoring cellular dynamics. The applications of such techniques already include biosensors and microchip-based biological systems for cell biological research, medical research and drug development. The state-of-the-art and future developments, e.g. miniaturization of planar interdigitated electrodes in order to achieve a gap/width size regime on the nanometer scale and thus considerable signal amplification, are summarized [9,10].

Electron transfer by thermally activated hopping through localized centers is an essential element for a broad variety of vital biological and technological processes. The use of electrode/self-assembled monolayer (SAM) assemblies [11] to explore fundamental aspects of long-and

INTECH
open science | open minds

© 2014 The Author(s). Licensee InTech. This chapter is distributed under the terms of the Creative Commons Attribution License (http://creativecommons.org/licenses/by/3.0), which permits unrestricted use, distribution, and reproduction in any medium, provided the original work is properly cited.

short-range electron exchange between electrodes and redox active molecules, such as proteins, are reviewed comprehensively in a chapter.

In the chapter we provide also a clear overview of the fundamentals and applications of nanoelectrochemistry in biology and medicine. The key concepts related to the double layer, mass transport and electrode kinetics and their dependence on the dimension and geometry of the electrode will be discussed. Next, various fabrication schemes utilized in making nano-sized electrodes are reviewed, along with the inherent challenges in characterizing them accurately [12].

Quantum-dot electrodes [13] and charging, and finite-size effects are also described. Recent advances in the electrochemistry-electrochemical scanning probe techniques used in the investigations of immobilized biomolecules are presented. Finally, a brief survey of the applications of nanoelectrodes in biosensors and biological systems is provided [14].

2. Cellular redox environment — monitoring

2.1. Cellular redox environment — the biological significance

It is now realized that the direction of many cellular processes depends on redox state. But at present the term redox state is not very well defined. The research in this area is mostly observational in that cells or tissues are subjected to an oxidative or reductive stress and then the effects are observed. In general, energy is required to maintain the ordered state of a living organism [15]. The energy comes from the movement of electrons from oxidizable organic molecules to oxygen. This results in an overall reducing environment in cells and tissues. Redox couples in cells are, of course, responsive to electron flow, that is, changes in the reducing/oxidizing environment. Some of these redox couples are linked to each other to form a set of related couples. Sets of couples can be independent from other sets if activation energies for reactions are high and there are no enzyme systems to link them kinetically. The redox environment of a cell is a reflection of the state of these couples [15].

The imbalance of the cellular redox environment is described as oxidative stress which can lead to the damage of biologically important molecules, such as proteins, lipids and DNA. Therefore, it is crucial to detect the cellular redox environment. To achieve this, besides detecting the reactive oxygen species (ROS) which are the indications of oxidative stress, it is also possible to detect the reducing capacity of the cells which plays an inevitable role in keeping the balance of the cellular redox environment.

The original works connected with cellular redox biochemistry were done by Bücher [16]. He developed approaches to determine the states of various redox couples in cells and was the first to estimate the actual cellular reduction potentials (*Ist-Potential*) for the $NAD^+/NADH$ and $NADP^+/NADPH$ couples.

These days the definition of redox environment should be more general; the redox environment of a linked set of redox couples as found in a biological fluid, organelle, cell, or tissue is the sum of the products of the reduction potential and reducing capacity of the linked redox couples present [15].

Reduction potential can be thought of as a voltage and reducing capacity would be total charge stored, that is, number of electrons available. Reducing capacity would be estimated by determining the concentration of the reduced species in a redox couple; the reduction potential can be estimated with the *Nernst equation*. In mathematical terms this could be represented by:

$$redox\,enviroment = \sum_{i=l}^{n(couple)} F_i X\,[reduced\,species]_i \tag{1}$$

where, E_i is the half-cell reduction potential for a given redox pair and *[reduced species]*$_i$ is the concentration of the reduced species in that redox pair.

It may be difficult (in practical point of view) to measure all linked redox couples present in biological settings to determine the redox environment. Alternatively, a representative redox couple could be used as indicator for changes in the redox environment. For example, in the cell the GSSG/2GSH (glutathione system) couple provides a very large pool of reducing equivalents. It is considered to be the cellular redox buffer. Therefore, the redox state of this couple could be used as an indicator for the redox environment of the cell [15].

Glutathione is considered to be the major thiol-disulfide redox buffer of the cell [17]. On average, the GSH concentration in the cytosol is 1–11 mM [18]. This is far higher than most other redox active compounds. Measurements of total GSH and/or GSSG levels have been used to estimate the redox environment of a cell. Many researchers estimate the redox state of the system by taking the ratio of [GSH]/[GSSG]. This is convenient as the units divide out, so it is not necessary to determine an absolute concentration. A measurement in mg/mg protein, arbitrary fluorescence units, or the area under an HPLC peak can be entered into the ratio and a useful estimate made. In contrast to the NADPH system however, the absolute concentrations of the components of the GSSG/2GSH redox pair have an impact on the reduction potential. The half-cell reaction is:

$$GSSG + 2H^+ + 2e^- \rightarrow 2GSH \tag{2}$$

Thus, the *Nernst equation* for the reduction potential of the GSSG/2GSH half-cell will have the form:

$$E_{hc} = -240 - (59.1/2)\,\log\left([GSH]^2/[GSSG]\right)\,mV \quad \text{at } 25°C,\ pH\ 7.0 \tag{3}$$

A good example for the importance of calculating E_{hc}, rather than estimating only the GSH/GSSG ratio, is provided by Kirlin et al. [19]. Colon cancer cells were differentiated with sodium butyrate and then exposed to benzene isothiocyanate, a compound that induces detoxifying enzymes. Sodium butyrate brought about a 3.6-fold decrease in the GSH to GSSG ratio; addition of benzene isothiocyanate decreased the ratio 3.7-fold. Sodium butyrate changed E_{hc} by 14 mV while benzene isothiocyanate changed E_{hc} by 40 mV. The difference of 26 mV in E_{hc} resulted because when cells were treated with benzene isothiocyanate they had a much smaller GSH pool due to the pretreatment with sodium butyrate.

Development of an advanced organism starts with the division of a single cell. The progeny of this cell must turn into a multi-cell, multi-tissue, and multi-functional entity. This development occurs in a relatively reducing environment.

In eukaryotes, cells divide when they attain a certain size or when triggered by extracellular stimuli such as growth factors or hormones. Proliferative signals flow through intracellular signaling pathways to activate the cell-cycle (Figure 1).

Stress signals from within or outside a cell oppose proliferation. Therefore, cells rely on proliferative signaling pathways, as well as stress surveillance systems (or checkpoints), to regulate entry into the cell-cycle [20, 21]. Stress signaling, particularly in response to DNA damage and infection, is vital for survival to ensure cells elicit appropriate defense or repair mechanisms [22].

The signaling cascades triggered by DNA damage and inflammatory responses have been well studied to this moment [20]. In addition to the primary damage, many stress situations induce reactive oxygen species (ROS). ROS are highly reactive radicals or molecules produced intracellularly from several compartments (mitochondria, endoplasmic reticulum, peroxisomes) [23]. However, they are also induced by external sources such as ionizing agents, vitamins, or herbicides [24].

ROS can interact with biomolecules resulting in oxidation of amino acyl residues in proteins, mutations in DNA, and lipid peroxidation producing more free radicals. Excess production of ROS in cells overwhelms cellular detoxifying systems, resulting in oxidative stress [24].

Recently, an increasing body of work has indicated that redox signaling is mediated by switching the redox state of certain cysteine residues in proteins to elicit an oxidative stress response in cells [25]. Although reactive cysteine residues have been identified in many cell-cycle regulators, their function in controlling cell division has only recently been elucidated in yeast [26,27].

Redox regulation by specific cysteine residues opens up the possibility of manipulating cell division by designing redox-active molecules to block the activity of cell-cycle regulators.

http://dx.doi.org/10.5772/58335

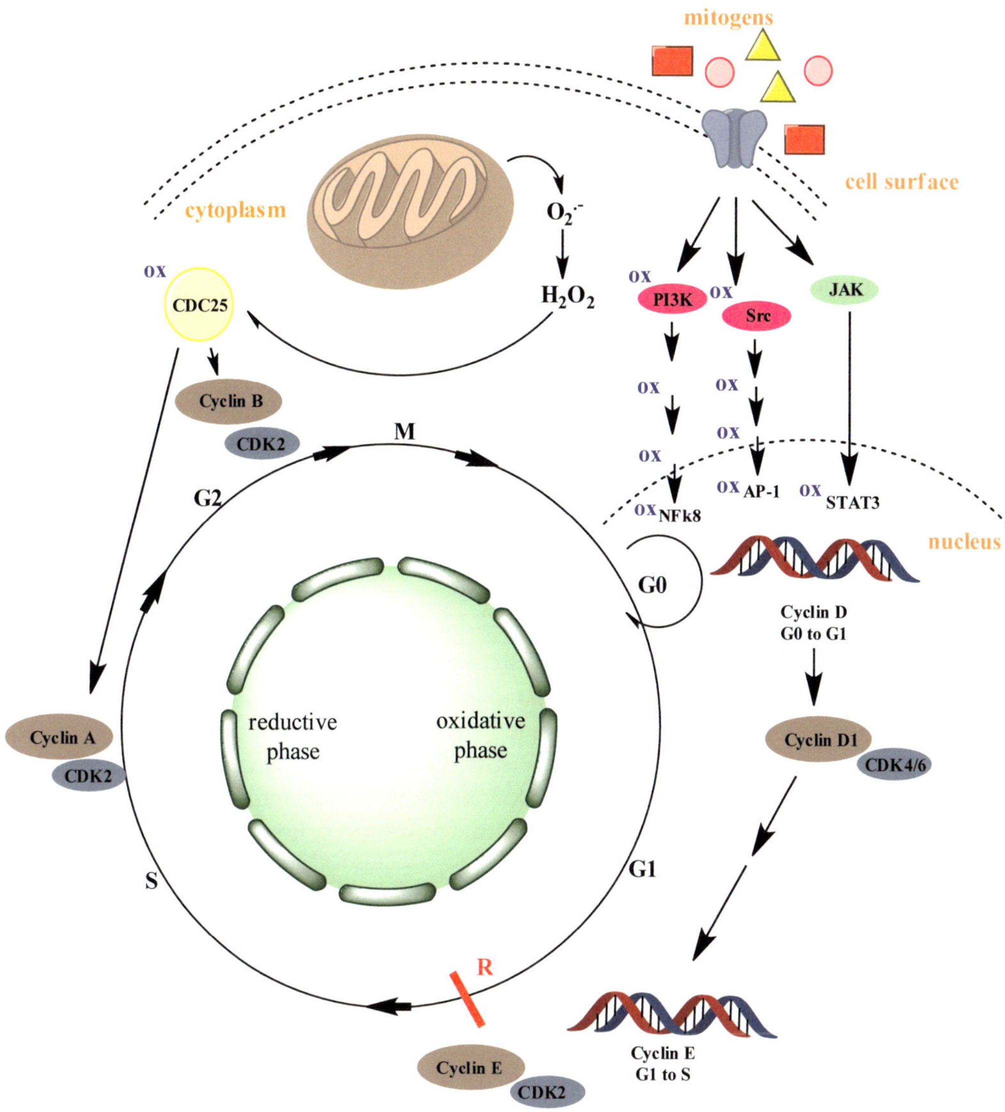

Figure 1. Overview of redox control of the eukaryotic cell-cycle. ROS (i.e. superoxide and H_2O_2) are produced as a result of oxidative phosphorylation in the mitochondria. Cell-cycle regulators with reactive cysteine residues are highlighted by a dark blue 'ox'. In response to mitogens, signaling pathway kinases convey proliferative signals to activate expression of cyclin D. Cyclin D complexes with Cdk4 or Cdk6 and phosphorylates to release its inhibitory effects driving cells to re-enter the cell-cycle from quiescence to G1. Once the restrictive point at late G1 (labeled as red R) is passed, cells are committed to cell division. Activation process leads to the transcription of cyclin E for transition from G1 to S phase. Subsequent expression of cyclin A and cyclin B leads to transition of S to G2 and G2 to M phases, respectively. The phosphatase Cdc25 activates cyclin A–Cdk1, cyclin B–Cdk1, cyclin E–Cdk2 for entry into M phase by removing the inhibitory phosphorylation on Cdk1 and Cdk2. The cell-cycle is synchronized with the metabolic cycle of the cells with S phase and M phase occurring only during the reductive phase of metabolism and G1 in the oxidative phase [according to 20].

2.2. Signaling through cysteine modification

Signals transduced by phosphorylation relays are amplified to produce a prolonged response in cells. This is beneficial during proliferation to ensure cellular activities are diverted to cell division and growth in response to growth stimuli [28,29].

Similarly, it would be advantageous for cells to amplify signals from harmful stimuli to remove relatively stable damage. A burst of ROS usually occurs in a localized manner and is short-lived. Therefore, redox signaling uses a different mechanism for sensing through reactions with specific residues in target proteins to sense and transmit signals [30].

In other words, the 'receptors' of ROS are specific reactive groups, mainly thiol groups of highly reactive cysteine residue(s) within targeted proteins [30]. Cysteine residues are widespread in proteins – the specificity of the 'ROS receptors' in redox signaling is determined by the susceptibility of cysteine residues to oxidation, which is determined by the pKa of their thiol groups [31]. In free cysteine, this pKa is approximately 8.2 [31], but cysteine residues located in proximity to positively charged basic residues, aromatic residues, or metal centers have lowered pKa (<6.5).

Depending also on the accessibility of a cysteine thiol to the solvent, its pKa can be influenced by small fluctuations in the local pH [32]. The micro environment of a cysteine residue therefore determines its reactivity. After oxidation, cysteine residues can be modified (Figure 2) by: formation of intramolecular or intermolecular disulfide bonds; formation of mixed disulfides with glutathione; and oxidation of the cysteine thiol (R-SH) to sulfenic acid (R-SOH) or to the more highly oxidized sulfinic (R-SO 2H) or sulfonic (R-SO3H) acids [23]. When a vicinal cysteine for disulfide bond formation is not available, sulfenic acid intermediate can also react with the main chain nitrogen of an adjacent residue forming a sulfonamide.

All these modifications have been observed in cell-cycle regulators in response to oxidative stress. Although oxidation of cell-cycle regulators has been detected *in vitro* in many cases [34], more recently, an increasing number have also been demonstrated *in vivo* [20, 36].

2.3. Monitoring of cellular redox environment-methods

The intracellular redox environment is a highly dynamic arrangement governed by the formation and degradation of many reactive species of oxygen and nitrogen. Under physiological conditions, the cytosol, the nucleus, and the mitochondrial matrix space maintain homeostatic conditions preferring a highly reducing environment [35]. Intracellular reducing conditions are largely maintained by mM concentrations of reduced glutathione and its enzymes that together constitute the glutathione system [36].

Recent methodological advances have made it possible to focus studies of pro-oxidative changes to specific redox couples within defined subcellular compartments [37], potentially granting greater specificity in mechanistic investigations of the oxidative effects of xenobiotic exposures. A new generation of genetically encoded fluorophores permits direct assessment of the oxidative effects of xenobiotic compounds in relation to the GSH/GSSG redox pair with unprecedented spatial and temporal resolution [37]. Redox-sensitive green fluorescent protein

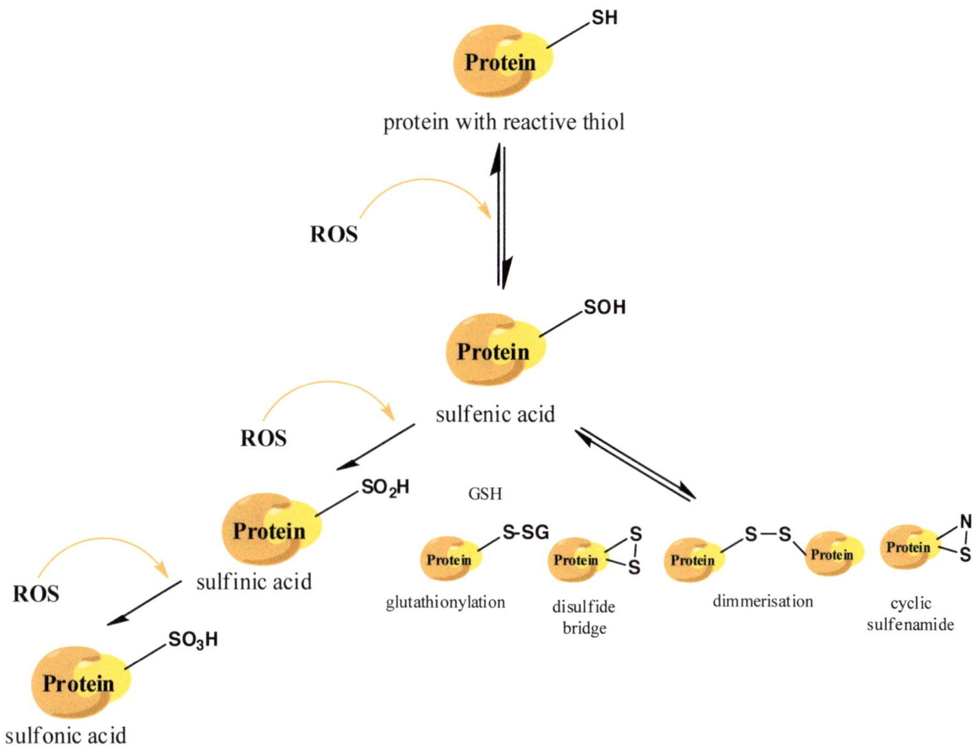

Figure 2. Oxidative modifications of reactive cysteine thiols. Oxidation of a reactive thiol to a sulfenic acid represents the first step for consecutive modification. Reversible modifications of sulfenic acid residues are shown that include: coupling with GSH (*S*-glutathionylation), disulfide formation with a neighboring cysteine (intramolecular) or with another protein (dimerisation), and covalent linkage to the nitrogen atom of a neighboring amino acid residue to form a sulfenamide. These modifications can be reversed by cellular reductants such as a glutathione (GSH). Sulfenic acid can also be irreversibly oxidized further to sulfinic or sulfonic acid species.

roGFP acts as a reporter of intracellular E_{GSH} by equilibrating with the GSH/GSSG redox pair [37]. In reaction that depends on catalysis by glutaredoxins, roGFP responds to oxidation of reduced glutathione (GSH) to its oxidized form (GSSG) *via* the internal formation of a disulfide bond [37,38] (Figure 3). The formation of the disulfide bond alters the spectral characteristics of the GFP fluorophore causing the intensity of the emitted green fluorescence (~520 nm) by excitation at 488 nm to decrease, while causing the emitted fluorescence after excitation at 405 nm to increase, thus making this sensing unit a ratiometric probe. Further efforts to improve the responsiveness of roGFP have led to the conjugation of pathway-specific enzymes to create a fusion of proteins operating as redox relays. In particular, the conjugation of glutaredoxin (Grx) to roGFP has been shown to enhance the kinetics of the roGFP response to the oxidation of glutathione [38].

Fluorescent probes for detection of the cellular redox environment are promising tools that can provide spatial and temporal information in living cells. An excellent probe is to couple the nitroxide with a fluorophore covalently (F–NO$^{\cdot}$). The fluorescence of the fluorophore can

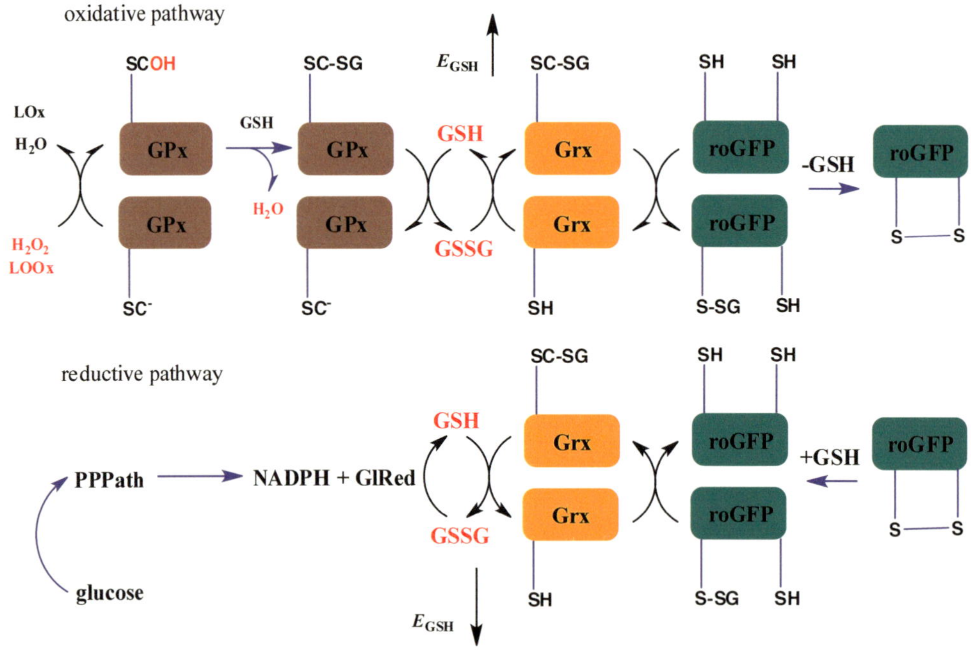

Figure 3. roGFP interactions with the glutathione system (according to [38]). Glutathione peroxidases (GPx) oxidize GSH to GSSG in response to peroxides, including H_2O_2 and lipid hydroperoxides (LOOx), thus increasing the glutathione redox potential (E_{GSH}). Abbreviations: LOx, reduced lipid oxide; SC^-, reduced selenocyteine; SCOH, oxidized selenocyteine; SCSG, glutathionylated selenocysteine. In response to the increase in GSSG, one of the engineered vicinal cysteines of roGFP becomes *S*-glutathionylated by glutaredoxin (Grx). Glutathionylation in turn causes disulfide bond formation and alteration of the spectral properties of the GFP fluorophore. In the reductive pathway, Grx catalyzes the reduction of roGFP disulfide bonds through deglutathionylation as GSSG levels decrease and normal levels of GSH are reestablished by glutathione reductase (GlRed), at the expense of NADPH, causing a renormalization of E_{GSH}. Glucose and the pentose-phosphate pathway (PPPath) create NADPH, which is used by GlRed to reduce GSSG to GSH.

be quenched efficiently by the nitroxide through electron exchanges but recovers significantly when converting to the corresponding diamagnetic derivatives by reduction or reaction with a free radical [39,40]. So it is usually called a profluorescent probe. The probes have been used to detect free radicals [41], ascorbic acid [42] and the redox environment of the cells [43]. The nitroxide reduction is mediated by various reductants, such as ascorbic acid, GSH (glutathione), NADPH and ubiquinol [34]. However, in cytosol, the reduction of nitroxide is primarily due to ascorbic acid, especially in hepatic cells where the concentration of ascorbic acid is higher than in the other cells [45]. So it is of potential to monitor the redox environment of hepatic cells by F–NO˙ probes based on the reduction of the nitroxide.

Expanding evidence suggests a close relation between glucose metabolism and cancerogenity. The tumour cells metabolize glucose more fiercely than normal cells and the apparent difference has led to attempts to develop novel cancer therapeutic strategies [46].

Despite the fact, the advantages of selective redox-regulating systems for cancer therapy have been in considerable attention, a method for testing the therapy effects of the regulators remains imperceptible, it is difficult to measure the intracellular redox environment over time.

http://dx.doi.org/10.5772/58335

One of the method probe to monitor the redox environment is synthesized by Cao et al. [47] a novel rhodamine nitroxide probe to observe the redox environment of human hepatoma cells and normal human liver cells by detecting the reducing capacity of the cells. The synthetic route and reduction mechanism are shown in Figure 4.

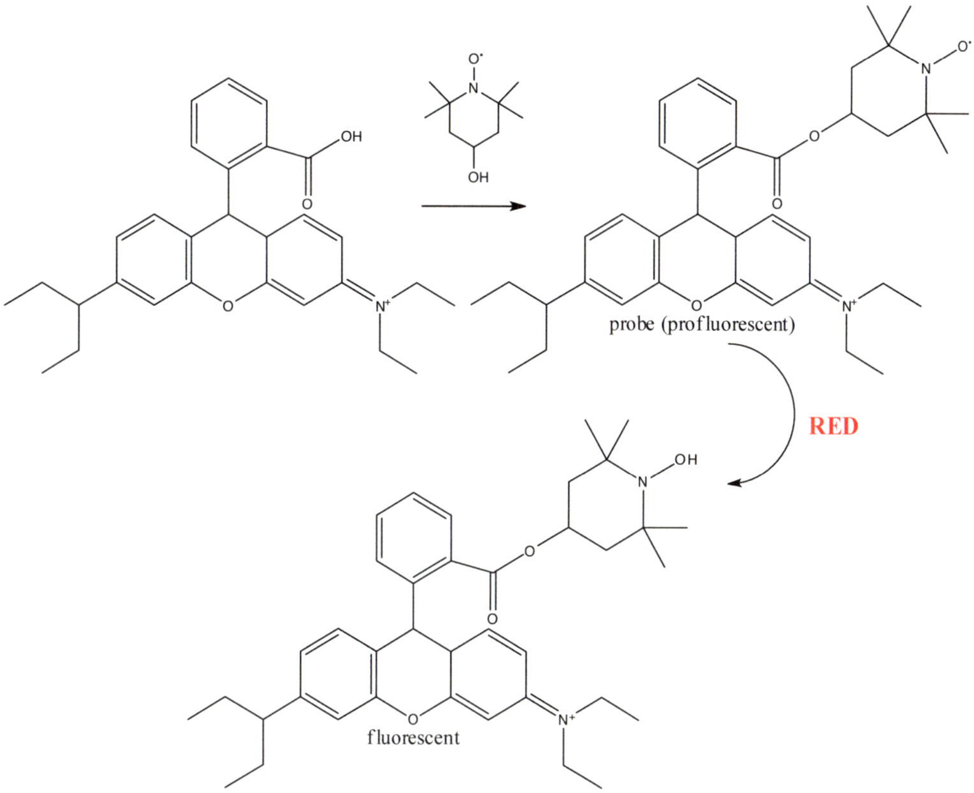

Figure 4. The synthetic route and reducing mechanism of the probe (novel rhodamine nitroxide)

Several other typical biological reducing compounds were also selected to verify the reduction properties of the probe for further application in cells. The addition of these reducing compounds, namely GSH, cysteine and uric acid, resulted in a small increase of the fluorescence.

3. Exocytosis — monitoring

3.1. Exocytosis — function

Exocytosis is a general term used to denote vesicle fusion at the plasma membrane, and it is the final step in the secretory pathway that typically begins in the endoplasmic reticulum, passes through the Golgi apparatus, and ends at the outside of the cell. Endocytosis refers to the recovery of vesicles from the plasma membrane (Fig. 4). Exocytotic vesicle fusion involves the coalescence of vesicle and plasma membranes [48] and allows the so-called fusion pore to form. The fusion pore is a channel that passes through the vesicle and plasma membranes and allows delivery of the vesicle contents to the extracellular compartment. Docking is the process by which the exocytotic vesicle is fixed beneath the plasma membrane before fusion. It is generally believed to involve molecular recognition between vesicle and plasma membrane and is therefore one aspect of vesicle targeting. Another kind of targeting can be provided by the cytoskeletal proteins that move vesicles around the cell [48].

A literature review indicated that there are at least three parallel routes from the Golgi apparatus to the cell surface in mammals and that these pathways operate to varying extents in different cell types [49]. There is no reason to suppose that in plants the situation is any less complex. It cannot be determined by inspection whether vesicles are exocytotic, endocytotic, or bound for destinations other than the plasma membrane. In many cases, however, particularly in green algae [50], it seems clear that there is more than one type of vesicle delivered to the plasma membrane. Electrophysiological data similarly indicate at least two vesicle populations in barley aleurone protoplasts [51].

The commonly held view is that secretory vesicles are delivered to their target membrane by the cytoskeleton. In animal cells, this delivery role is effected by microtubules [52], and vesicles are distributed from a centrally located Golgi apparatus to the plasma membrane. In plants, there are many Golgi apparatuses in each cell, and microfilaments play the major role in vesicle delivery, as is evidenced in highly polarized cells such as the pollen tube, where microfilaments must transport vesicles considerable distances [53]. Microtubules are also present but appear to be more concerned with nuclear migration and cytoplasmic area than vesicle transport [54]. Even in cells from the coleoptile and root, which exhibit less polarized growth, current evidence suggests a mainly microfilament-based vesicle transport mechanism [55].

Proteins that have a putative protective function often exemplify post-Golgi processing and maturation [56]. In many cases, these proteins accumulate within vacuoles and presumably are released by holocrine secretion, a crude type of exocytosis in as much as it is characterized by the release of vacuolar contents coupled with cell degeneration. As an example, a proteinase inhibitor that is believed to act against insects [57] is expressed in stigmatic cells of *Nicotiana alata* as a pentameric protein that is thought to undergo conversion to its active form within the vacuole [58]. On the assumption that the protein precursor is inactive, it has been suggested that such post-Golgi activation may provide a self-protection mechanism. Proteins with putative antimicrobial activity, such as 2S albumins, thionins, and chitin binding proteins, also accumulate within vacuoles, and their release can involve non-specific cell breakdown. Nevertheless, the release of vacuolar contents in response to invasion and other triggering processes may be more regulated [59].

http://dx.doi.org/10.5772/58335

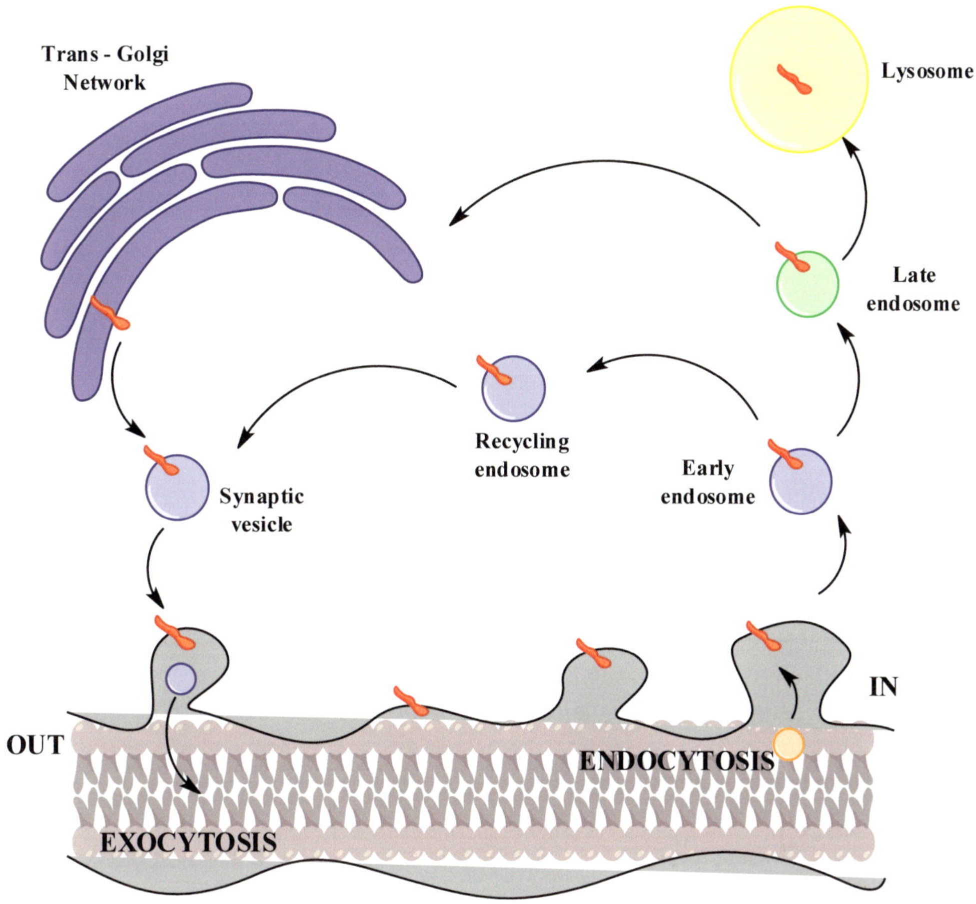

Figure 5. Scheme of exo-and endocytosis

3.2. Exocytosis — methods of monitoring — electrochemistry

Intercellular communication occurs through the release of chemical (or biochemical) messengers from an emitting cell to a target cell. This transmission is mostly achieved by vesicular exocytosis, such as in neurons, neuroendocrine cells, or for the control of hormonal fluxes in blood [60].The expulsion of the messengers from the emitting cell and their further diffusion toward the neighboring cells can be briefly shown. First of all, secretory vesicles located in the emitting cell cytoplasm are initially filled with the biochemical messengers [61]. Second, following an appropriate cell stimulation (which provokes a Ca^{2+}entry or increase), the available vesicles dock to the cell membrane through a step requiring multiple protein-protein interactions, that is, the formation of soluble *N*-ethylmaleimide sensitive fusion protein attachment receptors complexes [62]. To overcome the natural electrostatic repulsions between the cell and vesicular membranes, *N*-ethylmaleimide sensitive fusion protein attachment receptors complexes allow membranes to interact at molecular distances so that the electrical field carried by each electroporates them. According to local reorganization of phospholipidic bilayers, the formation of a fusion pore ensues through which the release of the vesicular content toward the extracellular medium onsets. In most cell models investigated experimentally, exocytosis implies "dense-core" vesicles, whose intrinsic properties make them a better analytical target. Thus, the cationic messengers are compacted into a matrix behaving like a polyelectrolytic gel and constituted of polyanionic proteins such as chromogranins, which fill the vesicle core. Because of the formation of the fusion pore, the first ionic exchanges between this matrix and the extracellular medium occur spontaneously by allowing partition of the cations. This situation provokes a local exchange with the cations of the external medium and thus a destructuration of the gel due to the different molecular and supramolecular interactions [60]. In a natural environment, each messenger cation is exchanged by a fully hydrated small cation (Na^+, H_3O^+), and a local matrix swelling is induced. This increases the *Laplace tension* over the membranes junction area, which may ultimately counterbalance the pore edge energy.

3.2.1. *Fast scan cyclic voltammetry and amperometry: respective advantages*

Aerobic living cells require oxygen-containing solutions to perform as close as possible to physiological conditions. Because O_2 is in high or comparable concentrations (0.23 mM) *vs* the released species flowing into the semi artificial synapse, reductive detection is generally prohibited [8]. Electrochemical measurements of a cell secretion then generally rely on the oxidation of the molecules released at the electrode surface located close to the cell membrane. The recorded currents evidence the nature of the molecules released and their quantities as well as the dynamics of the release itself. Two main electrochemical techniques are applied, amperometry and cyclic voltammetry [7,8].

In amperometry, the working electrode is held at a constant potential at which the molecules under investigation are oxidizable. It ensues a continuous recording of the oxidation current as a function of the time without severe contamination by capacitive current. However, the "potential information" is lost.

In cyclic voltammetry, the current is recorded as a function of the potential applied at the electrode (triangular voltage ramps), so discontinuous monitoring ensues, and capacitive currents may impede measurements, whenever they are not stable enough to be subtracted. In the absence of a semi-artificial synapse configuration, electrochemistry is coupled to another analytical technique (liquid chromatography or capillary electrophoresis) [8].

Unfortunately, although such measurements may be achieved at the single cell level, they merely report confirmations once the whole exocytotic process has ended [63]. As example, the catecholamines content of a single bovine chromaffin cell can be determined by analyzing the extracellular solution (resulting from the cell lysis or after exocytosis) through microcolumn liquid chromatography coupled with amperometry at a carbon fiber electrode [64]. This type of study has allowed the quantification of the amount of catecholamines per cell, and it has further evidence that chromaffin cells are classified in either a single type of catecholamines secreted or a mixture of species released. Moreover, the individual cells have been shown to deliver catecholamines in the same proportions in which they store catecholamines [65]. Similar studies (in which the coupled electrochemical method is voltammetry) were also performed on individual neurons [66]. Although these postanalytical methods can provide essential data concerning the vesicular content, no temporal and spatial information could be obtained during the very exocytotic release system. In the semi-artificial synapse mode, amperometry or voltammetry can be achieved along the occurrence of effective exocytotic events. As example, successive cyclic voltammograms performed on chromaffin cells at relatively low scan rates (10 Vs^{-1}) during exocytosis afforded a clear discrimination between adrenaline and nor-adrenaline releases. From the voltammogram of catecholamines under physiologically compatible conditions, an oxidation of the catechol into *o*-quinone during the anodic scan and subsequent reduction of the latter during the reverse cathodic scan can be observed [61]. An additional peak is detected on the reverse reduction scan for adrenaline, due to a fast cyclization reaction that eventually leads to the formation of a reducible adrenochrome, which is reduced [65].

After appropriate stimulation, vesicles that are primed to undergo exocytosis dock to the cell membrane by the mean of *N*-ethylmaleimide sensitive fusion protein attachment receptors complexes. The cell and vesicular membranes mix and evolve to the formation of a fusion pore, through which neurotransmitters begin to diffuse out the vesicular matrix. Resulting from the ionic exchanges (catecholamines cations vs Na^{+}or H_3O^{+}) and water entry between the intravesicular and the extracellular media, the vesicular matrix swells and provokes the fusion pore expansion.

Voltammetric studies performed during the exocytotic release by a single living cell could thus confirm the results obtained with chromatography, that is, at least three classes of chromaffin cells (one that only releases adrenaline, a second one that only releases nor-adrenaline, and a last one that releases a mixture of adrenaline and nor-adrenaline) may be implied during catecholamines secretion in adrenal glands [65].

It was found also that, the fast scan voltammetry (usually 800 Vs^{-1}) also allows the continuous differentiation of adrenaline and nor-adrenaline.

As compared to cyclic voltammetry measurements, complex features of the dynamics of the exocytotic events can be amperometrically obtained.

In general, four main parameters of the amperometric trace have been evidenced to reveal the features of secretion events at a single cell. The first one deals with the number of events detected per cell (frequency of release), and the others concern the characteristics of the individual amperometric spikes and provide individual and statistical data about the progress of the event itself [60].

4. Mechanisms of electron exchange between electrode and protein

Electrochemical methods offer several well-recognized, classical advantages over homogeneous approaches for elucidating charge transfer processes. Two of the more important advantages are:

- the ability to study a single redox-active component by using an electrode to act as the other reactant [67-68];
- and the ability to gradually differ the redox potential (electrode overvoltage); this trait allows for the gentle variation of the reaction free energy, ΔGo [67-68]. Standard heterogeneous pathways also have some obvious disadvantages. Noteworthy are the strong impact of the electrode/solution interfacial potential drop on the electron transfer process, especially in the free diffusion regime for which it complicates the over-all kinetic analysis [68,69]; and possible changes in the redox-active molecules upon adsorption to an electrode; in particular the adsorption of biomacromolecules, such as proteins, can lead to their inactivation [69].

The above mentioned disadvantages can be largely eliminated through self-assembly techniques that coat the electrode with a biocompatible surface. One common method uses thiolate groups, which form weak covalent bonds with surface-exposed atoms of metal electrodes; e.g., Au, Ag, Hg, etc. [70], and lead to the formation of almost perfect (defect-free) *quasi*-crystalline self-assembled monolayers – SAMs films. The chemical and physical properties of these films can be modified to provide additional control over the electron transfer. These include the variation of the chemical identity and composition (mixing/dilution) of the SAM terminal groups that are presented to the solution (electrolyte); e.g., they can be manipulated to control the binding position and orientation of biomolecules [69].

4.1. Freely diffusing redox couple — electron transfer

Systematic studies of electron exchange for freely diffusing redox couples with SAM-coated metal electrodes show a similar dependence of the rate constant on the SAM thickness. The most systematic studies of the couples $Fe(CN)_6^{3-/4-}$[71] and $Ru(NH_3)_6{}^{3+/2+}$, [72] at Au/SAM junctions show that the nature of the redox couple including the charge type and distribution within the ligand sphere plays an important role in determining whether water and counter-ions can penetrate into the SAM interior or be almost ideally excluded from the SAM (the case of $Fe(CN)_6{}^{3-/4}$).

The results [73] indicate that the kinetically fast redox couple $[Ru(NH_3)_6]^{3+/2+}$is better able to penetrate into unicomponent, CH_3-terminated SAMs of the general type: $S{-}(CH_2)n{-}CH_3$, with n running from 2 to 18,than is $Fe(CN)_6{}^{3-/4}$, and thus it can detect small defects in the SAM.

This interpretation has been substantiated by theoretical calculations of the charge distribution within the complex ions, which demonstrate that hexamine metal complexes have most of the excess charge located on the metal ion core, whereas hexacyano-metal complexes have it localized on the terminal nitrogen atoms of cyano-ligands [74]. This valid difference suggests that $[Ru(NH_3)_6]^{3+/2+}$species are able to penetrate more readily into SAMs and diffuse along the SAM chains, in the presence (or even absence) of probable collapsed sites, as compared to $[Fe(CN)_6]^{3-/4-}$. This difference could cause more variation in the effective charge-transfer distance, as compared to the ideal case with impermeable SAMs. Moreover, it can be suppose that, the SAM defects can be roughly classified as static and dynamic ones, ascribing the former to different kinds of pinholes and collapsed sites, whereas the latter corresponds to increased mobility, and may or may not be associated with particular structural defects [69].

These findings suggest that the dynamically controlled mechanism may be more pronounced in slowly relaxing media such as bio-molecules, ionic liquids or liquid crystals.

Indeed, the interior and the SAM-and solvent-adjacent layers of proteins can be viewed as an extremely viscous media with characteristic dielectric relaxation times similar to pure glycerol [75-76].

4.2. SAM — modified electrodes and immobilized species

The development of SAM-coated electrodes has enabled fundamental and applied studies of biomolecular electron transfer mechanisms, and offers many promising opportunities for creating arrays of redox active biomolecules. As methodologies to immobilize proteins [77-78] improve so that it is possible to better control over protein orientation and spatial placement on an electrode surface, the detail and precision of fundamental questions that can be addressed will improve and the applications of electrically wired protein arrays will increase. According to literature, applying of electrode/SAM assemblies to explore fundamental aspects of the electron exchange between electrodes and redox active molecules; most particularly, proteins is extremely interested field. These new developments identify a need to generalize modern charge transfer theory and bear on the use of electrochemical methods in modern bioengineering and biomedical applications [79]. Even the very first applications of SAM-based electrochemical systems enabled fundamental testing of basic features of contemporary charge transfer theory [69].

For example, studies of the electron transfer rate as a function of SAM thickness revealed that the nonaiabatic (diabatic) mechanism, or electron tunneling, is the rate controlling step for long range electron transfer. In this limiting case, theoretical considerations predict a distinct experimental signature, an exponential decay of the electron transfer rate constant, k_{et}, with the distance:

$$k_{ET(NA)} \propto \exp\left[-\beta(R_e - R_o)\right] \quad (4)$$

where *Re* is the electron transfer distance, *Ro* is the electron transfer distance at the closest approach of the redox-active couple to the electrode, and *β* is the decay parameter, normally of the order of *ca.* 1 $Å^{-1}$ [69].

Until recently, the mechanism(s) controlling short-range electron transfer, either at bare conductive electrodes (mostly implying small redox species), or SAM-modified metal electrodes (simple model systems as well as redox-active biomolecules), has remained more controversial. A generalized version of the classical electron transfer theory [69] accounts for a change in the inherent reaction mechanism with a gradual increase of electrode-reactant electronic coupling. The model systems allow the coupling to be tuned by changing the SAM thickness from thick (weak coupling/long range of electron transfer) to thin (strong coupling/short range of electron transfer) and in essence explore the connection between different electron transfer limits [80].

A number of examples of rigorously confirmed conformationally gated electron transfer have been reported [81] and it appears to be of considerable biological importance. Moreover, conformational gating should be considered as a special case compared to the more classical adiabatic electron transfer mechanism that has an essentially universal origin [79]. Unfortunately, as far as biological electron transfer is concerned, except of several cases limited to different cytochrome c assemblies [82,83] and azurin [84], the friction controlled mechanism has not been discussed in detail. In fact, viscosity-dependent kinetics (exhibiting a weak dependence of the rate on the SAM thickness) is often presented as arising from a conformational-gating mechanism without relevant experimental cross-testing or analysis to ascertain whether it could arise from a friction controlled electron transfer mechanism.

5. Nanomaterials and biosensors

The advantages and new possibilities offered by nanotechnology are varied. Materials exhibit new properties when scaled down from bulk material to nanometric dimensions. These properties can be precisely fine tuned, thus allowing for the fabrication of defined structures and materials optimized for a certain purpose.

Consequently, nanomaterials and concepts from nanotechnology have been much employed in biosensor development. Several reviews in this field [85-88] provide a distinct overview of the possibilities of nanotechnology in the field of biosensor research. The following summarizes the most important trends. The main challenges in the application of nanomaterials for biosensor designs are the definition of the material features, the reproducible synthesis of materials with suitable properties, and the meaningful application of nanotechnological concepts to biosensors. Instead, multidisciplinary effort will be necessary to obtain nanomaterials with properties as required for a novel biosensor design. The seemingly most challenging task of applying nanotechnology to biosensors is to really make use of "nano features" and not simply using nanomaterials without them adding value to the biosensor architecture. In the area of biosensor research some features of nanostructures become important in addition to pure material properties [1,2]. For example, in nanometric structures diffusion lengths become very short and hence mass transport is highly efficient. Since mass transport is valid in many biosensor designs, an increase or at least a change in sensor performance can be expected from using nanometric structures [8]. There are basically three broad categories of pathways towards nanobiosensors and in particular in electrochemical nanobiosensor development. The modification of a macroscopic transducer with nanomaterials is the first of these approaches. In electrochemical biosensors, this would translate into large electrodes modified with nanomaterials. The second approach is the miniaturization of the transducer, namely the use of nanoelectrodes [89] or other miniaturized circuitry of nanometric dimensions [7].

The modification of biomolecules with nanomaterials or coupling of biomolecules and nanomaterials is the third category of approach towards nanobiosensors. Of course, the lines between these approaches are blurred and some sensor designs may draw from more than one of these concepts.

5.1. Modification of biomolecules with nanomaterials

The oriented modification of biomolecules with nanomaterials resulting in biomolecule – nanomaterial hybrids offers crucial possibilities for biosensing. Gold nanoparticles can be used to immobilize biocatalysts or other bioelements on electrodes or other supports. However, in this case, the nanoparticles often just function as a linker and the sensor architecture does not benefit from a unique property due to nanometric dimensions. Unlike the unique optical properties of quantum dots (QDs) make these materials well suited as fluorescent labels in optical sensors [90,91] really taking advantage of a nanosized feature. The use of nanoparticle–enzyme hybrids has been recently reviewed [92,93] as has been the use of nanotechnology in the manipulation of redox systems at an earlier stage [94].

Such an establishment of electrical contact between a redox enzyme and an electrode can also be achieved by nanoparticles. Standing out in this field of research is the wiring of redox enzymes by gold nanoparticles [95]. Due to the fact, glucose oxidase (GOx) was reconstituted with a gold nanoparticle (1.4 nm in diameter, corresponding to the size of the redox center of the enzyme) that was functionalized with the enzyme's cofactor FAD [96].

Such enzyme – nanoparticle hybrids were assembled on gold electrodes leading to relevant good electrical contact between the protein redox center and the macroscopic electrode surface.

Nanotechnology has contributed significantly to recent developments in biosensor research. Modification of macroelectrodes with nanomaterials has resulted in new electron transfer properties and biocompatibility. Nanometric transducers have been used to obtain new classes of biosensor devices. Finally, biomolecule – nanoparticle conjugates show a suitable application potential in biodevice development. An aspect of nanotechnology that is rarely mentioned is the potential harm of nanomaterials towards health. Though a significant effort has been put into the research of this field [95,97-98], the consequences of the use of nanomaterials in everyday life are not yet clarified. Therefore, it appears, that the potential risks of nanomaterials are by far outweighed by the possibilities offered by nanotechnology.

5.2. Implanted biosensors in medicine

As shown in Figure 5, (electrochemical) biosensors are either placed in laboratory animals for fundamental physiological and neurochemical *in vivo* measurements or implanted in the human body for health and metabolite monitoring. In the field of *in vivo* medical research, biocatalyst-based analytical tools are often used for spatially confined measurements of their corresponding target species in preselected regions of living test subjects [1].

The vital challenges/problems in the field of implantable sensors are the stability of the device, the selectivity of the sensor, and the biocompatibility of the sensor. Most important, the sensor has not be rejected by the living organism. When implanted, the sensor should operate for a prolonged time to justify any surgical procedure necessary for the introduction of the sensor into the body. Even when these two challenges are ensured, the sensor has typically to deal with a very complex sample matrix, most commonly blood.

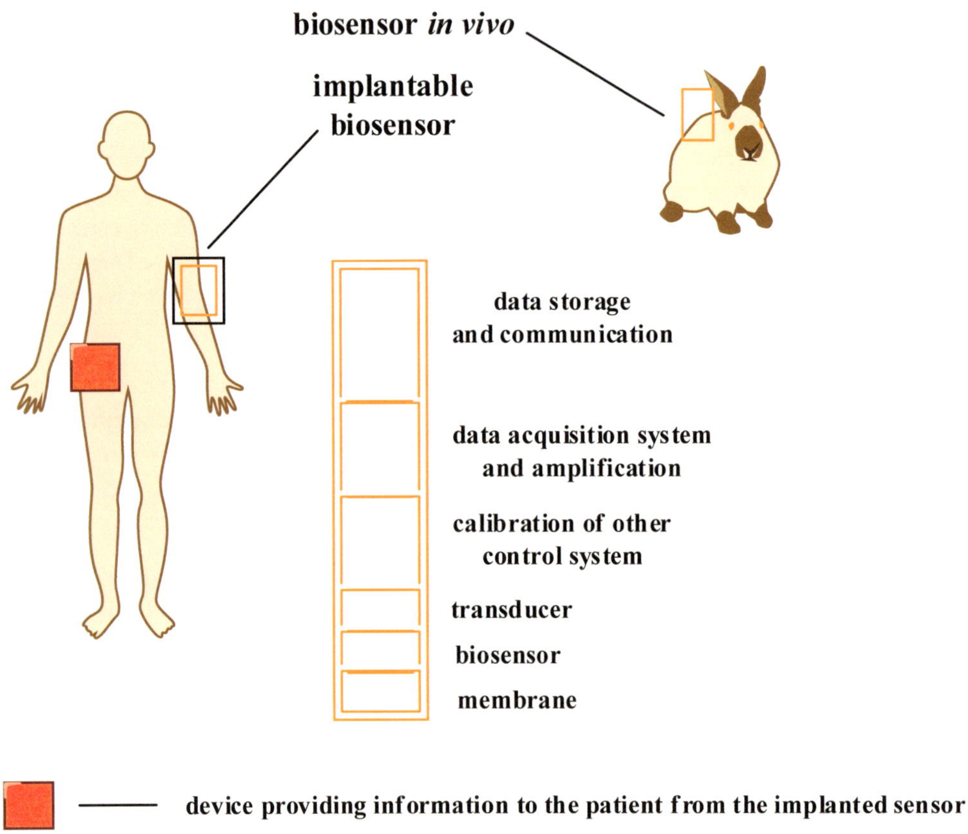

Figure 6. Analytical task of developing *in vivo* biosensor (according to [96])

In vitro sensors have to manage with the same demands in terms. Analytical task of developing and optimizing *in vivo* biosensors of their selectivity, stability is usually not such a critical point. Therefore, *in vitro* sensors also have to be biocompatible in such a way that their presence should not influence the biological medium. Currently, the highly heterogeneous rodent brain is probably most often addressed by *in vivo* and *in vitro* biosensors. Nevertheless, other parts of the rodent central nervous system, the many secretory glands of the regulatory endocrine system, or the tissue of muscles are sites of interest for implantable biosensors [8].

Fixed in a particular brain region for fundamental cognitive, pathological, and pharmacological investigations, the sensing tips of i.e., tapered voltammetric enzyme microbiosensors have demonstrated their ability to directly record up-and down-regulations of neurospecies that may appear in response to premeditated external stimuli such as feeding, drug administration, or gratification at the local level with high time resolution.

Biosensors that are implanted in human body, on the other side, are supposed to control the dynamics of the levels of metabolites related to mental or physical disease states [97]. To achieve this crucial goal they may be placed subcutaneously, just beneath the carrier's skin, or at deeper body locations close to target organs (kidney, liver, and pancreas or the muscles) [97]. Specific applications of biosensors in the human body include clinical point-of-care testing in hospital settings and personal diabetes management. Moreover, the main stream in this task of biosensor development is not on the description of a number of examples from *in vivo* and healthcare measurements but advancements that were reported in the last five years in terms of the design and quality of electrochemical biosensors for successful analysis in a firmly implanted configuration, be it in animals or humans [96]. For detailed information on specific examples of both classical *in vivo* (neurochemical) studies and standard human metabolite monitoring with implanted glucose, glutamate, lactate, acetylcholine, or peroxide biosensors the ones can be referred to several recently published comprehensive articles on the two subjects [98-102]. The clever involvement of new enzymes or adapted enzyme blends in the design of implantable biosensors was used to detect physiologically or pathologically relevant biochemical compounds. The release of the well-known purine ATP as potent extracellular signaling molecule was, for example, demonstrated *in vivo* for the Xenopus tadpole spinal cord during motor activity with implanted biosensors that had co-immobilized glycerol kinase, glycerol-3-phosphate oxidase, and phosphocreatine kinase [103]. Miniaturized carbon fiber-based biosensors for *in vivo* measurements of acetylcholine and choline has been prepared by means of a co-immobilization of acetylcholine esterase and choline oxidase [104]. The glio-transmittered-serine, well-known for a long time to modulate neurotransmission at the glutamatergic synapse, has been monitored in the rat brain striatal extracellular fluid with implanted biosensors employing mammalian *d*-aminoacid oxidase as the indicating biological recognition element [105].

The popular neurotransmitter dopamine is typically measured *in vivo* in particular brain sections with direct fast CV at the solid graphite disc of polished glass – epoxy insulated carbon-fiber microelectrodes [106,107]. In an attempt to improve the selectivity of local dopamine measurements in the complex extracellular matrix of brain fluid, an implantable enzyme-based dopamine microbiosensor has been constructed based on the immobilization of tyrosinase in a thin-film chitosan coating of carbon-fiber disc microelectrodes [108].

As with any classical electrochemical biosensor, an implanted biosensor should also have an exceptional selectivity and sensitivity for the target species, a low detection limit, and a fast response time that is well tailored to the time course of the expected dynamic changes in the concentration of the target analyte in the surrounding tissue. There are, however, important additional properties to look for when the purpose is for stable electrochemical biosensor performance in the complex matrix of the bodies of animals or humans. It is very important for *in vivo* brain biosensor analysis, but also crucial for other situations, to obtain a sensitive acquisition of a strongly localized signal from the molecule in question. Small sensor tip size will of course also be beneficial for placement with minimal (brain) cell and surrounding tissue damage. The exploration of glass-or polymer-insulated needle-type carbon or metal microelectrodes as diminutive precursor structures for biosensors offered an appropriate solution for this problem and no real innovation in this aspect arose in the period under consideration [96].

The second valid issue for success with biosensors in the chronically implanted configuration is sufficient sensor stability over the extended time of data acquisition throughout a trial. According to a lot of significant behavioral studies but basically in the general implantation case the desired period is days if not weeks of measuring time. The long-term quality of the sensor performance is of course impeded by the gradual loss of proper signal generation caused by the foreign body response and contaminating contact of functional sensor entities on the electrode surface, the immobilization matrix, and the immobilized biological recognition element with protein and lipid contents of the immediate physiological measuring environment [109-113].

Between the issues that can be adverse to long-lasting sensor functioning are (i) the fouling of the immobilization layer in the form of a delamination or loss of porosity which is essential for analyte diffusion, (ii) the degradation or denaturation of the biological recognition element, (iii) the passivation of the electrode surface by nonspecific adsorption of proteins and lipids, and (iv) the slow formation of a barrier for substrate diffusion through an ongoing fibrous encapsulation of the biosensor tip.

Several reports have dealt with the adaptation of redox hydrogels employed for the entrapment of the enzymes used *via* either a mild nondegrading biocompatible environment for the active macromolecule or the creation of hydrogel surfaces that are less prone to the adsorption of contaminating (protein) species. Suggestions include a self-cleaning nanocomposite hydrogel membrane [114], biomimetic hydrogels [115], the involvement of surfactants in the formation procedure of redox hydrogels [116], and hydrogels with optimized type and ratios of individual polymerizing components [117]. According to the fact that nitric oxide effectively inhibits platelet and bacterial surface adhesion, Shin and Schoenfisch proposed advanced biosensor interfaces with a high potential to resist biofouling *via* the implementation of an additional nitric oxide-releasing top coating made of *N*-diazenium diolate-modified polymers [118]. Self assembling polyelectrolyte – poly(ethylene glycol)-based nanofilm multilayers have been demonstrated on porous alumina species as an effective diffusion-controlling and protein adsorption-resistant coatings and were reported as optimized dual-function immobilization matrices for implanted biosensors [119]. Also recommended as surface modifications with promising biocompatibility properties were apparently low-fouling zwitterionic carboxybetaine methacrylate coatings [120], microporous collagen scaffolds that minimized unfavorable tissue reactions while stimulating angiogenesis in the vicinity of biosensor tips [121], new hydrophilic poly(ethylene glycol)-based redox copolymers bearing electrochemically active ferrocene and thiol/disulfide functionalities for anchoring to a gold electrode surface [122], and special nanoporous membranes [123,124].

This time, the effective blood glucose measurement and control is a top of analytical task in medical diagnostics and healthcare, respectively. Already prior to the period covered by this section, direct self-monitoring of internal glucose levels became routine in small-volume blood samples. The required commercial tools and information on both their technology and on the glucose meter marketplace is available, for example, in [100,101,125]. Glucose meters typically take advantage of sophisticated single-use screen-printed arrays of electrodes one of which is designed as the glucose sensor *via* specific immobilization of mostly GOx as the biological element and the involvement of artificial free-diffusing redox mediators. Upon placement of a droplet of whole blood, the electronics of the glucose measuring device assesses and digitally displays a glucose equivalent in reasonably short time [101]. Even if carried out several times a day, timed glucose monitoring with external sensors activated at user-chosen intervals obviously has the shortcomings that it fails to report irregular up and downs in between assessments and cannot utilize trends associated with daily habits of diabetics for an instant therapeutic action.

Electrochemical enzyme biosensors for *in vivo* studies and human body metabolite monitoring have in recent years been brought to quite an advanced level. A clear proof of the achievements is the good number of successful biosensor recordings of brain activity and the enterprise of marketable continuous blood glucose monitoring. Further improvements in the spatial and time resolution of *in vivo* measurements would need further sensor miniaturization and tapered nanobiosensors that should be similar to their microelectrode analogues in terms of the proper conductor embedment and resistance against sensor fouling. However, they should be equipped with a reduced total tip dimension for better positioning and nanometric sensing areas for fast and highly localized recording.

In vivo biosensor measurements at the single-cell level or at least a very small cell cluster level could then add novel information to the fundamental insights that were and still are gained through *in vitro* single-cell electrochemistry with isolated cells out of their native matrix [126]. Though tapered enzyme-based nanobiosensors with small total tip radii have already been reported [127] but not successfully operated *in vivo*. Here, there is proper space for future innovative research activity. Another area worth working on is the further extension of the lifetime of sensors for continuous blood glucose monitoring and the transfer of the principles of well-working GOx-based implanted biosensors to those incorpo-rating other enzyme systems for broadening the extent of target analytes. The related possible enhancements and expansion of vital health and disease marker monitoring could open up the individualized and portable medication and care plan that is envisioned by clinicians and patients around the world.

6. Summary

At present, scientific researches in the field of biosensing concern not only to the construction of miniaturized devices, faster, more efficient, low-cost, but also to the increase of integration of electronic and biological systems. The future development of biosensors and other bioelectronics of highly sensitive and specific properties demands the combination of interdisciplinary spheres like quantum chemistry and solid state physics as well as surface bioengineering, biology and medicine, electrical engineering, among others. Advances in any of these areas will have significant effect on the future of medical and environmental diagnositics where invention of more effective, real time monitoring procedures will be advantaged by biosensing technologies.

Acknowledgements

Authors are gratefully acknowledged for financial support of the Polish National Centre of Science Grant no. 2012/05/B/ST5/00749 and Wrocław University of Technology.

Author details

J. Sołoducho* and J. Cabaj

*Address all correspondence to: jadwiga.soloducho@pwr.wroc.pl

Wroclaw University of Technology, Faculty of Chemistry, Wrocław, Poland

References

[1] Logothetidis S. (ed.), Nanostructured Materials and Their Applications, Berlin Heidelberg, NanoScience and Technology, Springer-Verlag, 2012

[2] Priest S., Greenhalgh T., Kramer V., Risk perceptions starting to shift? U.S. citizens are forming opinions about nanotechnology. J. Nanopart. Res. 2010,12(1), 11–20

[3] Smith GB., Granqvist CG., Green Nanotechnology – Solutions for sustainability and energy in the built environment. Boca Ranton (FL), Taylor and Francis Group, 2011

[4] Choi H., Mody CCM., The Long History of Molecular Electronics: Microelectronics Origins of Nanotechnology. SSS, 2009, 39(1), 11–50

[5] Ibn-Elhaj M., Schadt M., Optical polymer thin films with isotropic and anisotropic nano-corrugated surface topologies. Nature, 2001, 410, 795-799

[6] Munger MA., Radwanski P., Hadlock GC., Stoddard G., Shaaban A., Falconer J., Grainger DW., Deering-Rice CE., In vivo human time-exposure study of orally dosed commercial silver nanoparticles. Nanomedicine: Nanotechnology, Biology, and Medicine, 2014, 10, 1–9

[7] Bio-Inspired and Nanoscale Integrated Computing, Ed. Eshaghian-Wilner M., Hoboken New Jersey, Wiley & Sons, Inc., 2009

[8] Application of Electrochemistry and Nanotechnology in Biology and Medicine I, Ed. Eliaz N., New York, Heidelberg, London, Springer, 2011

[9] Spegel C., Heiskanen A., Skjolding DLH.and Emnéus J. Chip Based Electroanalytical Systems for Cell Analysis, Electroanalysis, 2008, 20, 680-702

[10] Yakovleva J. and Emnéus J. Electrochemical Immunoassays in Bioelectrochemistry. Fundamentals, Experimental Techniques and Applications, Ed. Philip N. Bartlett. Chichester, Wiley, 2008

[11] Eckermann AL., Feld DJ., Shaw JA., Meade TJ., Electrochemistry of redox-active self-assembled monolayers. Coordination Chemistry Reviews, 2010, 254, 1769–1802

[12] Skourtis SS., Beratan DN., and Waldeck DH. Coherence in electron transfer pathways, Procedia Chemistry, 2011, 61, 461-485

[13] Shi CG., Shan X., Pan ZQ., Xu JJ., Lu C., Bao N., Gu HY. Quantum dot (QD)-modified carbon tape electrodes for reproducible electrochemiluminescence (ECL) emission on a paper-based platform. Anal Chem. 2012, 84(6), 3033-3038

[14] Rassaei L., Singh PS. and Lemay SG., Lithography-Based Nanoelectrochemistry, Analytical Chemistry 2011, 83, 3974 – 3980

[15] Schafer FQ. and Buettner GR. Redox Environment of the Cell as Viewed through the Redox State of the Glutathione Disulfide/Glutathione Couple, Free Radical Biology & Medicine 2001, 30(11) 1191–1212

[16] Bücher T.; Klingenberg M. Wege des Wasserstoffs in der lebendigen Organisation. Angewante Chemie 1958, 70, 552–570

[17] Gilbert HF. Molecular and cellular aspects of thiol-disulfide exchange. In: Meister A., ed. Advances in enzymology. New York: Wiley Interscience; 1990

[18] Smith CV., Jones DP., Guenthner TM., Lash, LH., Lauterburg BH. Contemporary issues in toxicology. Compartmentation of glutathione: implications for the study of toxicity and disease. Toxicol. Appl. Pharmacol. 1996, 140,1–12

[19] Kirlin WG., Cai J., Thompson SA., Diaz D., Kavanagh TJ., Jones DP. Glutathione redox potential in response to differentiation and enzyme inducers. Free Radic. Biol. Med. 1999, 27, 1208–1218

[20] Chiu J. and Dawes IW., Redox control of cell proliferation, Trends in Cell Biology 2012, 22(11) 592-601

[21] Zhang W. and Liu HT. MAPK signal pathways in the regulation of cell proliferation in mammalian cells. Cell Res. 2002, 12, 9–18

[22] Kim DH., Liberati NT., Mizuno T., Inoue H., Hisamoto N., Matsumoto K., Ausubel FM. Integration of Caenorhabditis elegans MAPK pathways mediating immunity and stress resistance by MEK-1 MAPK kinase and VHP-1 MAPK phosphatase. Proc. Natl. Acad. Sci. U. S. A. 2004, 101, 10990–10994

[23] Go YM. and Jones DP. Redox compartmentalization in eukaryotic cells. Biochim. Biophys. Acta 2008, 1780, 1273–1290

[24] Gutteridge JMC. and Halliwell B. Free Radicals in Biology and Medicine, Oxford University Press, 1999

[25] Janssen-Heininger YM., Mossman BT., Heintz NH., Forman HJ., Kalyanaraman B., Finkel T., Stamler JS., Rhee Sg., van der Vliet A. Redox-based regulation of signal transduction: principles, pitfalls, and promises. Free Radic. Biol. Med. 2008, 45, 1–17

[26] Chiu, J., Tactacan CM., Tan S-X., Lin RCY., Wouters MA., Dawes IW. Cell cycle sensing of oxidative stress in Saccharomyces cerevisiae by oxidation of a specific cysteine residue in the transcription factor Swi6p. J. Biol. Chem. 2011, 286, 5204–5214

[27] Tomko RJ., Jr and Lazo JS. Multimodal control of Cdc25A by nitrosative stress. Cancer Res. 2008, 68, 7457–7465

[28] Pearce AK., Humphrey TC., Integrating stress-response and cell-cycle checkpoint pathways, Trends Cell Biol. 2001, 11, 426–433

[29] Zhang W., Liu HT.MAPK signal pathways in the regulation of cell proliferation in mammalian cells. Cell Res., 2002, 12, 9–18

[30] Rudolph TK., Freeman BA., Transduction of redox signaling by electrophile-protein reactions. Sci. Signal. 2009, 2(90) re7

[31] Winterbourn CC., Hampton MB. Thiol chemistry and specificity in redox signaling Free Radic. Biol. Med. 2008, 45, 549–561

[32] Marino SM., Gladyshev VN. Analysis and functional prediction of reactive cysteine residues. J. Biol. Chem. 2012, 287, 4419–4425

[33] Dalle-Donne I., Milzani A., Gagliano N., Colombo R., Rossi R. S-glutathionylation in protein redox regulation. Free Radic. Biol. Med. 2007, 43, 883–898

[34] Whittal RM., Benz CC., Scott G., Semyonov J., Burlingame AL., Baldwin MA. Preferential oxidation of zinc finger 2 in estrogen receptor DNA-binding domain prevents dimerization and, hence, DNA binding. Biochemistry 2000, 39, 8406–8417

[35] Cannon MB., Remington SJ. Redox-sensitive green fluorescent protein: probes for dynamic intracellular redox responses. A review. Methods Mol Biol. 2008, 476, 51–65

[36] Anderson ME. Glutathione: an overview of biosynthesis and modulation. Chem. Biol. Interact 1998, 111–112, 1–14.

[37] Meyer AJ., Dick TP. Fluorescent protein-based redox probes. Antioxid Redox Signal 2010, 13, 621–650

[38] Gutscher M., Pauleau AL., Marty L., Brach T., Wabnitz GH., Samstag Y., Meyer AJ., Dick TP. Real-time imaging of the intracellular glutathione redox potential. Nat Methods 2008, 5, 553–559

[39] Yang XF. and Guo XQ. Investigation of the anthracene-nitroxide hybrid molecule as a probe for hydroxyl radicals. Analyst, 2001, 126, 1800-1804

[40] Jia M., Tang Y., Lam YF., Green SA. and Blough NV. Prefluorescent nitroxide probe for the highly sensitive determination of peroxyl and other radical oxidants. Anal. Chem. 2009, 81, 8033-8040

[41] Morris C., McMurtrie JC., Bottle SE. and Fairfull-Smith KE. Generation of Profluorescent Isoindoline Nitroxides Using Click Chemistry. J. Org. Chem. 2011, 76, 4964-4972

[42] Maki T., Soh N., Nakano K. and Imato T. Flow injection fluorometric determination of ascorbic acid using perylenebisimide-linked nitroxide. Talanta, 2011, 85, 1730-1733

[43] Ikeda M., Nakagawa H., Suzuki T. and Miyata N. Novel bisbenzimide-nitroxides for nuclear redox imaging in living cells Bioorg. Med. Chem. Lett., 2012, 22, 1949-1952

[44] Trnka J., Blaikie FH., Smith RAJ.and Murphy MP. A mitochondria-targeted nitroxide is reduced to its hydroxylamine by ubiquinol in mitochondria. Free Radical Biol. Med.,2008, 44, 1406-1419

[45] Banhegyi G., Braun L., Csala M., Puskas F. and Mandl J. Ascorbate metabolism and its regulation in Animals. Free Radical Biol. Med., 1997, 23, 793-803

[46] Coleman MC., Asbury CR., Daniels D., Du J., Aykin-Burns N. and Smith BJ., Li L., Spitz DR., Cullen JJ. 2-deoxy-D-glucose causes cytotoxicity, oxidative stress and radiosensitization in pancreatic cancer. Free Radical Biol. Med., 2008, 44, 322-331

[47] Cao L., Wu Q., Li Q., Shao S. and Guo Y. Visualizing the changes in the cellular redox environment using a novel profluorescent rhodamine nitroxide probe. New J. Chem., 2013, 37, 2991-2994

[48] Battey NH., James NC., Greenland AJ., Brownlee C., Exocytosis and Endocytosis. The Plant Cell, 1999, 11, 643-659

[49] Keller P., Simons K. Post-Golgi biosynthetic trafficking. J. Cell Sci. 1997, 110, 3001–3009

[50] Domozych DS. The Golgi apparatus and membrane trafficking in green algae. Int. Rev. Cytol. 1991, 131, 213–253

[51] Homann U., Tester M. Ca^{2+}-independent and Ca^{2+}/GTP-binding protein–controlled exocytosis in a plant cell. Proc. Natl. Acad. Sci. USA 1997, 94, 6565–6570

[52] Cole NB., Lippincott-Schwartz J. Organization of organelles and membrane traffic by microtubules. Curr. Opin. Cell Biol. 1995, 7, 55–64

[53] Taylor LP., Hepler PK. Pollen germination and tube growth. Annu. Rev. Plant Physiol. Plant Mol. Biol. 1997, 48, 461–491

[54] Cai G., Moscatelli A., Cresti M. Cytoskeletal organization and pollen tube growth. Trends Plant Sci. 1997, 2, 86–91

[55] Battey NH., Blackbourn HD. The control of exocytosis in plant cells. New Phytol. 1993, 125, 307–338

[56] Broekaert WF., Cammue BPA., De Bolle MFC., Thevissen K., De Samblanx GW., Osborn RW. Antimicrobial peptides from plants. Crit. Rev. Plant Sci. 1997, 16, 297–323

[57] Heath RL., McDonald G., Christeller JT., Lee M., Bateman K., West J., van Heeswijck R., Anderson MA. Proteinase inhibitors from Nicotiana alata enhance plant resistance to insect pests. J. Insect Physiol. 1997, 43, 833–842

[58] Atkinson AH., Heath RL., Simpson RJ., Clarke AE., Anderson MA. Proteinase inhibitors in Nicotiana alata stigmas are derived from a precursor protein which is processed into five homologous inhibitors. Plant Cell 1993, 5, 203–213

[59] Battey NH., James NC., Greenland AJ. and Brownlee C., Exocytosis and Endocytosis, Plant Cell, 1999, 11, 643-659

[60] Amatore C., Arbault S., Guille M., and Lemaıtre F., Electrochemical Monitoring of Single Cell Secretion Chemical Reviews, 2008, 108, 2585-2621

[61] Sulzer D., Pothos EN. Regulation of quantal size by presynaptic mechanisms. ReV. Neurosci. 2000, 11, 159-212

[62] Sorensen JB. SNARE complexes prepare for membrane fusion. Trends Neurosci.2005, 28, 453-455

[63] Chen GY., Ewing AG. Chemical analysis of single cells and exocytosis. Crit. ReV. Neurobiol. 1997, 11, 59-90

[64] Cooper BR.; Jankowski JA.; Leszczyszyn DJ.; Wightman RM.; Jorgenson JW. Quantitative-determination of catecholamines in individual bosine adrenomedullary cells by reversed-phase microcolumn liquid-chromatography with electrochemical detection. Anal. Chem. 1992, 64, 691-694

[65] Ciolkowski EL., Cooper BR., Jankowski JA., Jorgenson JW., Wightman, RM. Direct observation of epinephrine and norepinephrine cosecretion from individual adrenal medullary chromaffin cells. J. Am. Chem. Soc. 1992, 114, 2815-2821

[66] Kennedy RT., Oates MD., Cooper BR., Nickerson B., Jorgenson JW. Microcolumn separations and the analysis of single cells. Science 1989, 246, 57-63

[67] Electron Transfer from Isolated Molecules to Biomolecules. Advances in Chemical Physics. Jortner J. and Bixon M., eds. New York, Wiley, 1999

[68] Fawcett WR., Liquids, Solutions and Interfaces; Oxford University Press, 2004

[69] Waldeck DH. and Khoshtariya DE., Fundamental Studies of Long-and Short – Range Electron Exchange Mechanisms between Electrodes and Proteins, Modern Aspects of Electrochemistry. Ed. Eliaz N., New York, Springer, 2010

[70] Armstrong FA., and Wilson GS. Recent developments in faradaic bioelectrochemistry. Electrochim. Acta 2000, 45, 2623-2645

[71] Khoshtariya DE., Dolidze TD., Zusman LD., and Waldeck DH. Observation of the turnover between the solvent friction (overdamped) and tunneling (nonadiabatic)

charge-transfer mechanisms for a Au/Fe(CN)$_6^{3-/4-}$electrode process and evidence for a freezing out of the Marcus barrier. J. Phys. Chem. A, 2001, 105, 1818-1829

[72] Dolidze TD., Rondinini S., Vertova A., Longhi M. and Khoshtariya DE. Charge-Transfer Patterns for [Ru(NH$_3$)$_6$]$^{3+/2+}$at SAM Modified Gold Electrodes: Impact of the Permeability of a Redox Probe Open Phys. Chem. J. 2008, 2, 17-21

[73] Krysinski P., Moncelli MR., and Taadini-Buoninesegni F. A voltammetric study of monolayers and bilayers self-assembled on metal electrodes. Electrochim. Acta 2000, 45, 1885-1892

[74] Fowcett WR., Hromadova M., Tsirlina GA. and Nazmutdinov RR. The role of charge distribution in the reactant and product in double layer effects for simple heterogeneous redox reactions. J. Electroanal. Chem. 2001, 498, 93-104

[75] Gavish B., and Yedgar S. in: Protein –Solvent Interactions, Gregory RB., ed., Marcel Dekker, New York, 1995

[76] Kotelnikov AI., Ortega JM., Medvedev ES., Psikha BL., Garcia D. and Mathis P. Effect of protein relaxation on electron transfer from the cytochrome subunit to the bacteriochlorophyll dimer in Rps. sulfoviridis reaction centers within mixed adiabatic/nonadiabatic model. Bioelectrochemistry 2002, 56, 3-8

[77] Ulman A. An Introduction to Ultrathin Organic Films: From Langmuir-Blodgett to Self-Assembly. New York, Academic Press, 1991

[78] Chen S. and Smith LM. Photopatterned Thiol Surfaces for Biomolecule Immobilization. Langmuir 2009, 25, 12275-12282

[79] Kosloff R., and Ratner MA. Rate constant turnovers: energy spacings and mixings. J Phys. Chem. B 2002, 106, 8479-8483

[80] Finklea HO. in: Electroanalytical Chemistry, Bard AJ., and Rubinstein I., eds., New York, Marcel Dekker, 1996

[81] V. Davidson, Protein control of true, gated and coupled electron transfer reactions. Acc. Chem. Res. 2008, 41, 730-738

[82] Yue H., Khoshtariya DE., Waldeck DH., Grochol J., Hildebrandt P., and Murgida DH. On the electron transfer mechanism between cytochrome C and metal electrodes. Evidence for dynamic control at short distances. J. Phys. Chem. B 2006, 110, 19906-19913

[83] Dolidze D., Rondinini S., Vertova A., Waldeck DH., and Khoshtariya DE., Impact of self-assembly composition on the alternate interfacial electron transfer for electrostatically immobilized cytochrome c. Biopolymers 2007, 87, 68-73

[84] Khoshtariya DE., TD. Dolidze, M. Shushanyan, KL. Davis, DH. Waldeck, and R. van Eldik, Fundamental signatures of short-and long-range electron transfer for the blue

copper protein azurin at Au/SAM junctions.Proc. Natl. Acad. Sci. USA 2010, 107, 2757-2762

[85] Pumera M., Sanchez S., Ichinose I., and Tang J. Electrochemical nanobiosensors. Sensors and Actuators B, 2007, 123(2) 1195–1205

[86] Pandey P., Datta M., and Malhotra B.D. Prospects of nanomaterials in biosensors. Analytical Letters, 2008, 41(2) 159–209

[87] Zhang XQ., Guo Q., and Cui DX. Recent advances in nanotechnology applied to biosensors. Sensors, 2009, 9(2) 1033–1053

[88] Urban GA. Micro-and nanobiosensors: state of the art and trends. Measurement Science and Technology, 2009, 20(1) 12001

[89] Arrigan DWM. Nanoelectrodes, nanoelectrode arrays and their applications. The Analyst, 2004, 129(12) 1157–1165

[90] Medintz IL., Uyeda H.T., Goldman ER., and Mattoussi H. Quantum dot bioconjugates for imaging, labelling and sensing. Nature Materials, 2005, 4(6) 435–446

[91] Costa-Fernandez JM., Pereiro R., and Sanz-Medel A. The use of luminescent quantum dots for optical sensing. Trends in Analytical Chemistry: TRAC, 2006, 25(3) 207–218

[92] Willner I., Basnar B., and Willner B. () Nanoparticle – enzyme hysystems for nanobiotechnology. FEBS Journal, 2007, 274(2) 302–309

[93] Guo SJ. and Dong SJ. Biomolecule – nanoparticle hybrids for electrochemical biosensors. Trends in Analytical Chemistry: TRAC, 2009, 28(1) 96–109

[94] Gilardi G. and Fantuzzi A. Manipulating redox systems: application to nanotechnology. Trends in Biotechnology, 2001, 19(11) 468–47

[95] Xiao Y., Patolsky F., Katz E., Hainfeld JF. and Willner I. Plugging into enzymes: nanowiring of redox enzymes by a gold nanoparticle. Science, 2003, 299(5614) 1877–1881

[96] Alkire RC., Kolb DM., Lipkowski J. (eds), Bioelectrochemistry. Fundamentals, applications and Recent Developments, Weinheim, Wiley-VCH, 2011

[97] Nanobioelectrochemistry, Crespilho FN. (ed.), Berlin Heidelberg, Springer-Verlag 2013

[98] Wiesner MR., Lowry GV., Jones KL., Hochella MF., Di Giulio RT., Casman E., and Bernhardt ES. Decreasing uncertainties in assessing environmental exposure, risk, and ecological implications of nanomaterials. Environmental Science & Technology, 2009, 43(17) 6458–6462

[99] Shvedova AA. and Kagan VE. The role of nanotoxicology in realizing the "helping without harm" paradigm of nanomedicine: lessons from studies of pulmonary effects of single-walled carbon nanotubes. Journal of Internal Medicine, 2010, 267(1) 106–118

[100] Heller A. and Feldman B. Electrochemistry in diabetes management. Accounts of Chemical Research, 2010, 43(7) 963–973

[101] Yoo EH., and Lee SY. Glucose biosensors: an overview of use in clinical practice. Sensors, 2010, 10(5) 4558–4576

[102] D'Orazio P. Biosensors in clinical chemistry. Clinica Chimica Acta, 2003, 334(1–2) 41-69

[103] Llaudet E., Hatz S., Droniou M., and Dale N. Microelectrode biosensor for real-time measurement of ATP in biological tissue. Analytical Chemistry, 2005, 77(10), 3267–3273

[104] Schuvailo ON., Dzyadevych SV., Elskaya A., Gautier-Sauvigne S., Csoregi E., Cespuglio R., and Soldatkin AP. Carbon fibre-based microbiosensors for in vivo measurements of acetylcholine and choline. Biosensors & Bioelectronics, 2005, 21(1), 87–94

[105] Zain ZM., O'Neill RD., Lowry JP., Pierce KW., Tricklebank M., Dewa A., and Ab Ghani S. Development of an implantable D-serine biosensor for in vivo monitoring using mammalian D-amino acid oxidase on a poly (o-phenylenediamine) and Nafion-modified platinum-iridium disk electrode. Biosensors & Bioelectronics, 2010, 25(6) 1454–1459

[106] Kita JM. and Wightman RM. Microelectrodes for studying neurobiology. Current Opinion in Chemical Biology, 2008, 12(5), 491–496

[107] Robinson DL., Hermans A., Seipel AT., and Wightman RM. Monitoring rapid chemical communication in the brain. Chemical Reviews, 2008, 108(7) 2554–2584

[108] Njagi J., Chernov MM., Leiter JC., and Andreescu S. Amperometric detection of dopamine in vivo with an enzyme based carbon fiber microbiosensor. Analytical Chemistry, 2010, 82(3) 989–996

[109] Onuki Y., Bhardwaj U., Papadimitrakopoulos F., and Burgess DJ. A review of the biocompatibility of implantable devices: current challenges to overcome foreign body response. Journal of Diabetes Science and Technology, 2008, 2(6) 1003–1015

[110] Dungel P., Long N., Yu B., Moussy Y., and Moussy F. Study of the effects of tissue reactions on the function of implanted glucose sensors. Journal of Biomedical Materials Research Part A, 2008, 85A(3) 699–706

[111] Gifford R., Kehoe JJ., Barnes SL., Kornilayev BA., Alterman MA., and Wilson GS. Protein interactions with subcutaneously implanted biosensors. Biomaterials, 2006, 27(12) 2587–2598

[112] Khan AS. and Michael AC. () Invasive consequences of using micro-electrodes and microdialysis probes in the brain. Trends in Analytical Chemistry: TRAC, 2003, 22(9) 503–508

[113] Wisniewski N., Moussy F., and Reichert W. Characterization of implantable biosensor membrane biofouling. Fresenius Journal of Analytical Chemistry, 2000, 366(6–7) 611–621

[114] Gant RM., Abraham AA., Hou Y., Cummins BM., Grunlan MA., and Cote GL. () Design of a self-cleaning thermoresponsive nanocomposite hydrogel membrane for implantable biosensors. Acta Biomaterialia, 2010, 6(8), 2903–2910

[115] Justin G., Finley S., Rahman AR, and Guiseppi-Elie A. Biomimetic hydrogels for biosensor implant biocompatibility: electrochemical characterization using micro-disc electrode arrays (MDEAs) Biomedical Microdevices, 2009, 11(1) 103–115

[116] Mitala JJ. and Michael AC. Improving the performance of electrochemical microsensors based on enzymes entrapped in a redox hydrogel. Analytica Chimica Acta, 2006, 556(2) 326–332

[117] Oldenziel WH., Dijkstra G., Cremers T., and Westerink BH. Evaluationof hydrogel-coated glutamate microsensors. Analytical Chemistry, 2006, 78(10), 3366–3378

[118] Shin JH. and Schoenfisch MH. Improving the biocompatibility of in vivo sensors via nitric oxide release. The Analyst, 2006, 131(5) 609–615

[119] Park J. and McShane MJ. Dual-function nanofilm coatings with diffusion control and protein resistance. ACS Applied Materials & Interfaces, 2010, 2(4) 991–997

[120] Brault ND., Gao CL., Xue H., Piliarik M., Homola J., Jiang SY. and Yu QM. () Ultra-low fouling and functionalizable zwitterionic coatings grafted onto SiO 2via a biomimetic adhesive group for sensing and detection in complex media. Biosensors & Bioelectronics, 2010, 25(10) 2276–2282

[121] Ju YM., Yu BZ., West L., Moussy Y., and Moussy F. A novel porous collagen scaffold around an implantable biosensor for improving biocompatibility. II. Long-term in vitro/ in vivosensitivity characteristics of sensors with NDGA-or GA-crosslinked collagen scaffolds. Journal of Biomedical Materials Research Part A, 2010, 92A(2) 650–658

[122] Nagel B., Gajovic-Eichelmann N., Scheller FW., and Katterle M. Ionic topochemical tuned biosensor interface. Langmuir, 2010, 26(11) 9088–9093

[123] Narayan RJ., Jin CM., Menegazzo N., Mizaikoff B., Gerhardt RA., Andara M., Agarwal A., Shih CC., Shih CM., Lin SJ., and Su YY. Nanoporous hard carbon membranes for medical applications. Journal of Nanoscience and Nanotechnology, 2007, 7(4–5), 1486–1493

[124] Adiga SP., Jin CM., Curtiss LA., Monteiro-Riviere N.A., and Narayan RJ. Nanoporous membranes for medical and biological applications. Wiley Interdisciplinary Reviews: Nanomedicine and Nanobiotechnology, 2009, 1(5), 568–581

[125] Lee TM. Over-the-counter biosensors: past, present, and future. Sensors, 2008, 8(9) 5535–5559

[126] Schulte A. and Schuhmann W. Single-cell microelectrochemistry. Angewandte Chemie International Edition, 2007, 46(46) 8760–8777

[127] Fei JJ., Wu KB., Wang F., and Hu SS. Glucose nanosensors based on redox polymer/glucose oxidase modified carbon fiber nanoelectrodes. Talanta 2005, 65(4), 918–924

Chapter 2

Electrochemical Techniques for Characterization and Detection Application of Nanostructured Carbon Composite

Florica Manea

Additional information is available at the end of the chapter

http://dx.doi.org/10.5772/58633

1. Introduction

Nowadays, the introduction of nanomaterials within sensors composition has been shown a growing interest because they offer exciting new opportunities in the field of sensing based on their fascinating properties.

Nanostructured carbon in forms of carbon nanotubes and carbon nanofibers represents the most commonly studied nanostructured materials in sensor obtaining. Single-wall carbon nanotubes and multi-wall carbon nanotubes forms of carbon nanotubes have been studied more extensively in comparison with carbon nanofibers for the sensing applications based on their more sensitive properties. However, both forms of nanostructured carbon exhibit great potential for use as unmodified or modified electrode for the detection applications [1,2].

It is well-known that the carbon based sensors/electrodes are widely used in electroanalytical applications because of the properties of the carbon materials, *i.e.*, large potential window, low background current, low cost, versatile surface chemistry, suitability for various sensing applications [1-5]. Carbon-nanotubes and carbon nanofibers based electrodes are very promising for the detection of inorganic and organic compounds [1, 2, 6-18]. However, for certain electroanalytical requirements in relation with the detection of various target analytes as individual, selective or simultaneous detection the nanostructured carbon electrode should be modified with suitable nanoparticles, *e.g.*, silver and copper nanoparticles. Almost all of the electrochemical sensors made up of unmodified/modified nanostructured carbon materials could be applied for the detection of analytes ranging from chemical to biological molecules [6-26].

INTECH
open science | open minds

© 2014 The Author(s). Licensee InTech. This chapter is distributed under the terms of the Creative Commons Attribution License (http://creativecommons.org/licenses/by/3.0), which permits unrestricted use, distribution, and reproduction in any medium, provided the original work is properly cited.

Several detection applications using carbon nanotubes and nanofibers and silver/copper nanoparticles-modified carbon nanotubes electrodes are summarized in Table 1.

Electrode type	Target analyte	Reference
Carbon nanotubes	H_2O_2	[6]
	Bi (III)	[7]
	As(III), Pb (II)	[43]
	Valocyclovir	[10]
	Carbohydrates	[12], [13]
	Phenol derivates	[14], [15], [16]
	Butylparaben	[18]
Carbon nanofibers	H_2O_2	[2]
	Phenol derivates	[17]
Silver-modiefied carbon nanotubes	Ibuprofen	[23]
	Dopamine, ascorbic acid	[26]
Copper-modified carbon nanotubes	Glucose	[22], [24]
	Ammonia	[25]

Table 1. Various inorganic and organic species detected at carbon nanotubes and nanofibers and silver/copper nanoparticles-modified carbon nanotubes electrode materials.

This chapter is focused to present the unmodified/modified nanostructured carbon composite electrode correlated with different types of electrochemical techniques, i.e., cyclic voltammetry (CV), differential-pulsed voltammetry (DPV), square-wave voltammetry (SWV), chronoamperometry (CA) and pulsed amperometry (PA), applied for sensing application as voltammetric/amperometric sensors. The electrochemical detection performance is directly linked to the electrochemical properties of electrode surface in conjunction with the specific electrochemical techniques. The electroanalytical performance for voltammetric/amperometric detection applications of the nanostructured carbon composite (NCC) and metallic nanoparticle-modified nanostructured carbon composite (M-NCC) electrodes, *i.e.*, carbon nanotubes-epoxy (CNT), carbon nanofiber-epoxy (CNF), silver nanoparticles-electrodecorated carbon nanotubes-epoxy (AgCNT), silver nanoparticles-electrodecorated carbon nanofibers-epoxy (AgCNF), copper nanoparticles-electrodecorated carbon nanotubes-epoxy (CuCNT), copper nanoparticles-electrodecorated carbon nanofibers-epoxy (CuCNF), are discussed and presented in this chapter. CV technique is applied for the electrochemical characterization of composite electrodes based on the carbon nanotubes and nanofibers within epoxy matrix synthesised by two-roll mill method in order to use them for the detection applications. Also, M-NCC electrodes are characterised and tested comparatively to improve the electroanalytical performance in the detection applications.

2. Nanostructured carbon based composite (NCC) and metallic nanoparticles-modified nanostructured carbon composite (M-NCC) electrodes

Taking into consideration the main advantage of the composite electrode due to the specific behavior of ensemble/array of microelectrodes, which are characterized by significant higher useful currents due to the large surface area [27], the nanostructured carbon-based composite (NCC) electrodes represent the new generation with the enhanced properties. The most representative and often used of the nanostructured carbon composite materials are nanofibers and nanotubes [21, 28]. Also, metallic nanoparticles decorated on the composite electrode have been studied for various applications [22, 23].

Various detection and sensing applications of NCC electrodes have been extensively studied due to their interesting mechanical, electrical and electrochemical properties [4, 5, 9, 15, 29].

A major importance is given by the composite electrode obtaining, which is responsible for the NCC electrode properties and further detection/sensing applications.

The development of composites electrodes consisting of conductive nanostructured carbon phases dispersed within polymeric matrices has led to important advances in analytical electrochemistry and sensor devices. These new electrode materials combine the electrical properties of carbon with the ease of processing of resins and show attractive electrochemical, physical, mechanical, and economical features [19, 20, 30, 31]. The application of metallic nanoparticles has been shown a raised interest in the field of electrochemistry, in particular in electroanalysis due to their enhanced optical, electronic, magnetic and catalytic properties. Also, in terms of electroanalysis the metallic nanoparticles materials are appropriate to modify electrodes due to several advantages, *e.g.*, high active surface area, enhanced selectivity, and improved mass transport [32-34].

Based on the composite classification, the composite electrodes discussed in this chapter belong to randomized ensembles, in which nanostructured carbon is mixed with epoxy resin, after its dispersion in a suitable solvent (Figure 1). Tetrahydrofuran (THF) and N, N-dimethylformamide (DMF) are efficient solvents for a good dispersion of carbon nanofibers (CNFs) and carbon nanotubes (CNTs) to avoid their agglomeration [15, 22, 23]. Various obtaining methods have been developed to deposit metal nanoparticles on various substrates to obtain so-called chemically modified electrodes (CMEs) characterized by enhanced properties for detection/sensing applications. The most common metal nanoparticles used are copper, silver, gold, and platinum, and as well as the most frequently methods encountered in literature are in respect to the pathway of metallic nanoparticles deposition. Thus, the chemical synthesis is referred at the reduction with different reagents, UV light or electron-beam irradiation [35-37]. Also, the deposition of metallic nanoparticles can be performed by electrochemical techniques [38]. The electrochemical deposition method provides an easy and rapid alternative for the preparation of metallic nanoparticles based electrodes within a short period of time. In addition, this technique present some advances over chemical method such as high purity of the particles, higher control over the dimension, lower particle size distribution, more control over the density [39-41].

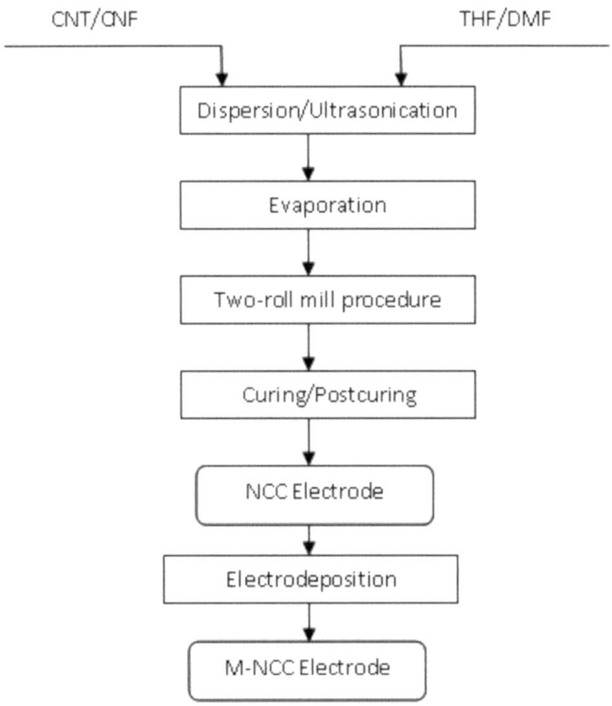

Figure 1. Schematic diagram of M-NCC-electrode preparation; metallic nanoparticles obtaining by electrodeposition.

Another main aspect is to distribute nanostructured carbon within epoxy matrix to reach the composite electrodes. Scanning electron miscroscopy (SEM) technique is very useful to evaluate the dispersion of the carbon filler within epoxy, which is responsible by the surface morphology and the structure and as consequence, the electrical and electrochemical properties of the NCC electrode (Figure 2). Also, the metallic nanoparticles can be observed and measured by this technique. Using two-roll mill procedure to obtain NCC electrode is very simple and assures a good dispersion and homogeneous distribution of nanostructured carbon filler within epoxy matrix (Figure 2 a, b). The electrochemical techniques used for the electrodeposition influenced also, the nanoparticles size and distribution on the conductive carbon filler. Thus, by maintaining the electrode potential at a certain value, e.g.,-0.4 V/SCE for Ag in $AgNO_3$ supporting electrolyte, which is assured by chronoamperometry (CA) technique (described in subchapter 3.1), a spontaneous formation of silver nanoparticles on NCC surface occurred, which are deposited in aggregated form. Ag particles are distributed randomly on the electrode surface and characterized by various sizes (see Figure 2c-e). Better distribution and homogeneous sizes of Ag nanoparticles are achieved if cyclic voltammetry (CV) (described in subchapter 3.4) is applied by succesive running within a certain potential range (Figure 2 f).

http://dx.doi.org/10.5772/58633

The electrode surface morphology, structure and nanoparticles sizes influence the electrical and electrochemical properties of the NCC electrode envisaging their detection applications.

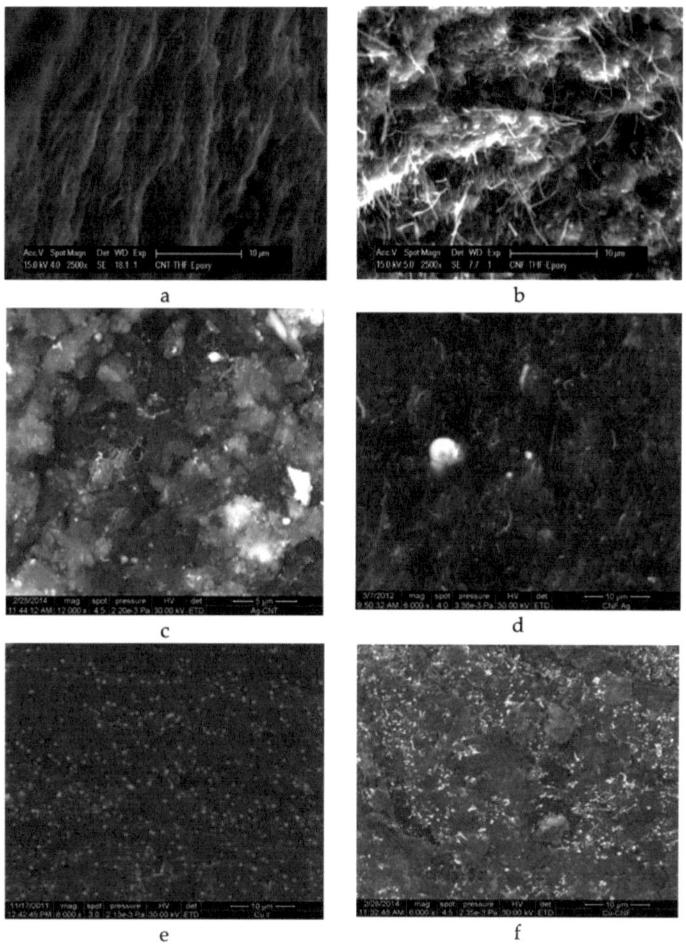

Figure 2. SEM image of NCC and M-NCC electrode surfaces: a)-CNT; b)-CNF; c)-AgCNT; d)-AgCNF; e)-CuCNT; f)-CuCNF.

2.1. Electrical and electrochemical properties of NCC electrodes

The electrical behavior of the NCC composite is an important property for the electrochemical detection application. For the composite electrode, the electrical property is given by the conductive filler loading within the insulating matrix and there is a percolation threshold loading beyond there is electrical conductivity. At low filler loading, the conductivity is very close to the pure insulating matrix conductivity because the fillers are dispersed individually or as small clusters in the matrix. Above the percolation threshold loading, independent conductive fillers tend to link together to form conductive networks, which leads to a significant increase of the electrical conductivity of the composite. Various percolation thresholds for carbon nanotubes-based composites have been reported by various research groups, from 0.0025 %, wt. for carbon nanotubes-epoxy composites up to 5%, wt. [42]. Above this loading, the electrical conductivity increased very much until the carbon nanotubes reached 20%, wt. and beyond this loading no significant increasing of the electrical conductivity occurred.

In general, the electrical conductivity is measured by four-point probe resistance measurements (FPP), and some values for various NCC electrode surface are given in Table 2.

Electrode	Electrical conductivity (Scm^{-1})
CNT	0.596
CNF	0.247
AgCNT	0.602
AgCNF	0.320
CuCNT	0.605
CuCNF	0.319

Table 2. The electrical conductivity of several NCC and M-NCC electrode materials.

The presence of metallic nanoparticles enhanced slightly the electrical conductivity for the electrode materials. All electrode materials exhibit the electrical conductivity suitable for the electrochemical applications.

The electrochemical properties related to the electroactive surface area are studied by cyclic voltammetry (CV). CV is a very versatile electrochemical technique that allows to identify the aspects regarding the mechanism of redox system and transport properties of an electroactive species in solution. This technique is the first electrochemical experiment performed in order to characterize an electrode material for every type of application. This technique provides rapid information on the thermodynamic redox processes, on the kinetics of heterogeneous electron-transfer reactions, and on coupled chemical reactions or adsorption processes. This is accomplished with a three-electrode arrangement whereby the potential relative to some reference electrode is scanned at a working electrode, while the resulting current flowing through a counter (or auxiliary) electrode is monitored in a supporting electrolyte. The technique is ideally suited for a quick search of redox couples present in a system, and once located a couple may be characterized by more careful analysis of the cyclic voltammogram. Usually, the potential is scanned back and forward linearly with time between two extreme values using a triangular potential waveform.

The most common method to determine the electroactive surface area is the ferro/ferricyanide redox couple method [15]. Ferro/ferricyanide redox system gives rise to a reversible redox system that involves one electron per molecule. Given the one electron involving and chemical reversibility of the redox system, the cyclic voltammetric analysis of the ferro/ferricyanide process envisaged the determination of the electroactive area of the NCC and M-NCC electrodes through the apparent diffusion coefficient of this redox system on these electrodes, calculated based on the Randles–Sevcik equation:

$$I_{pf} = (2.69 \times 10^5) n^{3/2} A D^{1/2} C^* v^{1/2} \qquad (1)$$

Where n is the number of electrons involved in the redox process, A is the active area of the working electrode (cm^2), D is the diffusion coefficient ($cm^2\ s^{-1}$), C* is the bulk concentration of the electroactive species (mol cm^{-3}), and v is the potential scan rate (V s^{-1}).

Thus, using cyclic voltammetry recorded at different scan rate in the presence of 4 mM $K_3Fe(CN)_6$, the electrochemical behavior of ferrocyanide system is studied after CV recording (Figure 3), which offers the opportunity to determine the characteristics of a cyclic voltammetric response originating from a reversible process. The reversibility of the system was estimated by the peak-to-peak separation (the separation between the anodic peak potential and the catholic peak potential, $\Delta E_p = E_{pa} - E_{pc}$). For a reversible couple the ΔE_p is egual to 0.059/ n V (n, the number of electron exchanged in the reaction, and in this case is equal with 1), and it is independent of the scan rate. For quasi-and irreversible conditions, the ΔE_p depends on the voltage scan rate. Another important parameter that is relates to the electrochemical reversibility of an electrode reaction is the peak current, and more specific, the ratio between the anodic peak current and the cathodic peak one (i_{pa}/i_{pc}), whose value is unity for a simple reversible couple.

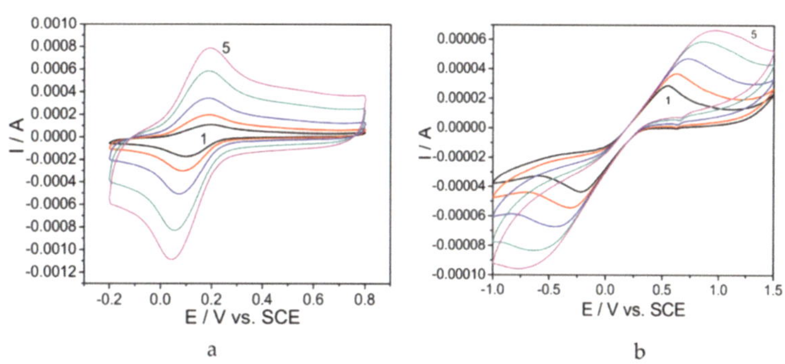

Figure 3. Cyclic voltammograms recorded at: a) CNT and b) Ag-CNT electrodes in the presence of 4 mM $K_3Fe(CN)_6$ and 1M KNO_3 supporting electrolyte; potential scan rate: 1-0.025, 2-0.05, 3-0.1, 4-0.2, 5-0.3 Vs^{-1}; potential range:-1 ÷ +1.5V/SCE.

By calculating diffusion coefficient based on the recorded voltammograms and comparing with theoretical diffusion coefficient ($6.7 \times 10^{-6}\ cm^2s^{-1}$), the values of the electroactive surface area can be determined. In general for all NCC and M-NCC the electroactive surface area are higher in comparison with the geometrical one. In addition, based on the reversibility parameters values, ΔE_p and i_{pa}/i_{pc}, some aspects regarding diffusion and adsorption process on the electrode surface can be elucidated, which must be correlated with the morphological, structural and electrical properties.

Also, other important aspects related to the electrochemical behaviour very important for electroanalysis are the potential window and the background currents. A large potential window and low background current are desired for the electrochemical detection applications. The background current is given by the capacitive component of the electrical double layer and it is desired that this component to be minimized. However, it is well-known that the electrode characterized by the electrocatalytic activity is characterized by high background current and low potential value for the oxygen evolution, which denotes a narrow potential window. Though, the electrodes characterized by the electrocatalytic activity are very useful for the detection application especial for the hard oxidizable or reduction species. In general, nanostructured carbon-based composite electrodes are characterized by low background current and large potential window (Figure 4a). However, it can be noticed that CNT exhibits higher background current and lower potential value for oxygen evolution, which can be attributed to a possible electrocatalytic effect towards this process. Metallic modified-carbon composite electrode exhibits the electrocatalytic activity towards oxygen evolution process, being characterized by higher background current and lower potential window (Figure 4b).

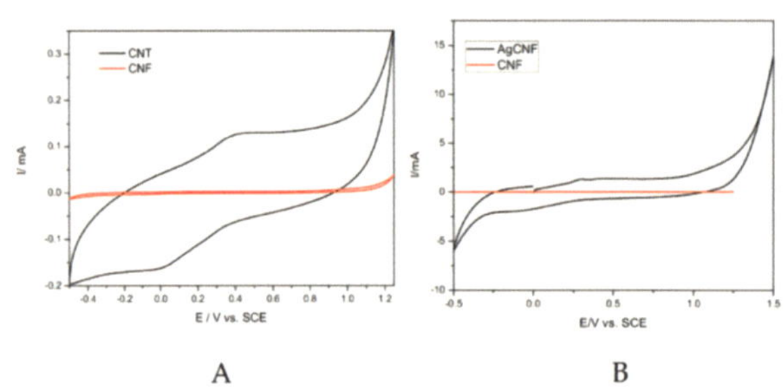

Figure 4. A. CVs recorded at a) CNF and b) CNT electrodes in 0.1 M Na_2SO_4 supporting electrolyte; scan rate $0.05Vs^{-1}$. **B.** CVs recorded at a) CNF and b) AgCNF electrodes in 0.1 M Na_2SO_4 supporting electrolyte; scan rate $0.05Vs^{-1}$.

3. Electrochemical techniques for the detection applications using NCC and M-NCC electrodes

3.1. Cyclic voltammetry

This technique is most useful technique to start the study of the detection applications for each type of the electrode material due to it provides information about the oxidation or reduction potential value, very important to optimize and simplify the detection procedure envisaging the batch injection analysis using chronoamperometry technique (BIA-CA), the easiest detection procedure. The information related to the electrochemical behaviours of NCC and M-NCC electrodes in the presence of various target analytes can be achieved by this technique, several examples will be given bellow. Also, some aspects about the mechanism elucidation related to the process control and adsorption aspects that are very useful for the detection application are provided by this technique operated at different scan rates.

http://dx.doi.org/10.5772/58633

3.1.1. Influence of scan rates

The oxidation or reduction processes of analytes on NCC and M-NCC electrodes are often very complex processes, involving both the adsorption of the reactant/intermediate or oxidation/ reduction products and the formation of passive, nonconductive layer on their surface by electropolymerization or adsorption. The mechanistic aspects of the overall oxidation or reduction processes of target analyte on the electrode surface are elucidated using the cyclic voltammograms recorded at various scan rates (0.01-0.2 Vs^{-1}) in the presence of a certain concentration of the target analyte. As example, Figure 5a shows CVs recorded at CNF electrode in 0.1 M Na_2SO_4 supporting electrolyte and the presence of 4 mM sulfide. The anodic current recorded at about+0.7 V vs. SCE increased linearly with the square root of the scan rate (Figure 5b) suggesting that the reaction is mass-transfer controlled. No zero intercept inform about the adsorption process that can not be neglected. Moreover, the starting potential of the oxidation peak shifted towards positive potential when increasing v indicating that the electrooxidation process of sulfide is irreversible (Figure 5c) that is proved also, by the lack of the cathodic corresponding peak. In addition, CVs shapes informed about two steps of the sulfide oxidation process to elemental sulfur, both steps being difussion-controlled.

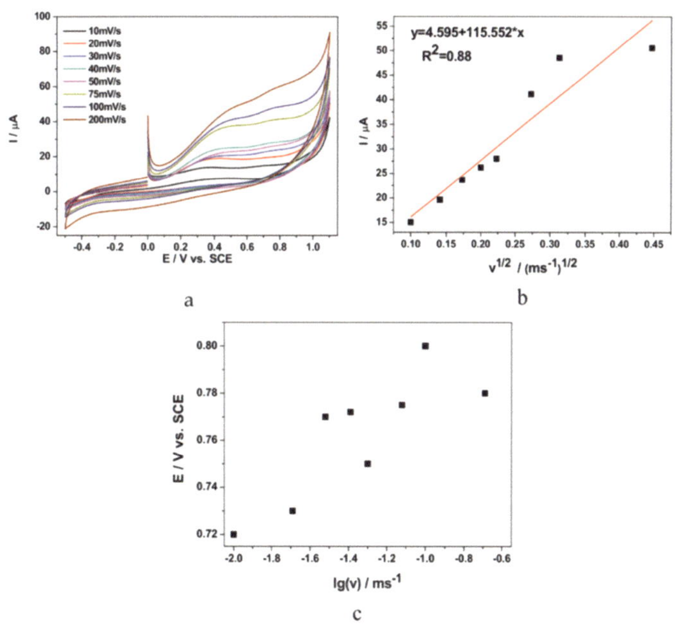

Figure 5. (a) CVs recorded at CNF electrode in 0.1 M Na_2SO_4 supporting electrolyte and 0.5 mM sulfide with different scan rates: 0.01, 0.02, 0.03, 0.04, 0.05, 0.07, 0.09, 0.1, 0.2 Vs^{-1} (curves 1-9); potential range: – 0.5 and+1.2 V/ SCE. (b) The anodic peak current recorded at+0.7 V/SCE vs. square root of scan rate; (c) The peak potential Ep vs. log (v).

Another example is given for the electrochemical behaviour of arsenic on AgCNF electrode (Figure 6) envisaging its detection. Arsenic (III) detection procedure involves anodic stripping voltammetry (ASV), which supposes two stages: first corresponding to reduction of arsenic

(III) at the electrode surface for a certain time followed by the second stage of electrochemically stripping from the electrode surface resulting a Faradaic response that is direct proportionally with arsenic concentration. The processes corresponding to these stages can be schematically represented by the following reactions [43]:

$$\text{Stage 1-Deposition: } As^{3+} + 3e^{-} \rightarrow As^{0} \tag{2}$$

$$\text{Stage 2-Anodic stripping: } As^{0} \rightarrow As^{3+} + 3e^{-} \tag{3}$$

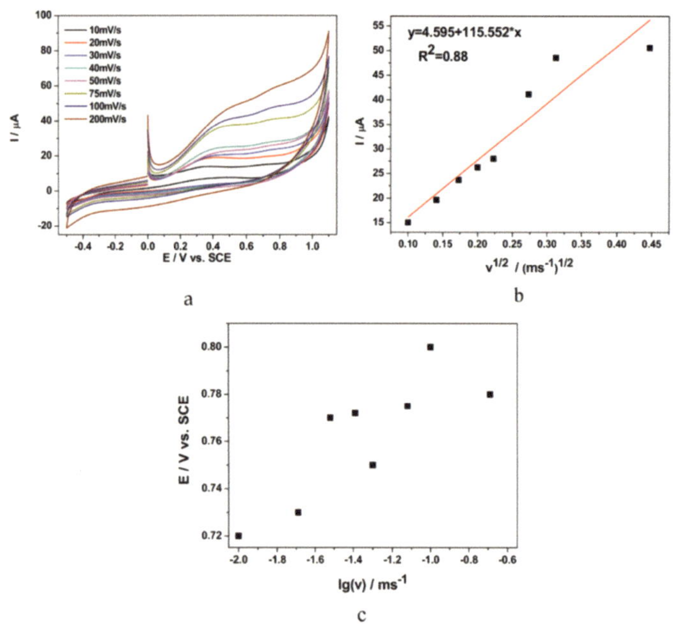

Figure 6. (a) CVs recorded at AgCNF electrode in 0.1 M Na_2SO_4 supporting electrolyte 0.09 M Na_2SO_4+0.01 M H_2SO_4 supporting electrolyte and in the presence of 3 mM As with different scan rates: 0.01, 0.02, 0.03, 0.04, 0.05, 0.07, 0.09, 0.1, 0.2 Vs^{-1} (curves 1-9); potential range: – 0.5 and+1.2 V/ SCE. (b) The anodic peak current recorded at+0.7 V/SCE vs. square root of scan rate; (c) The peak potential Ep vs. log (v).

The deposition stage for this procedure is achieved by maintaining the electrode potential at-0.4 V/SCE for 120 seconds. Based on the results regarding the CVs shapes, the liniar dependence of anodic peak height versus the square root of the scan rate, the oxidation potential shifting to positive direction with increasing scan rate and the presence of the cathodic peak corresponding to the anodic one, arsenic stripping voltammetric process is quasi-reversible and diffusion-controlled. These results are promising for the application of these oxidation processes for the detection purposes.

3.1.2. *CV for the detection application*

CV technique is very useful for the detection applications. Figure 7 presents the results of the application of CNT electrode using CV for sulphide detection in various supporting electrolytes envisaging different practical applications. 0.1 M Na_2SO_4 supporting electrolyte has been chosen due to it is very common for the detection application in environmental application. 0.1 M Na_2SO_4+1 M NaCl supporting electrolyte presents the composition very close to seawater and the question that raised was to check the possible interference of chloride anions. Real surface water was tested envisaging in-field detection application, for which no supporting electrolyte is added.

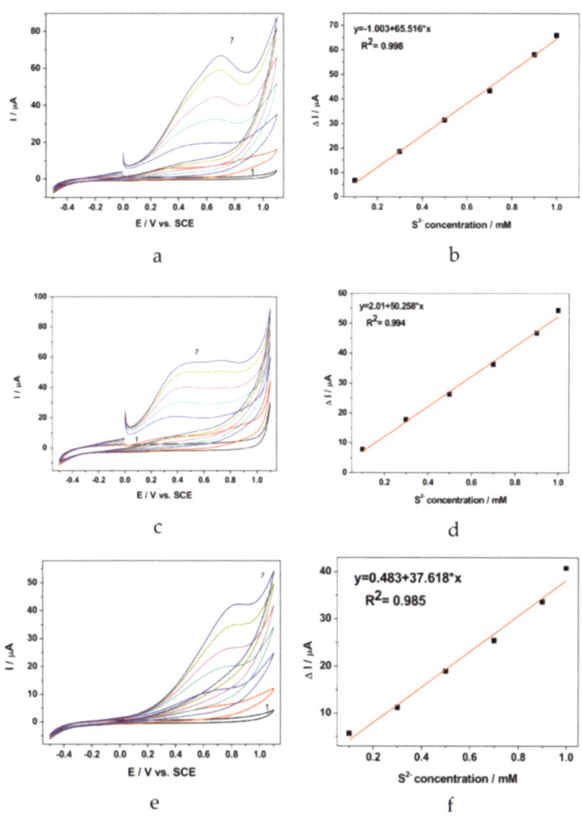

Figure 7. CVs recorded at CNF electrode in the presence of various sulfide concentrations 0.1; 0.3; 0.5; 0.7; 0.9; 1 mM in supporting electrolyte: 0.1 M Na_2SO_4 (a); 0.1 M Na_2SO_4+1M NaCl (c); real surface water without adding supporting electrolyte. Calibration plots of the anodic current recorded at+0.7 V/SCE vs. sulfide concentrations for each supporting electrolytes (b, d, f).

CV results presented in Figure 7b informed that high concentration of chloride does not affect anodic current response characteristics to the sulfide detection, because the chlorine evolution occurred at high potential value. This result is very promising for sulfide detection in the seawater.

The literature data has been reported about the electrochemical behavior of carbon-based composite electrodes characteristics to ordered (array) or randomized micro/nanoelectrode ensembles characterized by the heterogeneous higher electroactivity [5, 44-49]. Sometimes, the voltammetric responses of carbon-based composite electrodes are sometimes similarly to that found for „edge effect", which contribute significantly to the Faradaic current [5]. However, carbon-based composites obtained without a specific template are not rigorously arrays but are truly randomized ensembles, with the size, shape and inter-nano/micro carbon islands separation distributed over a wide range of insulating matrix. The random ensembles of microelectrodes can exhibit array behaviour in certain situation. Taking into consideration the distribution of the conductive filler islands within the insulator matrix, the main requirement is that the distance between individual conductive carbon microzones that should be much larger than its radius and mass transport is controlled by the spherical diffusion. The closely spaced carbon nano/microzones array will behave similar to a macroelectrode (linear diffusion-controlled mass transport) because of the diffusion layer overlap. Another important requirement is subjected to the insulating matrix to prevent current leakage, which resulted in the distortion of the cyclic voltammetry. Microelectrode arrays exhibit for the detection applications the advantages of single microeletrodes, e.g., reduced ohmic drop and charging current, the lower detection limit and better sensitivity. In the same time, their use overcomes the disadvantages of single microelectrode subjected to low current output and the high susceptibility to the electrochemical noise [47-49].

The above-presented CV results for the NCC and M-NCC obtained by two-roll mill procedure showed that these composite electrodes are not ordered (array) nano/microelectrodes, because it noticed the linear diffusion-controlled mass transport with a substantial hysteresis, explored by the scan rate influence of the CV shapes, which is characteristics to the macroelectrode behavior.

However, another practical micro/nanoelectrode array peculiarity is given by the ability to deliver the current responses in the absence of any supporting electrolyte, which is appropriate for in-situ or in-field detection applications. Even if the CV shapes recorded at NCC and M-NCC informed that these composites do not exhibit the array electrochemical behavior, however the similar anodic peak current for sulfide detection was noticed direct in surface water without supporting electrolyte at CNF composite electrode (Figure 7c). It can be seen that a slight lower sensitivity was found, and also, non considerable ohmic drop increasing was noticed in surface, which encourage the possible application for in-field sulfide detection.

3.2. Differential-pulsed voltammetry

Differential-pulsed voltammetry (DPV) is a very sensitive detection technique appropriate for trace levels of the analytes. The height of differential pulsed voltammogram is direct proportional to the concentration of analyte [27]:

$$i_p = \frac{nFAD^{1/2}C}{\sqrt{\pi t_m}}\left(\frac{1-\sigma}{1+\sigma}\right) \tag{4}$$

http://dx.doi.org/10.5772/58633

where $\sigma=\exp[(nF/RT)(\Delta E/2)]$, ΔE is the pulse amplitude.

The peak potential can identify the species especial for the simultaneous detection and the background current is much smaller in comparison with CV due to the charging current contribution is negligible. The optimization of the modulation amplitude (MA) and step potential (SP) operational parameters of DPV technique must be considered for each application in direct relation with the the sensitivity and the lowest limit of detection. Example of DPVs recorded on CNT in the presence of various pentachlorophenol (PCP) concentrations are shown comparatively with CV in Figure 8, and the electroanalytical parameters are gathered in Table 3. It can be noticed an improved detectability of PCP on CNT using DPV. Also, as we mentioned and expected, the detection peak potential is shifted to lower value, which is also very desired in electroanalysis.

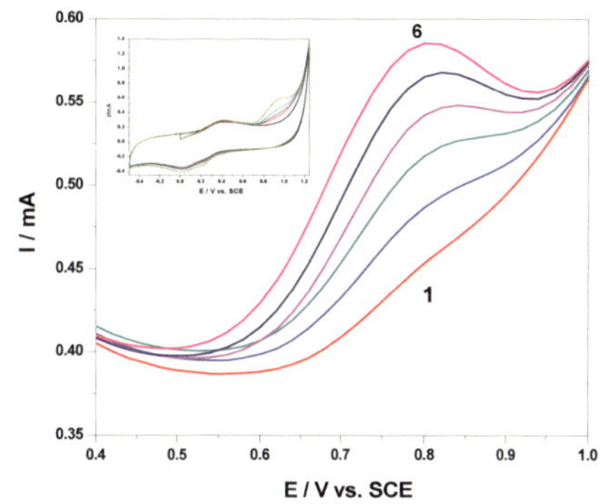

Figure 8. Differential-pulsed voltammograms recorded at CNT electrode (modulation amplitude 0.2 V, step potential 0.01 V) in 0.1 M Na_2SO_4 supporting electrolyte and in the presence of different pentachlorofenol concentration (20-120 μM) potential scan rate 0.05 Vs^{-1} and the potential range between 0 V and 1.25V vs. SCE

Peak potential	Technique used	Concentration range (μM)	Sensitivity (μA/ μM^{-1})	Correlation coefficient (R^2)	LOD (μM)
+ 0.97 V	CV	20-100	5.3	0.990	1.633
+ 0.80 V	DPV	20-120	11	0.984	0.801

Table 3. Electroanalytical performance of the CNT composite electrode for the detection of pentachlorophenol in 0.1M Na_2SO_4 supporting electrolyte.

The resolution for the simulatenous detection is improved by DPV measurements, sometimes being possible the detection of the analytes characterized by similar detection potential. An example of simultaneous detection of 4-chlorophenol (4-CP) and oxalic acid (OA) as its oxidation intermediar is given in Figure 9. The peak-shaped response coupled with the very low background current and the low detection potential values makes this technique involving NCC electrodes to be very useful for the simultaneous detection of the analytes.

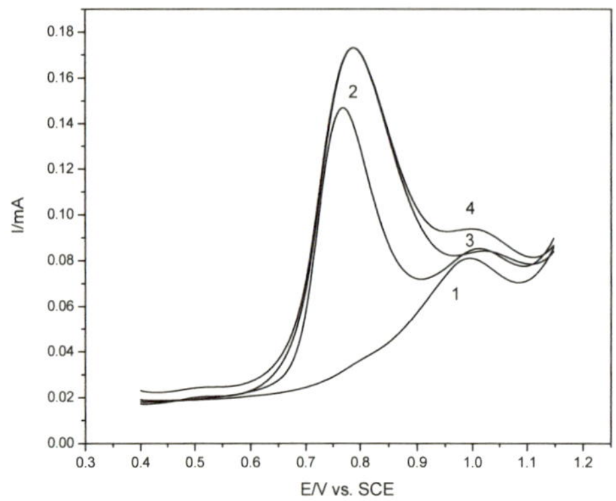

Figure 9. Simultaneous detection of 4-chlorophenol (4-CP) (E=+0.8V/SCE) and oxalic acid (OA)(E=+1 V/SCE) using CNF electrode: (1)-0.5 mM OA; (2)-0.5 mM OA+0.5 mM 4-CP; (3)-0.5 mM OA+0.6 mM 4-CP; (4)-0.6 mM OA+0.6 mM 4-CP.

3.3. Square-wave voltammetry

Square-wave voltammetry (SWV) is a large amplitude-differential technique characterized by a higher sensitivity than DPV, especial for the reversible system due to the reverse pulse cause the reverse reaction and the difference between the forward and reverse components are measured [27]. A very important advantage of this technique is the speed. The effective scan rate is given by fΔE, were f is frequency and ΔE is the step potential (SP). In comparison with DPV, for the detection of pentachlorophenol (PCP) the higher sensitivity was reached for 0.1 V MA, 0.01 V SP and 10 Hz frequency (see Table 4).

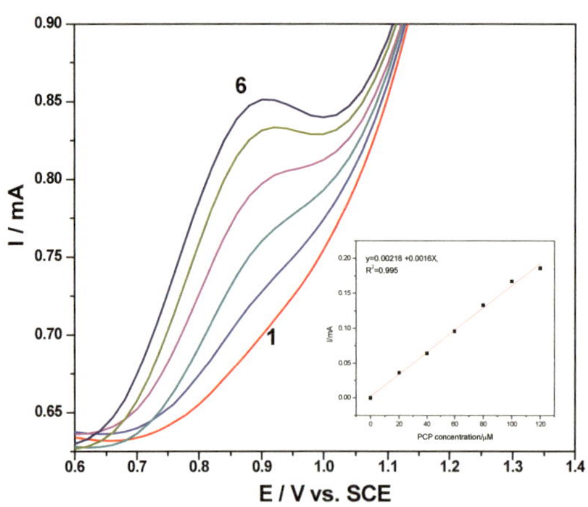

Figure 10. Square-wave voltammograms recorded at CNT electrode (modulation amplitude 0.1 V, step potential 0.01 V and frequency 10 Hz) in 0.1 M Na_2SO_4 supporting electrolyte and in the presence of different pentachlorophenol concentration (20-120 μM) potential scan rate 0.05 Vs^{-1} and the potential range between 0 V and 1.25V vs. SCE. Inset: Calibration plots of the anodic current recorded at E=+0.95 V/SCE vs. PCP concentration.

http://dx.doi.org/10.5772/58633

The voltammetric techniques can be applied in a preconcentration/voltammetric detection scheme, which is useful for the detection of the trace levels of the analytes in direct relation with the electrode material. The adsorption property of the nanostructured carbon-based electrode towards the target analyte is not generally desired, because of electrode fouling and the loss of the electrode activity. However, this aspect could be exploited however in a positive way to detect organics at the trace level by applying the preconcentration/voltammetric detection scheme. The extent of preconcentration is a function of accumulation time, which determines the degree of adsorption on the electrode surface. The accumulation time represents the time of maintaining the potential value at an open-circuit potential (OCP) value [50]. As example, the effect of accumulation time on the currents of the square-wave anodic peaks recorded at+0.90 V/SCE corresponding to PCP oxidation was investigated. The enhancement factor of 10 determined as ratio of the peak current recorded after 30 minutes accumulation times to that recorded without a preconcentration scheme was found for 8 μM PCP detection using CNT electrode. In Figure 11 is shown the SWVs recorded through the preconcentration-square-wave voltammetric technique. It can be noticed that preconcentration applying improved significantly the sensitivity (Table 4).

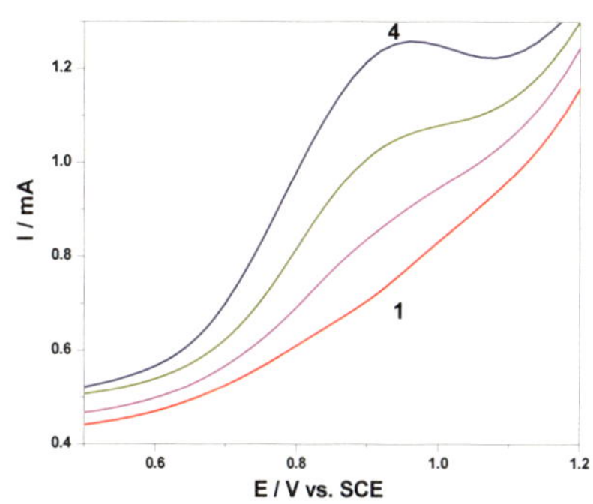

Figure 11. Square-wave voltammograms recorded at CNT electrode (modulation amplitude 0.1 V, step potential 0.01 V and frequency 10 Hz) in 0.1 M Na_2SO_4 supporting electrolyte and in the presence of different pentachlorophenol concentration (2-8 μM) potential scan rate 0.05 Vs^{-1} and the potential range between 0 V and 1.25V vs. SCE.

Peak potential	Technique used	Concentration range (μM)	Sensitivity (μA/ μM^{-1})	Correlation coefficient (R^2)	LOD (μM)
+ 0.80 V	DPV	20-120	11	0.984	0.801
+ 0.90 V	SWV	20-120	16	0.998	0.991
+ 0.95 V	Prec./SWV	2-8	138	0.952	0.033

Table 4. Electroanalytical performance of the CNT composite electrode using pulsed techniques for the detection of pentachlorophenol in 0.1M Na_2SO_4 supporting electrolyte.

3.4. Chronoamperometry

Chronoamperometry is the easiest electrochemical detection technique and it is appropriate for the practical detection applications. This technique is operated at the constant potential value, which is selected based on the existing well-established essential point of reference provided by the cyclic voltammograms, and the current-time dependence is monitorized. As mass transport is controlled by difussion, the current-time dependence reflects the change in the concentration gradient of the electrode surface vicinity. For macroelectrode characterized by linear difussion, this technique should be used for difussion coeficient determination and as consequence, for the specific electroactive surface area determination using Cotrell equation [27]:

$$i = \frac{nFACD^{1/2}}{\pi^{1/2} t^{1/2}} = kt^{-1/2} \tag{5}$$

For the detection application, the most common method is batch injection analysis (BIA), where the amperometric response recorded by chronoamperometry is obtained for successive and continuous addition of a certain concentration of the analyte. Figure 12 shows the amperometric response (BIA) recorded at AgCNF electrode in 0.1 M Na_2SO_4 supporting electrolyte by adding 2.4 μM ibuprofen (IBP) at an applied potential of 1.1 V vs. SCE. The response of the electrode is linear in the concentration range of 2.4 μM to 21.6 μM IBP (Inset of Fig. 12).

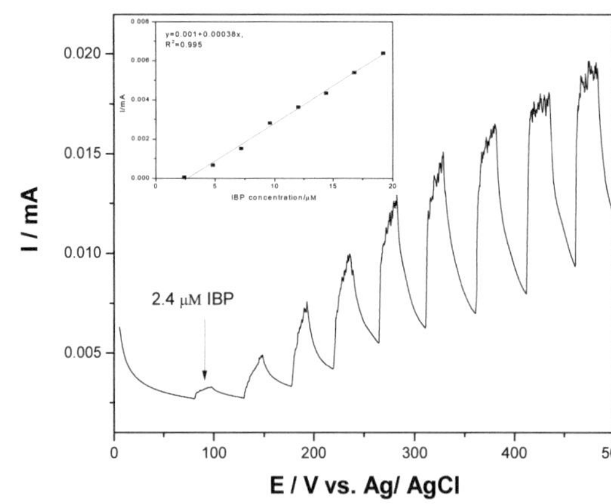

Figure 12. Amperometric response-batch system analysis (BIA) of the Ag-CNF electrode for the successive and continuous addition of 2.4 μM IBP. Applied potential: 1.1 V vs. SCE. Inset: calibration plots of useful signal vs. IBP.

3.5. Pulsed-amperometry

Pulsed-amperometry is an alternative for chronoamperometry to overcome the disadvantage of the electrode fouling during the detection application. In different working programs applied in practice of the pulsed amperometric detection (PAD) the measurement potential is kept for a short time (measurement pulse) after the electrode cleaning pulse and conditioning potential pulse [51]. This technique has proved to be extremely sensitive for the detection of organic molecules because of the advantage of *in-situ* cleaning and reactivating the electrode surface during the electroanalytical detection [52].

PA technique is very useful for the practical detection application especial for the detection of the analytes that can be very easily adsorbed on the electrode surface causing electrode fouling. Application of this technique for the detection of pentachlorophenol at CNT electrode as alternative for CA allowed achieving one hundred times better sensitivity under the operating conditions of three potential pulses, whose values were established on CV behaviour. The pulses were applied continuously using the following scheme [24]:

- *Pulse 1* operated at +0.97 V/SCE for a duration of 50 ms, where PCP is directly oxidized on the electrode surface,
- *Pulse 2* operated at +1.25 V/SCE for a duration of 50 ms, considered as cleaning potential because O_2 evolution occurred,
- *Pulse 3* operated at -0.1 V/SCE for a duration of 50 ms, where the electrode surface involved-reduction process occurred.

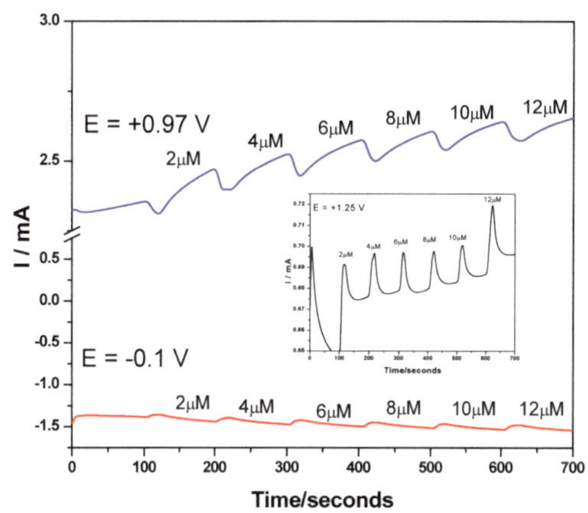

Figure 13. Pulsed amperograms (PAs) recorded at CNT electrode in 0.1 M Na_2SO_4 supporting electrolyte for the successive and continuous addition of 2 μM PCP recorded at E=+1.25 V;+0.97 and-0.1 V vs. SCE.

4. Conclusions

Electrochemical techniques should be regarded as suitable tools for the electrochemical characterization and the detection applications of the nanostructured carbon composite (NCC) and metallic nanoparticles-modified nanostructured carbon composite (M-NCC) electrodes. Cyclic voltammetry (CV) is very useful for the electrochemical characterization that provided mechanistic information based on which the operating conditions are optimized. In general, CV results are considered as reference for the all electrochemical technique applied in the detection application, especially for chronoamperometry (CA) and pulsed amperometry (PA). NCC and M-NCC electrodes obtained by two-roll mill procedure are very promising materials for the detection applications, due to fascinating properties of the nanostructured carbon and metallic nanoparticles. The electrode morphology affected the electrical and electrochemical behaviour of the electrode materials. Their electrochemical detection performances are linked to the electrochemical technique specificity, the best results being achieved using pulsed voltammetric techniques (DPV and SWV). These techniques are very effective also, for the simultaneous detection of multi-components. However, for the practical detection application, CA is most useful due to its simplicity but sometimes, for the specific process that occurred on the electrode surface during the detection, this technique favours the electrode fouling and as consequence, the loss of the electrode activity. PA overcomes the electrode fouling disadvantage since in-situ electrochemical cleaning can be assured by applying simultaneously the potential at which desired advanced oxidation/reduction as cleaning step and mild oxidation/reduction for electrode stabilizing besides the detection potential.

Acknowledgements

This work was supported by grants of the Romanian National Authority for Scientific

Research, CNCS – UEFISCDI-partially by the project: PN-II-ID-PCE 165/2011 and partially by the project: PNII-60/2012 (WATUSER).

Author details

Florica Manea*

Faculty of Industrial Chemistry and Environmental Engineering, Politehnica University of Timisoara, Romania

References

[1] Kirgoz ÜA, Timur S, Odaci D, Pérez B, Alegret S, Merkoci A (2007) Carbon Nanotube Composite as Novel Platform for Microbial Biosensor. Electroanal. 19 (7-8): 893–898.

[2] Li Z, Cui X, Zheng J, Wang Q, Lin Y(2007) Effects of Microstructure of Carbon Nanofibers for Amperometric Detection of Hydrogen Peroxide. Anal. chim. acta 597: 238–244.

[3] Manea F, Radovan C, Corb I, Pop A, Burtica G, Malchev P, Picken S, Schoonman J (2007) Electrochemical Oxidation and Determination of Oxalic Acid at an Exfoliated Graphite-Polystyrene Composite Electrode. Sensors. 7: 615-627.

[4] Sun D, Zhu L, Zhu, G (2006) Glassy Carbon Ceramic Composite Electrodes. Anal. chim. acta. 564: 243-247.

[5] Ramirez-Garcia S, Alegret S, Cespedes F, Forster RJ (2002) Carbon Composite Electrodes: Surface and Electrochemical Properties. Analyst. 127: 1512-1519.

[6] Kurusu F, Koide S, Karube I, Gotoh M (2006) Electrocatalytic Activity of Bamboo-Structured Carbon Nanotubes Paste Electrode Toward Hydrogen Peroxide, Anal. let. 39 (5): 903-911.

[7] Fathirad F, Afzali D, Mostafavi A, Shamspur T, Fozooni S (2013) Fabrication of a New Carbon Paste Electrode Modified with Multi-Walled Carbon Nanotube for Stripping Voltammetric Determination of Bismuth(III), Electrochim. acta. 30: 206–210.

[8] Rivas GA, Rubianes MD, Rodríguez MC, Ferreyra NF, Luque GL, Pedano ML, Miscoria S A, Parrado C (2007) Carbon Nanotubes for Electrochemical Biosensing. Talanta. 74: 291-307.

[9] Pumera M, Merkoci A, Alegret S (2006) Carbon Nanotube-Epoxy Composites for Electrochemical Sensing. Sens. actuators B-chem. 113: 617-622.

[10] Shah B, Lafleur T, Chen A (2013) Carbon Nanotube Based Electrochemical Sensor for the Sensitive Detection of Valacyclovir. Faraday discuss. 164: 135-146.

[11] Huang J, Liu Y, You T (2010) Carbon Nanofiber Based Electrochemical Biosensors: a Review, Anal. methods. 2: 202-211.

[12] Deo RP, Wang J (2004) Electrochemical Detection of Carbohydrates at Carbon-Nanotube Modified Glassy-Carbon Electrodes, Electrochem. commun. 6: 284–287.

[13] Pham X-H, Ngoc Bui M-P, Li CA, Han KN, Kim JH, Won H, Seong GH (2010) Electrochemical Characterization of a Single-Walled Carbon Nanotube Electrode for Detection of Glucose, Anal. chim. acta. 671: 36-40.

[14] Lamas-Ardisana PJ, Queipo P, Fanjul-Bolado P, Costa-García A (2008) Multiwalled Carbon Nanotube Modified Screen-Printed Electrodes for the Detection of *p*-Aminophenol: Optimisation and Application in Alkaline Phosphatase-Based Assays. Anal. chim. acta. 615: 30-38.

[15] Remes A, Pop A, Manea F, Baciu A, Picken S J, Schoonman J (2012) Electrochemical Determination of Pentachlorophenol in Water on a Multi-Wall Carbon Nanotubes-Epoxy Composite Electrode. Sensors. 12(6): 7033-7046.

[16] Arribas AS, Martínez-Fernández M, Moreno M, Bermejo E, Zapardiel A, Chicharro M (2013) Analysis of Total Polyphenols in Wines by FIA with Highly Stable Ampero-

metric Detection using Carbon Nanotube-Modified Electrodes. Food. chem. 136: 1183-1192.

[17] Baciu A, Manea F, Remes A, Motoc S, Burtica G, Pode R (2010) Anodic Determination of Pentachlorophenol from Water Using Carbon Nanofiber-Based Composite Electrode. Environ. eng. manag. j. 9 (11): 1555-1562.

[18] Lorenzo MÁ, Arribas AS, Moreno M, Bermejo E, Chicharro M, Zapardiel A (2013) Determination of Butylparaben by Adsorptive Stripping Voltammetry at Glassy Carbon Electrodes Modified with Multi-Wall Carbon Nanotubes. Microchem. j. 110: 510-516.

[19] Wang J (2005) Carbon-Nanotube Based Electrochemical Biosensors: a Review, Electroanal. 17 (1): 7-14.

[20] Balasubramanian K, Burghard M (2006) Biosensors Based on Carbon Nanotubes, Anal. bioanal. chem. 385: 452–468.

[21] Trojanowicz M (2006) Analytical Applications of Carbon Nanotubes: a Review. TrAC 25 (5): 480-489.

[22] Pop A, Manea F, Orha C, Motoc S, Ilinoiu E, Vaszilcsin N, Schoonman N (2012) Copper-Decorated Carbon Nanotubes-Based Composite Electrodes for Nonenzymatic Detection of Glucose. Nanoscale. res. lett. 7: 266-269.

[23] Manea F, Motoc S, Pop A, Remes A, Schoonman J (2012) Silver-Functionalized Carbon Nanofiber Composite Electrodes for Ibuprofen Detection. Nanoscale. res. lett. 7: 331-334.

[24] Kang X, Mai Z, Zou X, Cai P, Mo J (2007) A Sensitive Nonenzymatic Glucose Sensor in Alkaline Media with a Copper Nanocluster/Multiwall Carbon Nanotube-Modified Glassy Carbon Electrode. Anal. biochem. 363: 143-150.

[25] Valentini F, Biagiotti V, Lete C, Palleschi G, Wang J (2007) The Electrochemical Detection of Ammonia in Drinking Water Based on Multi-Walled Carbon Nanotube/ Copper Nanoparticle Composite Paste Electrodes. Sens. actuators B-chem. 128: 326-333.

[26] Chih YK, Yang MC (2014) Simultaneous Detection of Dopamine and Ascorbic Acid using Silver/Silver Sulfide Modified Carbon Nanotube Electrodes. J. Chin. inst. chem. cng. 45: 833-839.

[27] Wang J (2000) Analytical Electrochemistry. New York: Wiley-VCH. 222 p

[28] Husssain F, Hojjati M, Okamoto M, Gorga RE (2006) Polymer-Matrix Nanocomposites, Processing, Manufacturing, and Application: an Overview. J. composite mater. 40(17): 1151-1575.

[29] Uslu B, Ozkan SA (2007) Electroanalytical Application of Carbon Based Electrodes to the Pharmaceuticals. Anal. lett. 40: 817–853.

[30] Cespedes F, Martinez-Fabregas E, Alegret S (1996) New Materials for Electrochemical Sensing, I. Rigid Conducting Composites. TrAC, 15(7): 296-304.

[31] Merkoci A, Pumera M, Llopis X, Perez B, del Valle M, Alegret S (2005) New Materials for Electrochemical Sensing VI: Carbon Nanotubes. TrAC 24: 826-838.

[32] Chu H, Wei L, Cui R, Wang J, Li Y (2010) Carbon Nanotubes Combined with Inorganic Nanomaterials: Preparation and Application. Coordin. chem. rev. 254: 1117-1134.

[33] Oyama M (2010) Recent Nanoarchitectures in Metal Nanoparticle-Modified Electrodes for Electroanalysis. Anal. sci. 26: 1-12

[34] Welch CM, Compton RG (2006) The Use of Nanoparticles in Electroanalysis: a Review, Anal. bioanal. chem. 384 (3): 601-619.

[35] Suna L, Zhang Z, Dang H (2003) A Novel Method for Preparation of Silver Nanoparticles. Mater. lett. 57: 3874–3879.

[36] Fukushima M, Yanagi H, Hayashia S, Suganuma N, Taniguchi Y(2003) Fabrication of Gold Nanoparticles and Their Influence on Optical Properties of Dye-doped Sol-Gel Films. Thin. solid. films. 438: 39–43.

[37] Magdassi S, Grouchko M, Kamyshny A (2009) Colloidal Dispersion of Metallic Nanoparticles Formation and Functional Properties. In: Hidalgo-Alvarez R editor. Structure and Functional Properties of Colloidal Systems. CRC Press. pp. 339-365

[38] Renedo OD, Alonso-Lomillo MA, Martınez MJA (2007) Recent Developments in the Field of Screen-Printed Electrodes and Their Related Applications-Review. Talanta. 73: 202–219.

[39] Khaydarov RA, Khaydarov RR, Gapurova O, Estrin Y, Scheper T (2009) Electrochemical Method for the Synthesis of Silver Nanoparticles, J. nanopart. res. 11: 1193-1200.

[40] Hyde ME, Compton RG (2003) A Review of the Analysis of Multiple Nucleation with Diffusion Controlled Growth, J. electroanal. chem. 549: 1-12.

[41] Safavi A, Maleki N, Farjami E (2009) Electrodeposited Silver Nanoparticles on Carbon Ionic Liquid Electrode for Electrocatalytic Sensing of Hydrogen Peroxide. Electroanal. 21(13): 1533 – 1538.

[42] Bauhofer W, Kovacs JZ (2009) A Review and Analysis of Electrical Percolation in Carbon Nanotube Polymer Composites. Compos. sci. technol. 69: 1486-1498.

[43] Baciu A, Pop A, Manea F, Schoonman J (2014) Simultaneous Arsenic (III) and Lead (II) Detection from Aqueous Solution by Anodic Stripping Square-Wave Voltammetry. Environ. eng. manag. j. accepted for publication

[44] Ballarin B, Cordero-Rando MM, Blanco E, Hidalgo-Hidalgo De Cisneros JL, Seeber R, Tonelli D (2003) New Rigid Conducting Composites for Electrochemical Sensors, Collect. Czech. chem. commun. 68: 1420-1436.

[45] Stulik K, Amatore C, Holub K, Marecek V, Kutner W (2000) Microelectrodes. Definitions, Characterization, and Applications (Technical Report), Pure. appl. chem. 72: 1483-1492.

[46] Simm AO, Banks CE, Ward-Jones S, Davies TJ, Lawrence NS, Jones TGJ, Jiang L, Compton RG (2005) Boron-Doped Diamond Microdisc Arrays: Electrochemical Characterisation and Their Use as a Substrate for the Production of Microelectrode Arrays of Diverse Metals (Ag, Au, Cu)via Electrodeposition. Analyst. 130: 1303-1311.

[47] Manea F, Radovan C, Pop A, Corb I, Burtica G, Malchev P, Picken S, Schoonman J (2009) Carbon Composite Electrodes Applied for Electrochemical Sensors. In: Baraton MI editor. Sensors for Environment, Health and Security. NATO Science for Peace and Security Series C: Environmental Security. Springer. pp. 179-189.

[48] Cięciwa A, Wuthrich R, Comninellis C (2006) Electrochemical Characterization of Mechanically Implanted Boron-Doped Diamond Electrodes. Electrochem. commun. 8: 375-382.

[49] Feeney R, Kounaves SP (2000) Microfabricated Ultramicroelectrode Arrays: Developments, Advances, and Applications in Environmental Analysis. Electroanal. 12: 677-684.

[50] Pop A, Manea F, Radovan C, Malchev P, Bebeselea A, Proca C, Burtica G, Picken S, Schoonman J (2008) Amperometric Detection of 4-Chlorophenol on Two Types of Expanded Graphite Based Composite Electrodes. Electroanal. 20(22): 2460-2466.

[51] Charoenraks T, Chuanuwatanakul S, Honda K, Yamaguchi Y, Chailapakul O (2005) Analysis of Tetracycline Antibiotics Using HPLC with Pulsed Amperometric Detection. Anal. sci. 21: 241-245.

[52] Bebeselea A, Manea F, Burtica G, Nagy L, Nagy G (2010) The Electrochemical Determination of Phenolic Derivates using Multiple Pulsed Amperometry with Graphite Based Electrodes. Talanta. 80(3): 1068-1072.

Chapter 3

Electrochemical Scanning Tunneling Microscopy (ECSTM) – From Theory to Future Applications

Ajay Kumar Yagati, Junhong Min and
Jeong-Woo Choi

Additional information is available at the end of the chapter

http://dx.doi.org/10.5772/57236

1. Introduction

The development of scanning tunneling microscopy (STM) clearly forms the creation of a new research tool by innovative implementation of scientific and technological knowledge, thereby advancing further in the fundamental science and technology [1,2]. The quantum-mechanical phenomenon of electron tunneling had been known for a long time, but the use of this phenomenon for the imaging of a conductive surface at atomic level was realized only in 1982 when the first STM as built by Binnig et al [3]. STM has a resolution of a few Ångstrom in lateral directions and less than one Ångstrom in the direction perpendicular to the surface [4]. It consists of a scanning tip which images the surface by means of a tunnel current. Hence, the sample needs to be conductive [5]. At present, STM is a powerful tool for analyzing metallic and semiconductor surface. The real-space visualization of surface at atomic scale is one of the most important features [6]. The spatial variation of the tunneling current or the spatial variation of the tip height is converted in to the real space image. The tunneling current decreases exponentially with the increase in tip-sample distance. Thus, at any given location of tip over the sample surface, the electron transfer involves only one atom or few atoms at the tip apex and on the surface closest to them. This makes it possible to visualize the structures with sub-angstrom resolution and to detect atomic scale defects that are not possible with other spectroscopic techniques [7,8]. The STM not only provides the three-dimensional information about the topography of the sample, but it also gives the information about the spectroscopic properties and local variations of work functions. Further, as a nanofabrication tool, STM can be used for atom manipulation, local deposition and imaging of the molecules [9]. Moreover, STM can be used to operate in air, ultra high vacuum (UHV) and in liquid solutions for electrochemistry applications which involves the immersion of the STM probe into the liquid

INTECH
open science | open minds

© 2014 The Author(s). Licensee InTech. This chapter is distributed under the terms of the Creative Commons Attribution License (http://creativecommons.org/licenses/by/3.0), which permits unrestricted use, distribution, and reproduction in any medium, provided the original work is properly cited.

media and the corresponding electrochemical control (ECSTM) for *in situ* monitoring of redox processes on the sample electrode [10,11]. Having these advantages STM has become most widely accepted analytical measurement system in the current research works.

2. Fundamentals of scanning tunneling microscopy (STM)

2.1. Origin and operation mode:

The scanning tunnelling microscope was developed by Binnig, Rohrer, Gerber and Weibel [12]. Since STM can be used for imaging on at atomic scale level, this belongs to the most powerful experimental techniques of surface science. In STM, a sharp metallic tip is placed very close to the surface and a small bias voltage is applied between the tip and the sample. As a result, a current of electrons (I_T), flows between the electrodes through the vacuum gap. This process is a quantum mechanical phenomenon and is called as "tunnelling" effect [13] shown in Fig. 1. The electrons "tunnel" through this electrically insulating layer, giving rise to a measurable current which displays an exponential dependence on the distance between the two conducting electrodes [14]. The tunneling current flowing between the STM tip and the sample surface through the insulating gap (s) under an applied V_{bias} which can be explained in a simple analytical tunneling expression assuming a 3-dimensional (3D) metal-insulator-probe junction into a one-dimensional metal-insulator-metal contact. The derived equation that relates with the applied V_{bias} with I_T and s yields [15]:

$$I_T = I_o \frac{V_{bias}}{s} e^{-k\sqrt{\phi_b}S} \tag{1}$$

It is the exponential dependence of the tunneling current on the distance between the two conductors (STM tip and the underlying surface) that provides the sensitivity of the measured current that can be interpreted as the surface structure. As the equation shows, other factors influence the current as well, such as the electron band structure of the two conductors (φ).

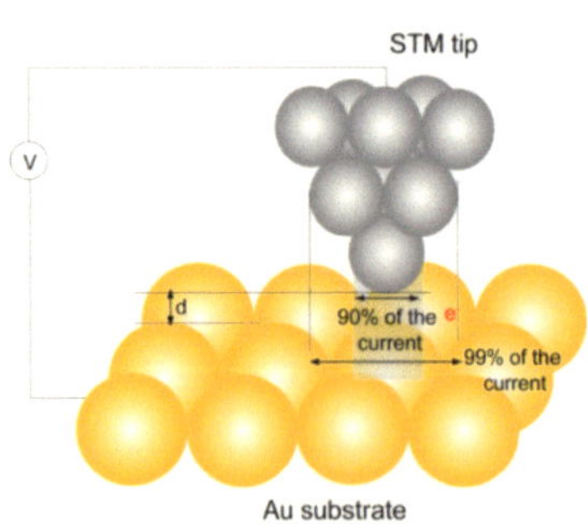

Figure 1. Schematic diagram of scanning tunneling microscopy depicts the tip sample interaction and the tunneling current between the tip and the sample surface.

http://dx.doi.org/10.5772/57236

2.2. Tunnelling methods

2.2.1. General modes of operation

Basically there are two modes in which STM can be operated. The first one being the constant current and the other one is termed as constant height. In the constant current mode, the tip is scanned over the surface at constant tunnel current, and the vertical tip position will be continuously changed to keep the tunnel current as a constant. In ideal conditions, at a homogenous surface constant current refers to constant interval between sample surface and the tip. In this mode, the height control mechanism will adjust the tip to move vertically up and down to keep the tunnel current to a constant value by the feedback voltage [16]. In the constant height mode which most effective for the investigation of atomically smooth surfaces (Z remains *const*). In this mode of operation, the tip moves above the surface at a distance of several Å, and the changes in the tunneling current are recorded as STM image [17]. Scanning may be done either with the feedback system switched off (here no topographic imaging is recorded), or at a speed exceeding the feedback reaction speed (only smooth changes of the surface topography are recorded). This method employs very high scan rates and fast STM images acquisition, allows observing the changes that occur on a surface in a real time.

2.2.2. The electrochemical STM (ECSTM) model

The electrochemical scanning tunneling microscopy (ECSTM) is an extended technique performed along with basic STM measurements for the study of electrode-electrolyte interfaces. Therefore, the same elements as in a standard STM experimental set up can be observed in the ECSTM set-up. The tunneling current flowing between a metallic tip and a conductive sample will be again used to obtain topographical information as well as the electronic structure of a determined electrode surface immersed in the corresponding electrolyte [18,19]. Even though, the fundamentals of both techniques are essentially the same, but two different elements were introduced in a conventional STM set up in order to control as an ECSTM: 1) a three electrode electrochemical cell in which the substrate used as a working electrode, a reference and a counter electrode completed the electrochemical cell. This configuration resembles as a normal three electrode configuration coupled to a potentiostat in which the potential of the working electrode can be controlled with respect to a high-impedance reference electrode, while the current is allowed to flow between the working and the counter electrode. 2) Development of suitable ECSTM probes. In this case, the tip is not an active electrode in the cell, but used only to image the surface morphology, even if it had to be under potential control in order to apply a voltage drop to drive the tunnelling current. The implementation of a liquid STM represented a great breakthrough for the in situ study of surface electrodes [20]. An improvement in the STM has been obtained with the introduction of the bipotentiostat approach, in which both tip and sample potentials are independently controlled with respect to a reference electrode in solution.

3. Electrochemistry in scanning probe microscopy: Basic concepts and applications

The electrochemical scanning tunneling microscope was the first tool for the investigation of solid-liquid interfaces that allowed in situ real space imaging of the underlying electrode surfaces at atomic level. Therefore ECSTM gained much importance and emerged as a prominent tool for the determination of the local surface structure as well as the dynamics of reactions/ processes that takes place at surfaces in an electrolytic environment. Although the fundamentals of both techniques are then essentially the same, but two different elements must be introduced in a conventional STM set up in order to operate as an ECSTM: an electrochemical cell consisting of two working electrodes with bipotentiostat approach and suitable ECSTM probes [21].

3.1. Preparation of reliable probes, tunneling in liquid environments at a bipotentiostat configuration

The general electrochemical experiment which is composed with standard potentiostat, the potential of the working electrode is controlled with respect to a reference electrode by the flow of current through a counter electrode. However, in ECSTM with *in-situ* electrochemical measurements, the same potentiostatic approach is able to control both the potential of the substrate and the potential of the tip that is present in solution with respect to a reference electrode in order to control electrochemical reactions taking place at its surface, which is practically not possible [22,23]. Generally in STM, it is expected that the current enters into the measuring system due to the charge transfer process is mainly due to tunneling current between the tip and the substrate. If the potential of one of these two electrodes is not controlled in the electrochemical cell, then electrochemical charge transfer mechanisms might become significant and contribute to the measured currents which leads to a strong source of noise in the STM control circuit [24,25]. Further, the absence of control on the electrochemical potential of one electrode in an electrochemical cell also gives rise to phase variation in the surface composition of the electrode. In this aspect, with bipotentiostatic approach an independent control on the potential of both the tip and the substrate with respect to a reference electrode in solution is established for the development of the *in-situ* STM.

The main purpose of the potentiostat is to control the potential of the electrode in an electrochemical cell from various impedances connecting to these electrodes. The controller maintains the potential of the reference with respect to the working electrode in such a way that the potential is exactly opposite to the controlled potential, which is free from fluctuations of the impedances [26]. Hence, the bipotentiostat controls the potential of two electrodes with respect to a reference electrode. The ECSTM setup is depicted in Fig. 2. Generally, the tip is virtually grounded and the tunneling current is measured by a high-gain current follower or it is fed to the STM control unit through a preamplifier. In all configurations, the tip and substrate potentials are controlled with respect to a current less reference electrode. Usually, the counter electrode in EC-STM setup is obtained by an Au or Pt wire which has enough stability [27]. Also, a good reference electrode is obtained when a metal is used in electrochemical equili-

http://dx.doi.org/10.5772/57236

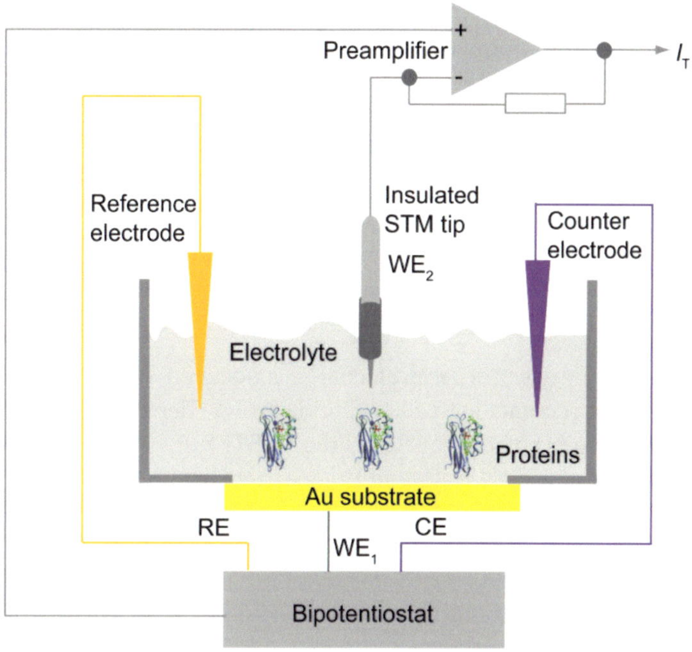

Figure 2. EC-STM configuration, the bipotentiostat controls potential of tip and sample with respect to reference electrode.

brium with the corresponding metal cation in solution. In the past decade, STM has been applied to study the solid-liquid interfaces, in which the STM probe is immersed into the liquid media and controlled by the electrochemical module of STM for *in-situ* monitoring of redox process on the sample electrode. ECSTM has advantages over classical or conventional STM operated either in air or in a vacuum. It allows precise and independent control of both tip potential and sample electrode potential through the use of a bipotentiostat and a quasi-reference electrode [28]. Generally, STM allows to perform electronic spectroscopy by recording the tunneling current (I_t) with the applied bias potential (V_b) known as scanning tunneling spectroscopy (STS) and the same can be applied in electrochemical environment, leading to electrochemical scanning tunneling microscopy (ECSTS) technique [29]. Even though ECSTS is a sophisticated and powerful tool it is not widely used because of complexity in tip preparation methods.

Generally in ECSTS, the current flowing through the tip has several components; a) STM tunneling current, b) faradaic current by the electrochemical reaction at the tip/electrolyte interface and c) charge-discharge process of the electrochemical double layer at the tip-electrolyte interface. So, if the faradaic currents and the charge-discharge processes are larger than the set-point tunneling current, the STM measurement will no longer be possible. Hence there should be an alternative to eliminate these two electrochemical contributions to the measured tip current. The most effective alternative is, taking the advantage of the fact that faradaic/capacitive currents are directly proportional to the exposed area of the electrode, coating the tip except for the very apex so that tunneling current can flow while faradaic and capacitive currents can be minimized [30,31]. Generally in STM, the tunneling current is usually set to between 1~10 nA. For accurate measurements, coated tips must yield faradaic and capacitive currents ≤ 0.1 nA at the end of the STS curve when ramping at the highest tip potential scan rate (up to 10 V/s) needed to minimize drift, whereas the STS curve is recorded without feedback. Various methods and materials have been proposed for coating purposes, such as apiezon wax, melt glass, copolymers, and a combination of glass and polymer, and in all these coating procedures, the faradaic and capacitative contributions are too large to be subtracted from the measured tip current, and thus, they do not allow STS spectra to be under electrochemical control [32].

As mentioned above, the electrochemical charge associated to the charging-discharging process of an electrode in contact with a specific electrolyte depends directly on the exposed area of the electrode itself. For this purpose, the ECSTM tips must be insulated from the electrolyte, in the way that just the very end tip apex remains in contact with the electrolyte. With this isolation process, the electrochemical current measured through the STM must be better than 10 % of the tunneling set point current, that is, typically ≤ 0.1 nA. Electrophoretic paints are increasingly used for coating Pt-It tips as they are chemically and electrochemically inert insulators [33]. However, among many protection methods available, apiezon wax is still used for ECSTM tip isolation, despite its disadvantages.

4. Characterizations with ECSTM

4.1. Understanding electrochemical processes such as corrosion, deposition, and adsorption

Applications of ECSTM in the field of corrosion have mainly focused on understanding the mechanism of corrosion initiation and the process of inhibition, including pitting initiation, surface dissolution, passive film formation, and the effect of inhibitors. ECSTM has been utilized to study a variety of materials, including Cu, Ni, and Fe, in many different corrosion environments [34]. In general, the term *corrosion* stands for material deterioration or surface damage in a liquid environment. In the case of metals, it is basically a chemical/electrochemical process that suggests an oxidation of a metal which transfer electrons to the electrolytic environment and undergoes a change of valence from zero to a positive value [35]. Normally, this initial process leads to a number of parallel electrochemical methods that result in material dissolution and/or eventual formation of secondary corrosion products. Basically the metal corrosion processes can be classified in two different forms: one is general corrosion [36] that symbolizes those corrosion processes involving the entire surface area of the material, and the second one localized corrosion [37] which mentions to a number of corrosion processes that are triggered at specifics sites on the material surface. The presence of a surface passive film prevents the metallic substrate from further oxidation, i.e., the surface becomes passivated. However, the passive state of a metal under certain conditions is susceptible to localized instabilities that prompt the corrosion of the metal electrode through the local dissolution of the passive layer. This process is called as pitting corrosion [38].

Pitting corrosion takes place at passivated metal surfaces and leads to a creation and growth of an active *pit* [39]. It is quite common that these pits can easily develop at defective sites on the passive film surface, on sharp edges or sites where the oxide film is thinner. The essence of the pitting process relays in its anodic nature (metal dissolution) compared with a passivated cathodic region surrounding it, thus allowing the continuous flow of electrons to take place [40]. The pit formation is depicted in Fig. 3 where peaks indicate the pitting corrosion. Most metal passive layers present a semiconducting or insulating electronic behavior and, therefore, it is expected to understand the initiation of the pitting corrosion process, this requires to embed semiconductor phenomena into an actual classical corrosion mechanisms.

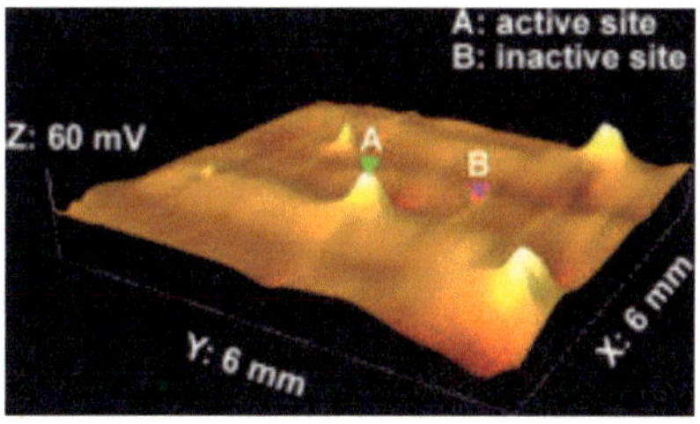

Figure 3. The typical 3D potential distribution image for the stainless steel in 10% (wt.%) $FeCl_3$ solution. Figure reproduced with permission from: ref. 43, © 2012 Elsevier

It is known that, charge injection always occurs at high electric fields in particularly under dc conditions. The presence of space charge in solid dielectrics will result in distortion of electric field distribution. Dielectric breakdown can began in the region where electric field enhancement takes place if the electric field exceeds the "threshold strength" of the material. That is, spontaneous generation of extra charge carriers in an electrically insulated region either by tunneling (Zener breakdown) or by collision (avalanche breakdown) [41]. Both these mechanisms occur when a sufficiently high electric field is applied to the oxide layer thus generating high currents passing through the interface.

The location of a carrier at the semiconductor solid surface is generally accounted as the lattice bond weakening that ultimately lead to the bond breaking and the corresponding electrochemical electrode corrosion (dissolution). This process may continue in different routes that comprise either the movement of electrons and holes to charged states [42]. The rate of these charge carriers (e^- and h^+) arriving to the particular charges state will then govern the dissolution rate and, consequently, the material corrosion. ECSTM can be used to analyze the pitting corrosion by measuring potential distribution image for the stainless steel in 10% (wt.%) $FeCl_3$ solution after 30 min of immersion [43]. Further, ECSTM is able to map the in-situ pitting process with micro and nano-spatial resolution and can locate the positions also the local activity of pitting corrosion in an early stage.

ECSTM has been a valuable tool for understanding electrochemical processes such as corrosion, deposition and adsorption. For example, a method to estimate the pitting corrosion of naked and self-assembled monolayer (SAM) of n-alkane thiol modified Au (111) surfaces in CN^- solutions was examined [44]. It is estimated that SAMs reduced the rate of corrosion, but fragile and contain too many defects. To estimate this corrosion, the applied potential was slowly changed from negative to positive values and it is observed that at small positive potential values leads the initial stage of corrosion, at slightly higher positive potentials pits and step edges occurs and even higher potentials the etching starts and surface becomes rough [45]. Overall, the SAM modified surface is more resistance to corrosion for longer periods of time and at more positive (etching) potentials than the naked Au surface. Hence, 1) the potential imaging technique is able to locate the positions and map local activity of the pit initiation. Hence ECSTM study becomes an alternative method for analyzing the pitting initiation at an early stage; 2) combined with potential imaging, *in situ* ECSTM can be useful for imaging the surface topographies that associated with dynamic process of local breakdown of passive layers and micropitting initiation; 3) a threshold potential can be defined as a critical potential to characterized the local breakdown of passive layers; 4) the pitting initiation is strongly depends on the surface conditions of passive film, concentration of the ions and pH in the solution.

4.2. Molecular resolution imaging of redox species in solution with ECSTM

In biological molecules, electron transfer (ET) in solid surface or liquid solution can occur between donor (D) and acceptor (A) separated by a long distance. To describe this process quantitatively, sophisticated models are required. Additionally, scanning tunneling and atomic force microscopy (STM and AFM, respectively) have opened an exciting new perspective for molecular imaging. STM imaging at the solid/air interface to molecular and occasionally sub molecular resolution also has been extended to biological macromolecules including DNA and a number of redox and non-redox proteins [46,47]. *In-situ* STM offers, on the other hand, new electrochemical spectroscopic probes in addition to current-bias voltage relations, particularly the relation between the tunnel current and the overvoltage of both the tip and substrate electrodes relative to a common reference electrode. Particularly, *in-situ* scanning tunneling microscopy (STM) of redox molecules, in aqueous solution, shows interesting analogies and differences compared with interfacial electrochemical electron transfer and also in homogeneous solution. With ECSTM, high resolution imaging and spectroscopy of adsorbed molecules can be achieved. It is understood that ECSTM combines electrochemical control and STM high-resolution profiles such as molecular imaging and scanning tunneling spectroscopy (STS) [48]. The possibility of using ECSTM to observe single-molecule charge transport was proposed in the early 1990s. The first ECSTM experiments was performed on iron porphyrin molecules adsorbed on highly ordered pyrolytic graphite (HOPG) surface, and the redox-tuned resonant tunneling effect was directly visualized by STM imaging [49]. Since then, it has become a powerful tool to study the interfacial electron transfer and molecular conductance of electro active species at single-molecule level in an electrochemical environment. For conductance imaging, with *in situ* ECSTM mapping was performed on redox molecules such as azurin (a redox protein) seen in reference [50]. To observe the electron transfer properties and its imaging, Azurin (Az) is a blue single-copper protein adsorbed on gold surface which functions as an electron carrier physiologically associated with oxidative stress responses in bacteria (e.g., *Pseudomonas aeruginosa*) and is a long-standing model for exploring electron tunneling through protein molecules was studied. The molecular assembly for the effective coupling of Azurin with gold surface can be examined with STM imaging or by examining the electrochemical property of the adsorbed Az on Au surface. Moreover, the combination of electrochemical and STM measurements thus provide a way to quantify one of the fundamental and long-lasting questions in adsorption chemistry of redox proteins: Such as mainly 1) what percentage of the protein molecules retains their biological activity in the immobilized state? 2) In addition to fast ET, stability is another key factor determining reproducibility and operation in applications of the system to molecular electronics. ECSTM enables to observe single-molecule current-voltage relations by tuning the over potential across the equilibrium redox potential. The energy state of both the substrate and the tip in ECSTM is under control by electrochemical potentials relative to a common reference electrode in an aqueous buffer environment [51,52]. This STM configuration is particularly suitable for *in situ* mapping of electronic properties of redox proteins during their biological action (e.g., ET or electrocatalysis), because the aqueous phase is essential for almost all biological processes in nature. STM imaging was performed to observe single molecules, for which high-resolution images of Az was obtained by keeping a constant bias voltage between the substrate and the tip, with the substrate potential set at the equilibrium redox potential (zero over potential) of azurin. Imaging was performed toward either positive or negative overpotentials by adjusting the substrate and tip potentials in parallel (i.e., at constant bias voltage) and finally was returned to the equilibrium potential [53]. The adsorbed azurin monolayer is sufficiently robust and can withstand repeated ECSTM imaging without loss of its activity. Hence, a series of STM images was obtained at various overpotentials. Fig. 4 shows typical images in which three molecules were targeted. Focus is on the central molecule; two molecules in the upper left region serve as a positioning reference. The single-molecule contrast is clearly tuned by the redox state of azurin, with a maximum around the equilibrium redox potential (Fig. 4C). The contrast decreases upon applying either positive (Fig. 4 A,B) or negative (Fig. 4 D, E) overpotentials, but the effects are not symmetric, with the decay being stronger at negative overpotentials.

4.2.1. Conductance images by potential variation

Potential-dependent EC-STM images of azurin was provided by an experiment in which azurin was chemisorbed on Au substrates studied by EC-STM [54]. *In-situ* cyclic voltammetry (CV) measurements were performed on monolayers of azurin to confirm that protein retained its redox activity and to examine the redox potentials. As proposed by Kim et. al., azurin adsorbed on Au surface has the redox peaks at with anodic redox wave at E_{pa}= 278 mV and cathodic wave at E_{pc}= 486 mV which corresponds the redox process of $Cu^{2+/1+}$ center in Azurin [55]. The open circuit potential (E_{oc}) of the cysteine-modified azurin on Au surface was found to be 80 mV. So, these three states were utilized to examine the images of azurin under different potentials. Typical bright spots are seen when tuning the substrate potential in a region, appear to be strongly potential-dependent of redox potentials. Such kind of behavior is consistent with a resonant nature of the current measured in STM experiments in the Au adsorbed azurin molecules.

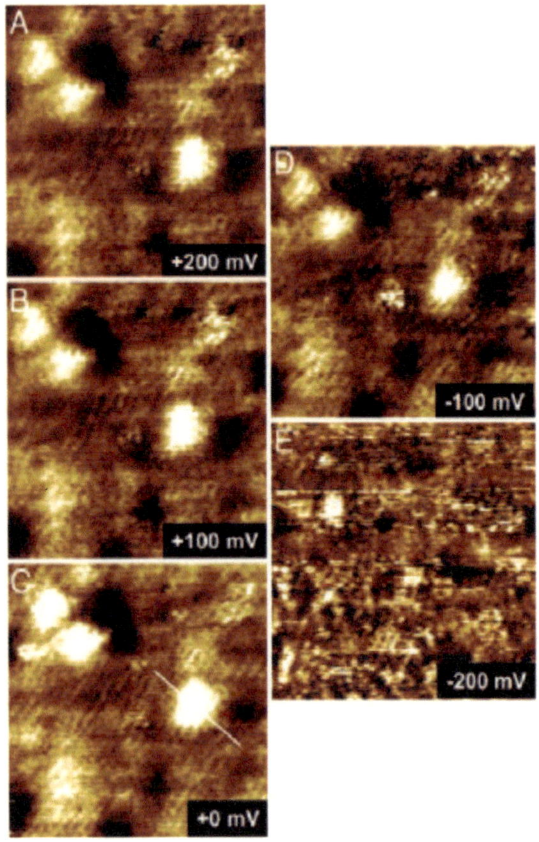

Figure 4. A series of STM images showing the *in-situ* observations of the redox gated electron tunneling resonance arising from the single azurin molecules. The images were obtained by using the azurin/thiol linker/Au at a fixed bias voltage of -0.2 V. by the substrate potential by (a) +200 (b) +100 (c) 0 (d) -100 and (e) -200 mV respectively. Scan area is 35x35 nm. Figure reproduced with permission from: ref. 27, © 2005 The National Academy of Sciences.

The single-molecule tunneling contrast is clearly measured by the redox state of the protein; the experimental observations represented in Fig. 5(a-c) can be explained by a two-step electron transfer mechanism in the STM redox process. The energy levels of the substrate electrode, the tip, and a redox molecule located in the substrate-tip gap may all be modified by changing the substrate potential, but the difference between the substrate and tip energy levels will remain constant if the bias voltage is fixed as the experimental conditions applied [56]. As a consequence, the redox level is shifted relative to the substrate and tip Fermi levels. The tunneling current (I_t) is gated by the molecular redox level, which is displayed directly by the changes in STM contrast. In other words, the contrast changes observed are due to the redox gated tunneling resonance. In contrast, the bias voltage will change with changing the substrate potential when the tip potential is fixed. The molecular energy levels can be mostly located either above or below the tip Femi level, resulting in no significant resonant tunneling tuned by the substrate potential.

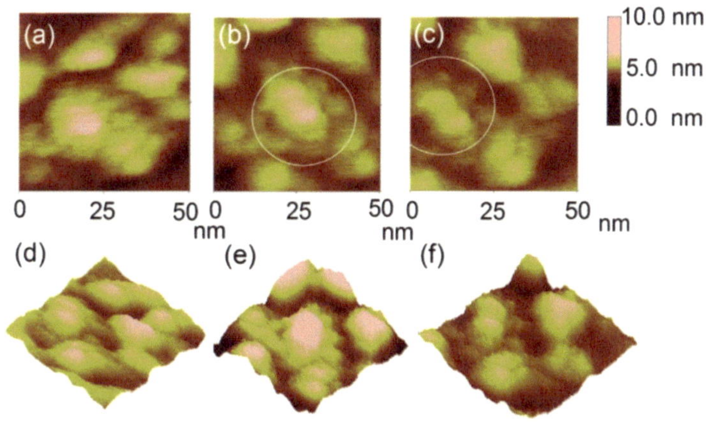

Figure 5. Sequence of ECSTM images obtained for immobilized azurin molecules on Au substrate for different bias potentials; (a) OCP (80 mV) (b) oxidation (486 mV) (c) reduction (278 mV) respectively. Scan size is 50 nm (d-f) is the 3-dimentional profile of (a-c) images respectively. Figure reproduced with permission from: ref. 30, © 2010 Elsevier

The tunneling current (I_t) dependence on the effective overpotential ($e\xi\eta$) and the bias voltage (eV_{bias}) can be analysed in steady-state electron transfer form in the following combination [50]

$$I_t = 2en\frac{k^{o/r}k^{r/o}}{k^{o/r}+k^{r/o}} \tag{2}$$

where $k^{o/r}$ and $k^{r/o}$ are the rate constants, for electron transfer between the tip and protein and between the protein and substrate respectively. n is the number of electrons transmitted in a single electron transfer event and e the electronic charge.

$$k^{o/r} = \kappa_t\rho_t\frac{\omega_{eff}2k_BT}{\alpha_t}\exp\left(-\frac{\left(\lambda - e\xi\eta - e\gamma V_{bias}\right)^2}{4\lambda k_BT}\right) \tag{3}$$

$$k^{r/o} = \kappa_s\rho_s\frac{\omega_{eff}2k_BT}{\alpha_s}\exp\left(-\frac{\left(\lambda - eV_{bias} + e\xi\eta + e\gamma V_{bias}\right)^2}{4\lambda k_BT}\right) \tag{4}$$

where κ_t and κ_s are electronic transmission coefficients for electron transfer between the tip and the protein, and between the protein and the substrate, respectively; ρ_t and ρ_s are the electronic level densities of the tip and the substrate; ω_{eff} is the effective nuclear vibrational frequency, α_t and α_s are the transfer coefficients for electron transfer between the tip and the protein and between the protein and the substrate, respectively; λ is the reorganization free energy; ξ is the fraction of the substrate-solution potential drop, η is the overpotential; γ is the fraction of the bias voltage drop at the site of the molecular redox center. It should be noted that, the parameters ξ and γ represent the electric potential distribution in the ECSTM tunneling gap.

4.2.2. Current-voltage characteristics and transition voltage spectroscopy of individual redox proteins

The measurement of I-V plots is essential for electrical characterization of devices, particularly in redox proteins by electrochemical scanning tunneling microscopy (ECSTM) is essential for the development of bioelectronic devices and biosensors. Further transition voltage spectroscopy (TVS), through which molecular level positions can be examined in molecular devices without applying higher voltages, was utilized for organic monolayer and also for non-redox proteins. For the first time, see [57] TVS spectroscopy for redox proteins which is useful for understanding voltage dependence of molecular conductance is specifically important for the mechanism of ET in redox-active molecule.

ECSTM was employed to measure the I-V characteristics of the redox protein azurin covalently bound to Au substrate in an electrochemical cell under bipotentiostatic control for the probe and sample electrodes versus Ag/AgCl reference electrode on both the reduced and oxidized azurin molecules. I-V measurements were carried out in the tunneling configuration (better known that there is no physical contact between the STM probe and the protein) and the wired configuration (where the probe is in contact with the protein). In tunneling configuration of ECSTM, like normal STS spectroscopy, the I-V curves were obtained by positioning the probe over a region with a high protein surface concentration once the imaging is finished. The I-V curves for the reduced azurin depicts two distinct behavior one relatively linear and the other more rectifying behavior which is not observed on bare Au electrode.

This rectifying behavior is due to the applied bias voltage such as the reduced and oxidized states of azurin. From the I–V characteristics, the conductance (G) of the tunneling gap in the presence of azurin was calculated from the relation $G = I/V$ to be between $10^{-6}\ G_0$ and $10^{-5}\ G_0$. To get better understanding of the obtained I-V curves, TV value that can be used to describe the electrochemical potential dependence of the azurin conductance in the framework of TVS. Individual I-V curves and plotted $\ln(I/V^2)$ versus $1/V$ displayed a minima in the curve, called transition voltage(TV) value, Fig. 6, which is not visible in bare Au samples. Several ET studies of redox molecules have shown that a transition in conductance occurs when the application of an external potential results in the arrangement of the molecular energy levels and the Fermi level of the electrodes. Here, the low TV value found for azurin suggests that the effective barrier for tunneling through a solution is lower than the barrier observed in pure tunneling processes is in agreement with experimental and theoretical works on tunneling through an electrochemical environment [58,59].

In particular, in the context of two-step ET in a redox molecule, a transition was predicted by theory to occur in the range where the effective voltage in the redox center is higher than the reorganization energy of the molecule.

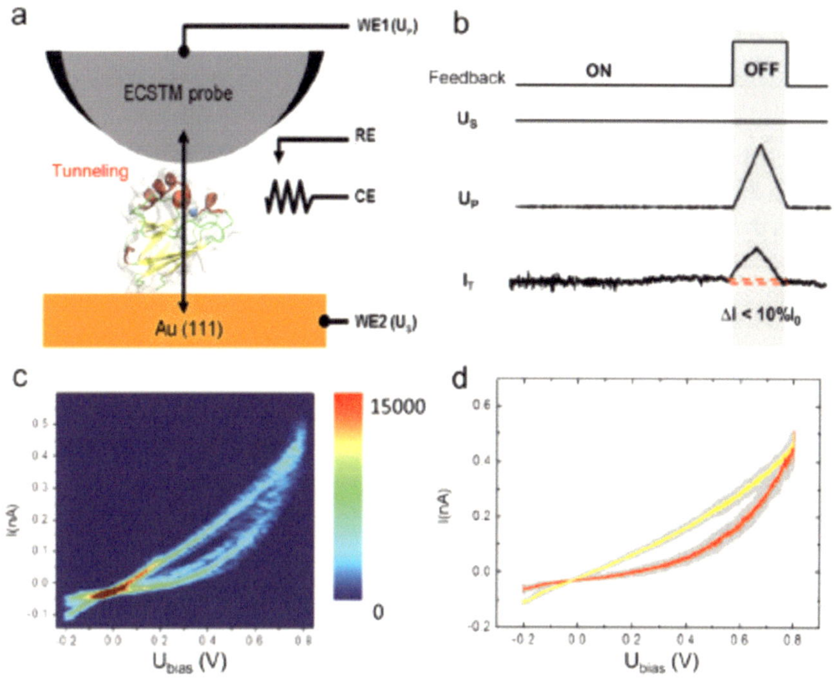

Figure 6. (a) Experimental ECSTM set-up for the azurin immobilized Au structure, WE, working electrode, RE, reference electrode, CE, counter electrode. (b) I-V curves, a triangular ramp was applied to the probe (WE_1) while the feedback is off. The current signal (I_T) was recorded at constant potential (WE_2). (c) 2D I-V histogram showing two populations of curves in azurin on Au. (d) Average of two I-V curves obtained in (c) corresponding the azurin (red curve) and bare Au (yellow curve). Gray error bars indicate the standard deviations. Figure reproduced with permission from: ref. 31, © 2012 ACS

For the determination of TV in the conductance of the redox protein, the tunneling current for the two-step electron transfer mechanism, I_T is given by [60],

$$I_T = e\kappa\rho(eU_{bias})\frac{\omega}{2\pi}\left\{\exp\left[\frac{e}{4\lambda kT}\left(\lambda+\xi\eta+\gamma U_{bias}\right)^2\right]+\exp\left[\frac{e}{4\lambda kT}\left(\lambda+U_{bias}-\xi\eta-\gamma U_{bias}\right)^2\right]\right\}^{-1} \quad (5)$$

in which κ is the electronic transmission coefficient; ρ is the density of states in the metal near the Fermi level; ω is the nuclear vibration frequency; k is Boltzmann's constant; T is the temperature; $U_{bias} = U_P - U_S$ is the potential difference between probe and sample electrodes; λ is the reorganization energy; η is the overpotential, given by $\eta = U_S - U_{Az}$, where U_{Az} is the redox potential of an azurin molecule; and γ and ξ are two model parameters describing the shifts in U_{bias} and η at the redox center, respectively. The parameters γ and ξ are related to the electronic coupling of the molecule with the probe and substrate, respectively

To obtain values of the parameters for azurin, the experimental I-V curves has been fitted using Pobelov and Wandlowski's numerical equation,[61]

$$I_T = 1820\kappa U_{bias}\left\{\exp\left[\frac{9.73}{\lambda}\left(\lambda+\xi\eta+\gamma U_{bias}\right)^2\right]+\exp\left[\frac{9.73}{\lambda}\left(\lambda+U_{bias}-\xi\eta-\gamma U_{bias}\right)^2\right]\right\}^{-1} \quad (6)$$

in which I_T is expressed in nA, potentials are in V, and λ is in eV. In this expression, typical values for ω in a liquid and ρ in a metal were used, and ξ, γ, λ, and γ were left as free model parameters. In the wired junctions, a TVS spectrum is also determined but a negative low value was determined. This minimum is due to the stronger coupling with the probe electrode, which lowers the energy barrier between the levels of the STM probe electrode and the molecule. Thus, in wired junctions, the TV is related to the contact resistance, as commonly found for single-molecule junctions. Hence, these measurements help in characterizing redox proteins and understanding their performance in biological ET chains and molecular electronic devices.

5. Applications of ECSTM

5.1. Combined instrument for electrochemical scanning tunneling microscopy (ECSTM) and scanning electrochemical microscopy (SECM)

Until now, we have analyzed the advantages of ECSTM but there are many versatile techniques such as scanning electrochemical microscopy (SECM) which is also a useful tool to analyze interfacial physicochemical processes. The main reason, ECSTM is able to localize electrochemical reactivity only if this is accompanied by changes in sample topography, e.g. for deposition, adsorption or dissolution reactions. Parallel to the advance of ECSTM, scanning electrochemical microscopy (SECM) was developed as a new tool for localized electrochemistry [62,63]. SECM uses an ultramicroelectrode (UME) as a probe and by either amperometric/potentiometrically utilized to investigate the activity and/or topography of an interface on a localized scale. The tip moves much longer distance than the tunneling distance and measures faradic currents that results from an electron transfer reaction at UME [64,65]. Several operational modes of SECM have been developed which allows the investigation of local chemical properties of interfaces. To study biomolecular interfaces these are most important methods which can be utilized for the analysis.

a. *Tip detection (or collection) method:* Local variations in concentrations about an interface can be mapped with both potentiometric and amperometric probes. In this method, the detector probe is generally assumed to be passive (i.e. non-perturbing to the interfacial process) [66]. Tip detection measurements have demonstrated particularly powerful in identifying localized transport pathways in synthetic membranes and biological tissues. Also it is useful for the investigation of the activity of immobilized enzymes [67]. However, mass transport between the tip and the surface, under tip detection conditions, is complicated.

b. *Feedback mode:* SECM involves in the usage of the tip to locally perturb an interfacial process, by electrolysis or ion transfer, and determine the kinetic effect from the resulting tip current. A mediator is added (concentration in mM) to the supporting electrolyte that is converted at the UME under diffusion controlled conditions which produces a steady-state current, $I_{T,\infty}$, in the bulk phase of the solution. If the UME is brought close to an inert and insulating surface, it blocks the diffusion of the mediator to the UME and the UME current I_T decreases below $I_{T,\infty}$ (called negative feedback) [68]. If the UME is brought above a conductive surface or a catalytically active surface, I_T increases above the value found for an inert and insulating surface. The magnitude of this increase depends on the local reactivity of the sample. A diffusion-controlled reaction at the sample and the tip constitute an important limiting and is called as positive feedback [69].

c. *Advanced tip positioning:* SECM tip usually needs to be positioned close to an interface with high precision. Accurate positioning is achieved by attaching the tip to piezoelectric translators. However, this still leaves the problem of determining the exact separation of the tip electrode and the surface commonly known as *'distance of closest approach'* of the electrode with the surface [70]. One can use the amperometric response of the tip electrode in some instances for many systems it might be difficult to add a redox-active species to the solution, without affecting the process or the viability of the sample. Also there are challenges that include in low analytes concentrations or background processes in biological media which means it is hard to measure the distance accurately from the amperometric response [71]. Hence, much effort has been directed towards the development of alternate procedures for tip positioning and distance determination.

Shear force modulation is one method to achieve the control of tip-sample separation by shaking the electrode through a small oscillation in the *x-y* plane. As the electrode is brought close to a surface, the oscillation is damped, to a degree which depends on the tip-substrate separation. Images are usually acquired at constant damping amplitude, which resembles to a constant distance between the tip and substrate; thus, the tip follows the surface contours [72]. Further, Tip position modulation SECM refers to an operation where an amperometric tip is oscillated in a sinusoidal motion perpendicular to the surface. The resulting current varies with the frequency of the driving oscillation. The amplitude and phase of the oscillating current enable one to deconvolute the activity and topography of the surface [73]. The phase of the current is the same as the phase of the tip-surface separation when the probe is oscillated above an inert surface, whereas they are entirely out of phase above a conducting surface (in positive feedback mode).

d. *SECM with ECSTM*: SECM coupled ECSTM has the dual benefits of nanometer scale resolution imaging along with the ability of the electrochemical measurements. For example see [74]. In order to perform measurements with SECM at nanometer scale many number of technical hurdles has to overcome. Some of them are 1) preparation of suitable nanoscale electrodes with insulator coatings 2) scanning and positioning the probe above the sample surface with a sample distance of 10 nm which is much larger than the tunneling distance but smaller than some electrode radii of the probe, while avoiding mechanical contact between sample and probe. In order to observe electrochemical reactivity at individual nanometer-sized features, such features have to be prepared on the sample surface with such a large distance that SECM can determine the signals of individual features. But it is very difficult to avoid the possibility of mechanical contact between the probe and the surface structures. Hence a method is adopted in which the probe is used in ECSTM mode over the protruding regions of the sample surface and retracted from there [75]. A novel instrumentation has been developed for the imaging with ECSTM and SECM with Pt/Ir wire coated with paint is used as the tip. The tip is kept at a working distance of 20 nm which is much larger than the tunneling distance but enough for feedback imaging for SECM. Probes for the operation of ECSTM and SECM is the most important factor so selecting a desired probe is crucial for better imaging and electrochemical analysis. Tips can be distinguished by the geometric shape of the active electrode area and the insulating sheath for different applications.

Disk-shaped microelectrodes which are the preferred shape for quantitative SECM experiments in the micrometer range, which consists of a Pt wire, sealed in glass or quartz with a laser-heated capillary puller (Fig. 7a) [76]. Because of the large insulating sheath these electrodes are unlikely to function as ECSTM probes. Ring-disk electrodes have been approximated as modified cantilevers for combined scanning force microscopy (SFM)/SECM experiments (Fig. 7b) [77]. The probes have been produced by modification of SFM cantilevers and shaping individual cantilevers by fast ion bombardment. The outer electrode was square-shaped with a side length of 1.5 mm.

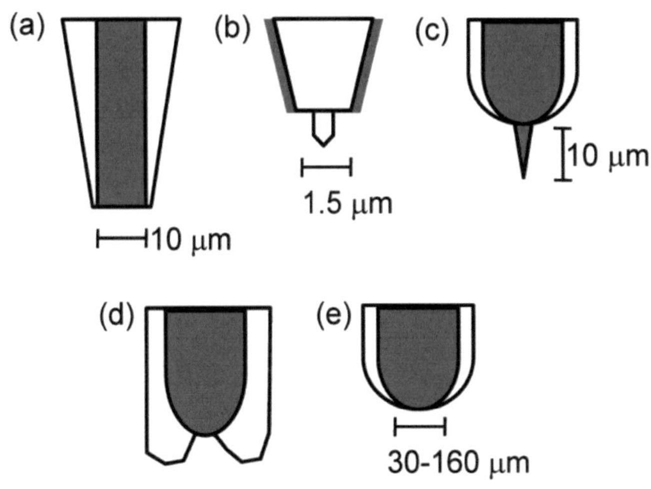

Figure 7. Schematics of the shapes of different microelectrodes used in scanning probe microscopy. Figure reproduced with permission from: ref. 40, © 2010 Elsevier

Electrodes shown in Fig. 7c are generally used for high-resolution ECSTM experiments. The pointed tip allows atomic resolution in tunneling experiments. However, the active electrode area is decreased by an insulating coating which leaves the pointed area of about 10 mm length open. This is sufficient because the potentials for the imaging experiments are selected such that no Faradic reactions proceed at the tip potential. The tunneling current can be as high as 1 nA so that other currents do not interfere significantly with the experiments. Such electrodes are, however, not useful for SECM feedback experiments. Mediators can access the UME by

http://dx.doi.org/10.5772/57236

diffusing parallel to the sample so that no diffusional blocking occurs above passivated samples [78]. The positive feedback would be insignificant compared with the large background current from mediator conversion at such electrodes. Electrodes with a slightly recessed active electrode area (Fig. 7d) were used for amperometric single molecule detection. No images were recorded with such probes. For the combined ECSTM/SECM operation the electrodes such as in Fig. 7e are required. Although they are not as pointed as the one in Fig. 7c, they should allow ECSTM experiments of flat samples although not with atomic resolution. Apart from that, the shallow cone would still provide a blocking of the mediator diffusion above passivated samples [79,80].

Operating principle:

The combined ECSTM/SECM operation is based on the sequential acquisition of ECSTM and SECM [74] data shown in Fig. 8. Once the probe is brought in tunneling contact, an ECSTM image is recorded in the constant current mode. It provides topographic data of the sample. After completion of the ECSTM scan, the electronic feedback loop is switched off and the probe is retracted 20 nm from the working point of the ECSTM scan. This distance is much larger than the tunneling distance. At the same time the potentials at the probe are switched for a desired range to obtain the redox properties of the adsorbed molecule on the surface. After completing the SECM scan, the probe is brought back into tunneling contact with the sample and a step perpendicular to the high-frequency scan axis is performed. From there the sequence is repeated until a full image frame is recorded.

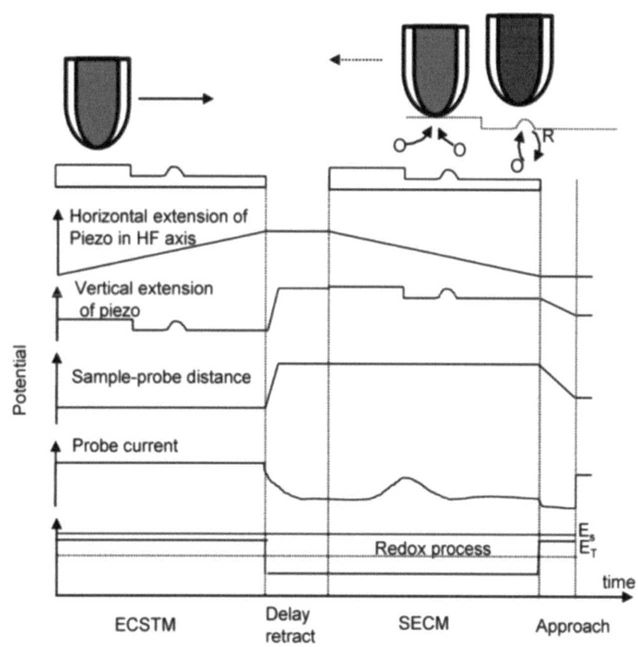

Figure 8. Schematic diagram for the combined operation of ECSTM-SECM performance. Figure reproduced with permission from: ref. 40, © 2010 Elsevier

5.2. Limitations of ECSTM

ECSTM has several limitations and drawbacks, generally these limitations resides in the difficulty to perform bias-dependent measurements such as when potential of the substrate is fixed during electrochemical process then the tip is adjusted to maximize the faradic current and to optimize STM measurements, with the tunnel voltage is no longer adjusted over long range. Hence in STS measurements, voltage dependent imaging is not possible to measure in ECSTM mode [81]. Limitations are also arises from the possible interference of the tip with the electrochemical process at the working electrode. The close proximity of the tip causes shielding effects for reaction at the sample-solution interface.

Further, corrosion studies by ECSTM are also facing problems as high rate of mass transport with respect to time for acquisition of images thus leads to low resolution images. Apart from these limitations, ECSTM is an effective tool which can provide valuable information related to molecular structures of the adsorbed molecules at nanoscale and also provide information relating to corrosion process like adsorption, dissolution and localized corrosion of metallic materials.

5.2.1. Instability and drift

Repeated electrochemical experiments on the all the coated tips leads two typical instability behaviors. A progressive increase of the maximum current is observed on aging, which is associated with dissolution of the coated layer that leaves more and more metal surface uncoated in time. Secondly, all the current levels at each potential drift in time toward higher values, which means that the effective electrical resistance at the tip due to coating decreases. Probably, the solution was leaking into the interface between metal and coating [82].

Mechanical drift is a critical parameter in high resolution imaging caused by the differential thermal expansion of individual instrument components. Sources of drift in the X-Y plane consist of these thermal effects and hysteresis in the piezo-scanner. The Z-direction presents a more complex situation due to the contributions of temperature, sample tilt in the X-Y plane, and hysteresis. Further, in the cases of *in-situ* and electrochemical experiments where drift is typically higher [83], the effects of time and evaporation of the solvent must be carefully balanced. The influence of drift in the X-Y plane detrimentally affects two main aspects of the imaging process. First, the ability to image a given region or feature over time is limited. Second, high-resolution imaging of atomic and molecular lattices demonstrates curvature in the observed periodic structure. The analysis of drift in the X-Y plane is carried out by the collection of sequential images [84].

The major effects of drift in the Z-direction are most disruptive during surface spectroscopic measurements. In the STM application, tunnel current versus distance curves (I-S) collected at a constant bias voltage provide a direct measurement of the effective local barrier height involved in the tunneling process. This value is generally unknown and directly influences the more common bias-dependent (I-V) spectroscopy. It is therefore desirable to minimize differential drift in the Z-direction between the tip and sample. Measurement of drift in the Z-direction was quantified through long-term acquisition of the Z-signal in the tunneling condition.

5.2.2. Poorly resolved images

High-resolution imaging has been the primary feature that attracted the researcher's attention to scanning probe microscopy yet there are still a number of outstanding questions regarding this function of scanning tunneling microscopes or coupled with other microscopic techniques. Some types of proteins may adhere to the tip. This will reduce resolution giving "fuzzy" images. If tip contamination is suspected to be a problem, it will be necessary to protect the tip against such contamination [85]. The ECSTM images become unclear and noisy at potentials more negative potentials due to the perturbation of the tip caused by severe reactions. A great quantity of spots will be observed on all over the surface. In addition, it is know that on the surface of graphite substrate the alkane molecules are assembled in flat-lying lamellar structures. The alkane adsorbates on graphite were analyzed by means of a droplet of saturated alkane solution that is deposited on graphite surface and the metallic tip penetrates this droplet and a molecular adsorbate at the liquid-solid interface until it detects a tunneling current. At these conditions the tip is scanning over the ordered molecular layer in immediate vicinity of the substrate. A specific feature of the STM imaging at the liquid-solid interface is that the probe is surrounded by the alkane saturated solution [86]. Any instability of the imaging and the use of low tunneling gap resistance cause a mechanical damage of the alkane order, and the probe might record the image of the underlying graphite. If the gap is increased again the alkane order is restored. It is difficult to get STM images of "dry" alkane layers on graphite because an occasional damage of the layer is not repairable. This progress relies on instrumental improvements (better signal-to-noise characteristics, low thermal drift, improved detection and control of the tip-sample forces, etc.) and the use of sharp probes [87].

The other issue is related to the better understanding of the nature of atomic-scale resolution in STM. In some cases, imaging provides only the lattice resolution in the contrast with true atomic resolution where a detection of such defects is expected. The imaging of the periodical lattices with the defects can be demonstrated with the results of the computer simulation which revealed that visualization of the defects does not necessarily mean that the surrounding molecular order is correctly reproduced in the images. These findings emphasize a need of a thorough interaction between the experiment and theory in the analysis of the atomic scale data.

6. Future trends and applications

The invention of the scanning tunneling microscope had a revolutionary influence on the development of material characterization at nanometer scale. The future of STM with electrochemical control depends on the development of appropriate probes for implementing the electrochemically controlled current sensing atomic force microscopy (ECAFM) [88]. In comparison of ECAFM and ECSTM, it can be said that both methods can measure the electronic properties of single molecules but ECAFM allows the control of force applied by the tip on the molecules adsorbed on the surface, thus allowing measure the mechanical properties of the single molecules. In comparison with STM, Atomic force microscopy (AFM) doesn't require conductive samples and tips. AFM has the advantages of measuring the local forces between the tip and the sample surface, including van der Waals, Born repulsion, electrostatic and magnetic forces, friction and adhesion [89]. In electrochemical applications, ECAFM is often preferred because it is easier to set-up and the obtained topographic information is independent of the conductivity of corrosion materials. Moreover, ECAFM is most often used at submicron level, i.e., at a lower level of spatial resolution that does not require preparation of atomically smooth surfaces as for ECSTM studies. Further, ECAFM can also be combined with a variety of other techniques to analyze corrosion and optimize corrosion protection properties [90]. However in comparison with ECSTM, even though ECAFM doesn't require atomically smooth surfaces like ECSTM and also in ECAFM the cantilever-tip assembly doesn't constitute a fourth electrode through which the electrochemical current that would flow, but there is a disadvantage that in the resonant contact mode, the fluid medium tends to damp the normal resonance frequency of the cantilever which is complicated. Further, the noncontact mode is impractical because the van der waals forces are even smaller making it as a big drawback for biological applications. Such instrument would enable better molecular investigation of electron transfer properties thereby maintaining the molecular conformation. To achieve this spectroscopy, specially designed probes such as insulated conductive probes along with bipotentiostat setup essential. The capability of recording atomic scale features with STM has increased the scope of research to develop other types of STM which provide information about the topography and mechanical, magnetic, electrical properties of the surfaces [91]. The development of these new applications is the design of specific probes, having improved spatial resolution, sensing the desired sample properties and operating in different environments. Developing a highly flexible, compact microscopic system can be easily interfaced with commercial control electronics and integrated with inverted optical microscopy. Simple design architecture enables simple exchange of the probe tip, sample, and scanner while enhancing system resonances and resistance to thermal drift. In recent years, the demand of the applications of ECSTM has greatly increased in numerous fields [92]. ECSTM in combination with SECM has the ability to perform local reactivity imaging simultaneously with ECSTM imaging as well as to induce local electrochemical surface modification in the same setup opens up new perspectives for the investigation of heterogeneous reactions in electrocatalysis at metal clusters and in corrosion processes in a new size regime. Hence with the help of STM, complemented by other characterization techniques, it is reasonable to believe that new advance in building specific and functional surfaces can be achieved in the future.

Author details

Ajay Kumar Yagati[1], Junhong Min[3] and Jeong-Woo Choi[1,2*]

*Address all correspondence to: jwchoi@sogang.ac.kr

1 Research Center for Integrated Biotechnology, Sogang University, Seoul, Republic of Korea

2 Department of Chemical and Biomolecular Engineering, Sogang University, Seoul, Republic of Korea

3 School of Integrative Engineering, Chung-Ang University, Seoul, Republic of Korea

References

[1] Eigler D.M, Schweizer E.K., Positioning single atoms with a scanning tunnelling microscope. Nature 1990; 344, 524-526.

[2] Stroscio J.A, Eigler D.M., Atomic and Molecular Manipulation with the Scanning Tunneling Microscope. Science 1991; 254, 1319-1326.

[3] Binnig G, Rohrer H. Scanning tunneling microscopy. Surface science 1983; 126(1-3) 236-127.

[4] Stolyarova E, Rim K.T, Ryu S, Maultzsch J, Kim P, Brus L.E, Heinz T.F, Hybertsen M.S, Flynn G.W, High-resolution scanning tunneling microscopy imaging of mesoscopic graphene sheets on an insulating surface. PNAS 2007; 104, 9209-9212.

[5] Heim, M, Eschrich, R, Hillebrand, A, Knapp, H.F, Guckenberger, R, Cevc, G. Scanning tunneling microscopy based on the conductivity of surface adsorbed water. Charge transfer between tip and sample via electrochemistry in a water meniscus or via tunneling? Journal of Vacuum Science & Technology B: Microelectronics and Nanometer Structures, 1996; 14, 1498-1502.

[6] Cui, X. D, Primak A, Zarate X, Tomfohr J, Sankey O.F, Moore A.L, Gust D, Harris G, Lindsay S.M, Reproducible Measurement of Single-Molecule Conductivity. Science 2001; 294, 571-574.

[7] Hallmark, V.M, Chiang s, Rabolt, J.F, Swalen, J.D, Wilson R.J. Observation of Atomic Corrugation on Au(111) by Scanning Tunneling Microscopy. Phys. Rev. Lett. 1987; 59, 2879-2882

[8] Bonnell D.A. Scanning tunneling microscopy. Encyclopedia of Materials: Science and Technology 2001, 8269-8281.

[9] Garcia R, Yuqiu J, Schabtach E, Bustamante C. Deposition and imaging of metal-coated biomolecules with the STM. Ultramicroscopy 1992; 42-44, 1250-1254.

[10] Itaya K. In situ scanning tunneling microscopy in electrolyte solutions. Progress in Surface Science 1998; 58, 121-247.

[11] Wang E. Electrochemical scanning tunneling microscopy. Analytical sciences 1994, 10 (1) 155-156.

[12] Binnig G, Rohrer H, Gerber Ch, Weibel E. (111) facets as the origin of reconstructed Au(110) surfaces. Surface Science letters 131 (1) L379–L384.

[13] Sobczyk S, Donarini A, Grifoni M. Theory of STM junctions for π-conjugated molecules on thin insulating films Phys. Rev. B 85, 2012; 205408 (17pp)

[14] Van de leemput, L E C, van Kempen H. Scanning tunnelling microscopy, Rep. Prog. Phys.1992; 55, 1165-1240.

[15] Frommer, J. Scanning tunneling microscopy and atomic force microscopy in organic chemistry. Angew. Chem.1992; 31(10) 1298-1328.

[16] Gross L, Mohn F, Moll N, meye G, Ebel R, Abdel-Mageed W.M, Jaspers M. Organic structure determination using atomic-resolution scanning probe microscopy. Nature Chemistry 2010, 2, 821-825.

[17] Kim Y.-T., Bard, A. J. Imaging and etching of self-assembled n-octadecanethiol layers on gold with the scanning tunneling microscope, Langmuir 1992; 8, 1096-1102.

[18] Halbritter J, Repphun G, Vinzelberg S, Staikov G, Lorenz W. J. Tunneling mechanisms in electrochemical STM -distance and voltage tunneling spectroscopy. Electrochimica Acta, 1995; 40, 1385-1394.

[19] Esplandiu M.J, Carot M.L, Cometto F.P, Macagno V.A, Patrito E.M. Electrochemical STM investigation of 1,8-octanedithiol monolayers on Au(1 1 1).: Experimental and theoretical study. Surface science 2006; 600(1) 155-172.

[20] Gitting D. I, Bethell D, Schiffrin D. J, Nichols R. J. A nanometre-scale electronic switch consisting of a metal cluster and redox-addressable groups. Nature 2000; 408 67-69.

[21] Dieluweit S, Giesen M. Determination of step and kink energies on Au(100) electrodes in sulfuric acid solutions by island studies with electrochemical STM. J. Electroanal. Chem. 2002; 524-525, 194-200.

[22] Funtikov, A.M, Linke U, Stimming U, Vogel R. An in-situ STM study of anion adsorption on Pt(111) from sulfuric acid solutions. Surface Science 1995; 324, L343-L348.

[23] Kunze, J, Strehblow H.-H, Staikov G. In situ STM study of the initial stages of electrochemical oxide formation at the Ag(1 1 1)/0.1 M NaOH(aq) interface. Electrochem. Commun. 2004; 6, 132-137.

[24] Eckhard K, Chen X, Turcu F, Schuhmann W. Redox competition mode of scanning electrochemical microscopy (RC-SECM) for visualisation of local catalytic activity. Phys. Chem. Chem. Phys., 2006; 8, 5359-5365.

[25] Zhang J, Christensen H.E.M, Ooi B.L, Ulstrup J. In situ STM imaging and direct electrochemistry of Pyrococcus furiosus ferredoxin assembled on thiolate-modified Au(111) surfaces. Langmuir 2004; 20(23) 10200-10207.

[26] Pan J, Jing T.W, Lindsay S.M. Tunneling Barriers in Electrochemical Scanning Tunneling Microscopy. J. Phys. Chem., 1994; 98, 4205-4208.

[27] Pobelov I. V, Li Z, Wandlowski T. Electrolyte gating in redox-active tunneling junctions-An electrochemical STM approach. J. Am. Chem. Soc., 2008; 130(47) 16045–16054.

[28] Albrecht T, Moth-poulsen K, Christensen J.B, Hjelm J, Bjornholm T, Ulstrup J. Scanning tunneling spectroscopy in an Ionic Liquid. J. Am. Chem. Soc., 2006; 128 (20), 6574-6575.

[29] Alessandrini A, Facci P. Electrochemical Scanning Tunneling Microscopy and Spectroscopy for Single-Molecule Investigation, Cellular and Subcellular Nanotechnology 2013; 261-273

[30] Zhu L, Claude-Montigny B, gattrell M. Insulating method using cataphoretic paint for tungsten tips for electrochemical scanning tunnelling microscopy (ECSTM). Applied Surface Science 2005; 252, 1833-1845.

[31] Zhang B, Wang E. Fabrication of STM tips with controlled geometry by electrochemical etching and ECSTM tips coated with paraffin. Electrochimica Acta 1994; 39, 103-106.

[32] Guell A. G. Biez-perz I, Gorostiza P, Sanz F. Preparation of reliable probes for electrochemical tunneling spectroscopy. Anal. Chem., 2004; 76 (17), 5218-5222.

[33] Hudson J. E., Abruna h. D. STM and ECSTM Study of the Formation and Structure of Self-Assembling Osmium Complexes on Pt(111). J. Phys. Chem., 1996, 100 (3), 1036-1042.

[34] Díez-Péreza, I, Sanza, F, Gorostizab, P. In situ studies of metal passive films. Curr. Opin. Solid State Mater. Sci. 2006; 10, 144-152.

[35] Ghali E, Dietzel W, Kainer K.-U. General and localized corrosion of magnesium alloys: A critical review. Journal of Materials Engineering and Performance 2004; 13(1) 7-23.

[36] Maurice V, Klein L. H. Marcus P. Atomic Structure of Metastable Pits Formed on Nickel. Electrochem. Solid-State Lett. 2001; 4, B1-B3.

[37] Bolzoni F, Contreras G, Fumagalli G, Lazzari L, Re G. Localized Corrosion: An Empirical Approach to the Study of Passive Film Breakdown Rates. Corrosion2013; 69, 352-363.

[38] Burstein, G. T, Liu, C, Souto, R. M, Vines, S. P. Origins of pitting corrosion. Corrosion Engineering, Science and Technology, 2004; 39(1) 25-30.

[39] Szklarska-Smialowska, Z. Pitting corrosion of aluminum. Corrosion science 1999; 41, 1743-1767.

[40] Kabasakaloğlu K, Kiyak T, Sendil O, Asan A. Electrochemical behavior of brass in 0.1 M NaCl. Appl. Surf. Sci. 2002; 193(1-4) 167-174.

[41] Szklarska-Smialowska Z. Mechanism of pit nucleation by electrical breakdown of the passive film. Corrosion Science 2002; 44(5) 1143-1149.

[42] Rangel C.M, Silva T.M, da Cunha Belo M. Semiconductor electrochemistry approach to passivity and stress corrosion cracking susceptibility of stainless steels. Electrochimica Acta 2005; 50(25-26) 5076-5082.

[43] Ye, C-Q., Hu, R.-G, Li, Y., Lin, C.-J., Probing the vertical profiles of potential in a thin layer of solution closed to electrode surface during localized corrosion of stainless steel. Corrosion Science 2012; 61, 242-245.

[44] Zamborni F. P, crooks R. M. Corrosion passivation of gold by n-alkanethiol self-assembled monolayers: Effect of chain length and end group. Langmuir, 1998; 14(12), 3279-3286.

[45] Marcus P. Surface science approach of corrosion phenomena. Electrochim. Acta 1998; 43, 109-118.

[46] Yagati A.K, Jung M,Kim S.-U, Min J, Choi J.-W. Nanoscaled redox active protein adsorption on Au-dot arrays: An electrochemical scanning probe microscopic investigation for application in nano-biodevices. Thin Solid films 2009; 518(2) 634-637.

[47] Tang L, Li X, Cammarata R.C, Friesen C, Sieradzki K. Electrochemical stability of elemental metal nanoparticles. J. Am. Chem. Soc., 2010, 132 (33), 11722–11726.

[48] Sripirom J, Kuhn S, jung U, Magnussen O, Schulte A. Pointed carbon fiber ultramicroelectrodes: A new probe option for electrochemical scanning tunneling microscopy. Anal. Chem., 2013, 85 (2), 837-842

[49] Tao N.J, Cardenas G, Cunha F, Shi Z. In Situ STM and AFM study of protoporphyrin and Iron (III) and Zinc (II) protoporphyrins adsorbed on graphite in aqueous solutions. Langmuir, 1995, 11(11), 4445-4448.

[50] Chi Q, Farver O, Ulstrup J. Long-range protein electron transfer observed at the single-molecule level: In situ mapping of redox-gated tunneling resonance. PNAS 2005; 102(45) 16203-16208.

[51] Yagati A.K, Lee T, min J, Choi J.-W. A robust nanoscale biomemory device composed of recombinant azurin on hexagonally packed Au-nano array. Biosensors and Bioelectronics 2013; 40(1) 283-290.

[52] Artes, J.M, Diez-perez I, Sanz F, Gorostiza P. Direct measurement of electron transfer distance decay constants of single redox proteins by electrochemical tunneling spectroscopy. ACS Nano, 2011, 5 (3), 2060-2066.

[53] Alessandrini A, Corni S, Facci P. Unravelling single metalloprotein electron transfer by scanning probe techniques. Phys. Chem. Chem. Phys., 2006; 8, 4383-4397.

[54] Alessandrini A, Gerunda M, Canter, G.W, Verbeet, M.Ph., Facci, P. Electron tunnelling through azurin is mediated by the active site Cu ion. Chem. Phys. Lett., 2003; 376, 625-630.

[55] Kim S.U, Yagati A.K, Min J, Choi J.-W. Nanoscale protein-based memory device composed of recombinant azurin. Biomaterials 2010; 31(6) 1293-1298.

[56] Petrangolini P, Alessandrini A, Berti L, Facci P. An Electrochemical Scanning Tunneling Microscopy Study of 2-(6-Mercaptoalkyl)hydroquinone Molecules on Au(111). J. Am. Chem. Soc., 2010; 132, 7445-7453.

[57] Artes J.M, Lopez-Martinez M, Giraudet A, Diez-perz I, Sanz F, Gorostiza P. Current-voltage characteristics and transition voltage spectroscopy of individual redox proteins. J. Am. Chem. Soc., 2012, 134 (50), 20218-20221.

[58] Nagy G, Wandlowski T. Double Layer Properties of Au (111)/H2SO4 (Cl) + Cu^{2+} from Distance Tunneling Spectroscopy. Langmuir, 2003; 19, 10271-10280

[59] Wigginton, N.S, Rosso, K. M, Stack A.G. Hochella Jr, M.F. Long-Range Electron Transfer across Cytochrome–Hematite (α-Fe_2O_3) Interfaces. J. Phys. Chem. C, 2009; 113, 2096-2103.

[60] Kuznetsov A.M, Ulstrup J. Theory of electron transfer at electrified interfaces. Electrochimica Acta 2000; 45(15-16) 2339-2361.

[61] Pobolev I.V, Li Z, Wandlowski T. Electrolyte gating in redox-active tunneling junctions-An electrochemical STM approach. J. Am. Chem. Soc., 2008, 130 (47), 16045-16054.

[62] Mirkin M.V. Horrocks B.R Electroanalytical measurements using the scanning electrochemical microscope. Anal.Chim. Acta 2000; 406(2) 119-146.

[63] Wittstock G. Modification and characterization of artificially patterned enzymatically active surfaces by scanning electrochemical microscopy. Fresenius' Journal of Analytical Chemistry 2001; 370 (4) 303-315.

[64] Bollo S, Jara-Ulloa P. Finger S, Nunez-Vergara, L.J, Squella, J.A. Scanning electrochemical microscopy (SECM) study of superoxide generation and its reactivity with 1,4-dihydropyridines. J. Electroanal. Chem. 2005; 577, 235-242.

[65] Bauermann, L. P, Schuhmann, W, Schulte, A. An advanced biological scanning electrochemical microscope (Bio-SECM) for studying individual living cells. Phys. Chem. Chem. Phys., 2004; 6, 4003-4008.

[66] Macpherson J.V. Unwin P.R. Scanning electrochemical microscopy as an in vitro technique for measuring convective flow rates across dentine and the efficacy of surface blocking treatments. Electroanalysis 2005;17(3) 197-204.

[67] Wang, Y. Kececi, K. Velmurugan, J. Mirkin, M.V. Electron transfer/ion transfer mode of scanning electrochemical microscopy (SECM): a new tool for imaging and kinetic studies. Chem. Sci., 2013; 4, 3606-3616.

[68] Macpherson, J.V., Slevin, C.J., Unwin, P.R. Probing the oxidative etching kinetics of metals with the feedback mode of the scanning electrochemical microscope. J. Chem. Soc., Faraday Trans., 1996; 92, 3799-3805.

[69] Turcu F, Schulte A, Hartwich G, Schuhmann W. Imaging immobilised ssDNA and detecting DNA hybridisation by means of the repelling mode of scanning electrochemical microscopy (SECM). Biosensors and Bioelectronics 2004; 20(5) 925-932.

[70] Lu, X., Wang, Q., Liu, X. Review: Recent applications of scanning electrochemical microscopy to the study of charge transfer kinetics. Anal. Chim. Acta. 2007; 601, 10-25.

[71] Gonsalves M, Barker A L, Macpherson J V, Unwin P R, O'Hare D, Winlove C P. Scanning electrochemical microscopy as a local probe of oxygen permeability in cartilage. Biophys. J. 2000; 78(3) 1578-1588.

[72] Katemann B.B, Schulte A, Schuhmann W, Constant-Distance Mode Scanning Electrochemical Microscopy (SECM)—Part I: Adaptation of a Non-Optical Shear-Force-Based Positioning Mode for SECM Tips. Chem. Eur. J.,2003; 9: 2025-2033.

[73] nebel, M., Eckhard, K., Erichen T., Schulte, A., Schuhmann, W. 4D Shearforce-Based Constant-Distance Mode Scanning Electrochemical Microscopy. Anal. Chem., 2010, 82, 7842–7848.

[74] Treutler T.H, Wittstock G. Combination of an electrochemical tunneling microscope (ECSTM) and a scanning electrochemical microscope (SECM): application for tip-induced modification of self-assembled monolayers. Electrochimica Acta 2003; 48(20-22) 2923-2932.

[75] Lugstein, A., Bertagnolli, E., Kranz, C. Kueng, A., Mizaikoff, B. Integrating micro- and nanoelectrodes into atomic force microscopy cantilevers using focused ion beam techniques. Appl. Phys. Lett. 2002; 81, 349 (3pp).

[76] Katemann B.B, Schuhmann W. Fabrication and characterization of needle-type. Electroanalysis, 2002;14: 22-28.

[77] Kranz C, Friedbacher G, mizaikoff B. Integrating an ultramicroelectrode in an afm cantilever: combined technology for enhanced information. Anal. Chem., 2001; 73 (11) 2491-2500.

[78] Neufeld, A.K., O'Mullane, A.P. Effect of the mediator in feedback mode-based SECM interrogation of indium tin-oxide and boron-doped diamond electrodes. J Solid State Electrochem. 2006; 10, 808-816.

[79] Szamocki R, Velichko A, Holzapfel C, mucklich F, Ravaine S, Garrigue P, Sojic N, Hempelmann R, Kuhn A. Macroporous ultramicroelectrodes for improved electroanalytical measurements. Anal. Chem., 2007; 79(2), 533-539.

[80] Hermans A, Wightman M. Conical tungsten tips as substrates for the preparation of ultramicroelectrodes. Langmuir, 2006; 22 (25), 10348-10353.4

[81] Sachs C, Hildebrand M, Volkening S, Ertl G, Spatiotemporal self-organization in a surface reaction: from the atomic to the mesoscopic scale. Science 2001; 293 (5535) 1635-1638.

[82] Zamborini F.P, Crooks, R.M. In-situ electrochemical scanning tunneling microscopy (ECSTM) study of cyanide-induced corrosion of naked and hexadecyl mercaptan-passivated Au(111). Langmuir 1997; 13(2) 122-126.

[83] Papadantonakis, K.M., Brunschwig, B.S., Lewis, N.S. Use of Alkane Monolayer Templates To Modify the Structure of Alkyl Ether Monolayers on Highly Ordered Pyrolytic Graphite. Langmuir, 2008; 24, 857–861.

[84] Stieg A. Z, Rasool H. I, Gimzewski J.K. A flexible, highly stable electrochemical scanning probe microscope for nanoscale studies at the solid-liquid interface. Review of Scientific Instruments 2008; 79 (10) 103701-7.

[85] Stoll E, Marti O. Restoration of scanning-tunneling-microscope data blurred by limited resolution, and hampered by 1/f like noise. Surface science 1987; 181(1-2) 222-229.

[86] Eigler D.M, Schweizer E.K. Positioning single atoms with a scanning tunnelling microscope. Nature 1990; 344, 524-526.

[87] Reiss G, Bruckl H, Vancea J, Lecheler R, Hastreiter E. Scanning tunneling microscopy on rough surfaces-quantitative image analysis. J. Appl. Phys.1991; 70, 523-525.

[88] Endres, F., Borisenko, N., Abedin, S. Z. E., Hayes, R., Atkin, R. The interface ionic liquid(s)/electrode(s): In situ STM and AFM measurements Faraday Discuss., 2012; 154, 221-233.

[89] Atkin, R., Abedin, S. Z. E., Hayes, R., Gasparotto, L.H.s., Borisenko, N., Endres, F. AFM and STM Studies on the Surface Interaction of [BMP]TFSA and [EMIm]TFSA Ionic Liquids with Au(111). J. Phys. Chem. C, 2009, 113 (30), 13266–13272.

[90] Macpherson, J.V., Unwin, P.R., Combined Scanning Electrochemical−Atomic Force Microscopy. Anal. Chem., 2000, 72 (2), 276–285.

[91] Hai, N.T.M., Huynh, T.M.T., Fluegel, A., Mayer, D., Broekmann, P. Adsorption behavior of redox-active suppressor additives: Combined electrochemical and STM studies. Electrochimica Acta. 2011; 56, 7361-7370.

[92] Gewirth, A.A., Niece, B.K. Electrochemical Applications of in Situ Scanning Probe Microscopy, Chem. Rev. 1997; 97, 1129–1162.

Chapter 4

Recent Trends in the Development of Electrochemical Biosensors for Organophosphorus Pesticides Determination

Margarita Stoytcheva and Roumen Zlatev

Additional information is available at the end of the chapter

http://dx.doi.org/10.5772/58310

1. Introduction

The environmental and public concerns provoked by the indiscriminate use of organophosphorus pesticides (OPs) and the adopted safety standards [1-6] incited the development of new sensitive methods enabling their determination in the nanomole-picomole range. Such analytical performances offer the nanostructured electrochemical biosensors.

The nanotechnological approach to electrochemical biosensing [7-16], due to the electrocatalytical properties of the nanostructures, their action as electron transfer mediators or electrical wires, large surface to volume ratio, structural robustness, and biocompatibility leads to electrode potential lowering, enhancement of the electron transfer rate with no electrode surface fouling, sensitivity increase, stability improvement, and interface functionalization.

In this review are presented the recent trends in the development of nanomaterials based electrochemical biosensors for organophosphorus pesticides determination. Their performance characteristics such as sensitivity, linear range, detection limits, and stability are compared and discussed.

2. OPs determination applying electrochemical biosensors

The electrochemical biosensors, because of the high sensitivity of the determinations, the simplicity of the operational procedure, the availability and the affordable cost of the equipment, are considered as an alternative to the expensive, time-consuming, and sophisticated chromatographic techniques currently applied for OPs quantification [17].

INTECH
open science | open minds

© 2014 The Author(s). Licensee InTech. This chapter is distributed under the terms of the Creative Commons Attribution License (http://creativecommons.org/licenses/by/3.0), which permits unrestricted use, distribution, and reproduction in any medium, provided the original work is properly cited.

The main processes involved in the electrochemical biosensors for OPs quantification are: cholinesterases activity inhibition by OPs or OPs hydrolysis catalyzed by organophosphorus hydrolase, both followed by the conversion of the signal produced by the interaction between the biorecognition element and the analyte into electrical one.

2.1. Inhibition based electrochemical biosensors for OPs quantification

The electrochemical biosensors which take advantage of the inhibitory effect of the OPs on cholinesterases activity have been extensively investigated [18-26]. The first generation of inhibition based electrochemical biosensors involves the following reactions:

$$\text{R-choline} + H_2O \xrightarrow{\text{ChE}} \text{choline} + \text{R-COOH} \tag{1}$$

$$\text{choline} + 2O_2 + H_2O \xrightarrow{\text{ChO}} \text{betaine} + 2H_2O_2 \tag{2}$$

$$2H_2O_2 \rightarrow O_2 + 2H^+ + 2e^- \tag{3}$$

or

$$O_2 + 4e^- + 2H_2O \rightarrow 4OH^- \tag{4}$$

where ChE is acylcholinesterase and ChO is choline oxidase.

The acylcholinesterases catalyzed R-choline hydrolysis (Eq. 1) affected by the enzyme activity inhibition with OPs, is coupled with the choline oxidase catalyzed betaine oxidation (Eq. 2). The current of the oxidation of the produced H_2O_2 (Eq. 3) or the current of the reduction of the consumed O_2 (Eq. 4) is registered as a sensor response and is correlated to the OPs concentration.

Nevertheless, the electrochemical biosensors for OPs determination of first generation present some intrinsic deficiencies: sophisticated design as two enzymes have to be integrated, multistep protocol, and possible interferences at the potential of H_2O_2 oxidation (+0.6 V/SCE), among other. These drawbacks imposed the development of the electrochemical biosensors for OPs determination of second generation. They quantify the cholinesterases inhibition applying a simpler measurement principle, consisting in the monitoring of the electroactive thiocholine formed upon enzymatic hydrolysis of acylthiocholine (Eq. 5):

$$\text{R-thiocholine} + H_2O \xrightarrow{\text{ChE}} \text{thiocholine} + \text{R-COOH} \tag{5}$$

The response generating reaction is the direct thiocholine oxidation (Eq. 6) at a potential of+0.8 V/SCE at conventional metal and graphite transducers [27-31]:

$$2\,\text{thiocholine} \rightarrow \text{dithio-bis-choline} + 2H^{+} + 2e^{-} \tag{6}$$

or the mediated thiocholine oxidation (Eqs. 7 and 8) at lower potentials using chemically modified electrodes [32-40], thus avoiding the interferences:

$$2\,\text{thiocholine} + M_{ox} \rightarrow \text{dithio-bis-choline} + M_{red} \tag{7}$$

$$M_{red} \rightarrow M_{ox} + 2e^{-} \tag{8}$$

where M_{ox} and M_{red} are the oxidized and the reduced forms of the mediator M.

The biosensors based on the inhibitory effects of OPs on cholinesterases activity are very sensitive, but the indirect sensing mechanism they use is associated with some shortcomings such as poor selectivity, an irreversible response, etc. The non-ideal behavior of the enzyme inhibition-based biosensors and biosensing systems for OPs determination is exhaustively commented by Luque de Castro [41].

2.2. Substrate based electrochemical biosensors for OPs quantification

The nitro phenyl-substituted OPs (paraoxon, parathion, methyl parathion, fenitrothion, etc.), and some chemical warfare agents (sarin, soman, tabun, VX, etc.) act as substrates for the enzyme organophosphorus hydrolase (OPH) [42-44]. The enzyme catalyzed hydrolysis of these substances yields p-nitrophenol (PNP). The PNP oxidation current, which is the sensor signal, measured at a fixed-potential is proportional to the OPs concentration. The occurring reactions, selecting paraoxon as a model, are the following:

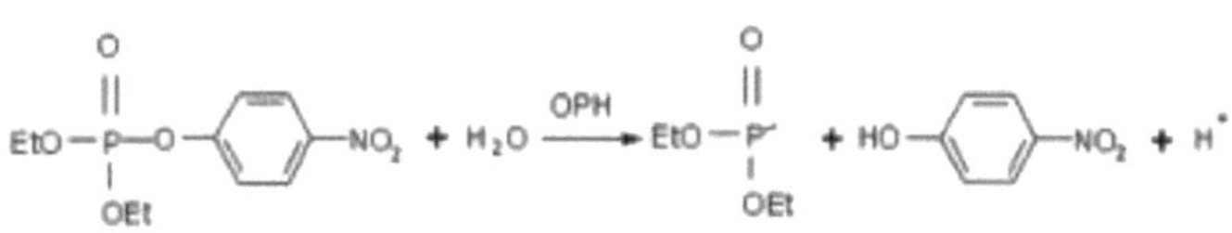

(9)

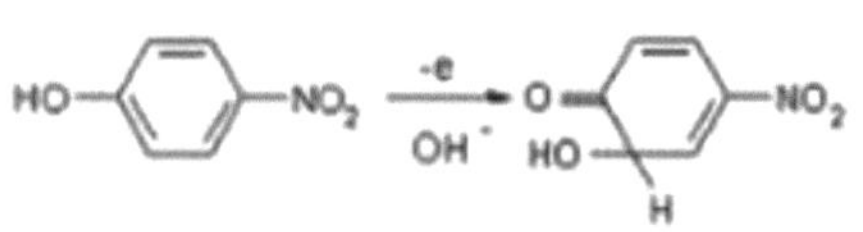

(10)

Thus, the use of OPH is extremely attractive for the direct and selective biosensing of OPs [19, 21, 22, 45-48]. Nevertheless, the sensitivity of the OPH based electrochemical sensors is lower and their detection limits are much higher than those of the inhibition based ones [20, 49]. The PNP oxidation that produces phenoxy radicals which couple to form an insulating polymeric film fouling the electrodes surfaces and inhibiting further phenols oxidation [50-59] and the complex, long-lasting, and expensive procedure for OPH extraction and purification, performed in specialized microbiological laboratories (this enzyme is not commercially available) [21] create additional problems.

3. Nanostructured electrochemical biosensors for OPs quantification

The bibliographical survey covering the period 2010-2013 demonstrated that the predominant part of the recently developed nanostructured electrochemical biosensors for OPs quantification make use of carbon nanotubes (CNTs) or gold nanoparticles (GNPs). Previous studies are revised and reported in the comprehensive reviews of Liu et al. [60] and Periasami et al. [61].

3.1. Carbon nanotubes

Carbon nanotubes are widely used in electrochemical biosensors because of their chemical stability, mechanical strength and stiffness [62, 63], and improved electron transfer properties attributed to the changes in the energy bands close to the Fermi level [64].

The single-walled carbon nanotubes (SWCNTs) display higher surface area and low electrical percolation thresholds in comparison to the multi-walled carbon nanotubes (MWCNTs). Nevertheless, their higher cost and poorer dispersibility limit their application. Moreover, SWCNTs form less regular layers onto the electrodes with a higher deviation of the signal measured. Nevertheless, data reported in the literature [65] demonstrate that OPH covalently immobilized on SWNTs conserves much higher activity than OPH conjugated to MWNTs. This was attributed to the more uniform deposition of OPH on the SWNTs and the formation of a SWNTs network. The dynamic concentration range for paraoxon determination applying SWNT-OPH sensor was found to be in the range 0.5-8.5 μmol L^{-1} with a detection limit of 0.01μg mL^{-1}. In addition, the SWNTs sensor with covalently immobilized enzyme exhibited enhanced solution-storage and operational stability: 25% signal loss over 7 months.

Earlier studies has also shown that the CNT surface modification could be useful for improving the sensitivity and stability of oxidative measurement of phenolic compounds, produced upon OPH catalyzed hydrolysis of OPs. The evaluation of the performances of the SWCNTs and of MWCNTs prepared by chemical vapor deposition (SWCNT-CVD and MWNT-CVD), and by the ARC discharge method (MWNT-ARC) demonstrates that both the SW-and MW-CVD-CNT coated surfaces offer sensitivity enhancement compared to the ARC-CNT and bare electrodes [66]. It was considered that the higher sensitivity of the CVD-CNT-modified electrode reflects differences in the density of edge-plane-like defects that leads to higher electrochemical reactivity [66].

SWCNTs were also used for the development of an inhibition based biosensor for OPs determination, applying a simple protocol. It includes the one-stage deposition of SWCNs and Co phtalocyanine followed by carbodiimide binding of acetylcholinesterase, providing directed coordination of the protein molecule at the terminal carboxylic groups of the oxidized SWCNTs [67]. The biosensor made it possible the detection of 5-50 ppb of paraoxon and 2-50 ppb of malaoxon with detection limits of 3 and 2 ppb, respectively. The amperometric measurements were performed at low potential (0.050 V/Ag, AgCl), thus avoiding the interferences.

The use of pristine MWCNTs and of MWCNTs modified with metal nanoparticles has attracted much attention. Au, Pt, Pd, and Ni nanoparticles are applied to enhance the performances of the CNTs modified electrodes, because of their high catalytic activity, biocompatibility, and increased surface area. Increase of enzyme loading, promotion of electron transfer, and synergistic effect in the biosensing of methyl parathion were observed using a nanocomposite sensing film prepared via the formation of gold nanoparticles on silica particles, mixing with multiwall carbon nanotubes and subsequent covalent immobilization of methyl parathion hydrolase. A linear response was obtained in the range from 0.001 μg mL^{-1} to 5.0 μg mL^{-1} with a detection limit of 0.3 ng mL^{-1} [68].

The one-step fabrication of MWCNTs-GNPs composite could be performed by in situ reduction of $HAuCl_4$ by $NaBH_4$. The self-coating of the GNPs on the MWCNTs produced a uniform nanocomposite. It was used for the fabrication of an acetylcholinesterase based electrochemical sensor for malathion determination [69]. The detection limit was found to be 0.6 ng mL^{-1}.

3.2. Gold nanoparticles

GNPs are extensively used in biosensors application, for the reason of their biocompatibility, catalytic activity, excellent conductivity, and high surface area [70, 71].

Various materials modified with GNPs were tested as enzyme immobilization matrices. Marinov et al. [72] suggest the use of GNPs loaded chemically modified poly(acrylonitrile-methylmethacrylate-sodium vinylsulfonate) membranes (PAN) as supports for acetylcholinesterase immobilization. Since PAN is not electroconductive, GNPs acted as electron transfer "wires". The high enzyme loading and the presence of GNPs resulted in high sensitivity. The detection limit for paraoxon determination was found to be 0.074 ng L^{-1} and a linear response was obtained in the range 0.1-100 ng L^{-1}. Important advantage of the developed biosensor is the substitutability of the enzyme membrane, as the enzyme carrier is a separate element that could be incubated in a pesticide solution and reactivated in PAM solution afterwards aside from the working electrode, which is hence available to be assembled with another enzyme membrane and used for further pesticide measurements.

Simple and efficient strategy for acetylcholinesterase immobilization onto a composite film of Au-polypyrrole nanowires was proposed by Gong et al. [73]. It is assumed that the three-dimensional interlaced polypyrrole nanowires network provides a favorable microenvironment to maintain the bioactivity of the enzyme, while the GNPs distributed in the network-structured matrix facilitate the electron transfer. The inhibition of methyl parathion was found to be proportional to its concentration ranging from 0.005 to 0.12 and 0.5 to 4.5μg mL^{-1} with a detection limit of 2 ng mL^{-1}. After 30-days storage the sensor retained 60% of its initial current response.

A fast method for the preparation of acetylcholinesterase-GNPs-$CaCO_3$ bioconjugates was suggested by Chauhan et al. [74]. The procedure includes the preparation of the hybrid GNPs-$CaCO_3$ material by $CaCO_3$ dispersion into Au colloid solution, followed by enzyme adsorption and encapsulation of the bioconjugates on the electrode surface using silica sol. The electrochemical measurements were performed at a potential of+0.2 V/Ag, AgCl, this avoiding the interferences. It was demonstrated that malathion and chlorpyrifos could be detected in the range of 0.1-100 nM and 0.1-70 nM, respectively, with a detection limit of 0.1 nM for both. The response current of the sensor decreased to 60% after 60 days.

A stable and sensitive inhibition based sensor for OPs quantification was fabricated by Sun et al. [75], using hollow gold nanospheres (HGNs). The protocol comprised glassy carbon electrode modification with chitosan, hollow gold nanospheres adsorption onto the surface of chitosan through electrostatic interactions, L-cysteine assemblage on HGNs through Au-S bond, and acetylcholinesterase covalent immobilization via the-COOH groups of L-cysteine. The inhibition rates of pesticides were found to be proportional to their concentrations in the range of 0.1-150 and 0.1-200 μg L^{-1} for chlorpyrifos and carbofuran, with detection limits of 0.06μg L^{-1} and 0.08 μg L^{-1}, respectively. After 40-days of storage, the sensor retained 85.4 % of its initial current response.

3.3. Other nanomaterials

Effective devices for OPs determination were developed using functionalized graphene structures. It has been demonstrated that the acetylcholinesterase sensors based on graphene oxide, GNP-graphene oxide, and nanoparticles (NiO, Pt, SnO_2)-graphene nanocomposites show high electron mobility, catalytic activity, and sensitivity [76-80]. They were successfully applied for methylparathion, chlorpyrifos, malathion, and dichlorvos quantification. Another sensitive acetylcholinesterase sensor was fabricated using oxidized exfoliated graphite nanoplatelet (xGnPs)-chitosan cross-linked composite [81]. It was used for chloropyrifos determination with a detection limit of 1.58×10^{-10} M.

Other carbonaceous materials used in OPs biosensing are the mesoporous carbons and carbon black [82]. The well-ordered nanopores, many edge-plane-like defective sites, and high surface area of the mesoporous carbon resulted in increased sensitivity, and allowed for nanomolar-range detection of the analyte paraoxon using an OPH-based sensor. The detection limit achieved was of 0.12μM (36 ppb).

The potential of the magnetic nanoparticles was exploited for the construction of a disposable acetylcholinesterase-coated Fe_3O_4/Au magnetic nanoparticles (GMP-AChE) sensor [83]. The GMP-AChE were absorbed on the surface of a screen printed carbon electrode modified by carbon nanotubes (CNTs)/nano-ZrO_2/prussian blue(PB)/Nafion (Nf) composite membrane by an external magnetic field. The biosensor exhibited a fast response, wide linear detection range and high sensitivity to OPs due to the conductive Fe_3O_4/Au NPs providing a large electrode surface. The other advantages of the biosensor are associated with the specific adsorption of OPs by the ZrO_2 NPs, and the easy removal and replacement of the Fe_3O_4/Au/AChE by applying an external magnetic field. The biosensor was used for dimethoate determination in the range 1.0×10^{-3}-10 ng mL^{-1} with a detection limit of 5.6×10^{-4} ng mL^{-1}.

An OPs biosensor fabricated by covalent acetylcholinesterase immobilization onto iron oxide nanoparticles and carboxylated multi walled carbon nanotubes modified Au electrode was reported by Chauhan et al. [84]. The synergistic action of Fe_3O_4 NP and MWCNT showed excellent electrocatalytic activity at low potential (+0.4 V). Under optimum conditions, the inhibition rates of OPs were proportional to their concentrations in the range of 0.1-40 nM, 0.1-50 nM, 1-50 nM and 10-100 nM for malathion, chlorpyrifos, monocrotophos and endosulfan, respectively. The detection limits were 0.1 nM for malathion and chlorpyrifos, 1 nM for monocrotophos, and 10 nM for endosulfan. The biosensor was stable (2 months) and reusable (more than 50 times).

4. Analytical performances of the recently developed nanostructured electrochemical biosensors for OPs quantification

The analytical performances of the recently developed nanostructures electrochemical biosensors for OPs quantification are summarized in Table 1.

Electrode/matrix	Technique	Immobilization method	LOD, μM	Linearity, μM	Analyte	Storage stability	Ref.
SWCNT/Co phtalocyanine	amperometric	covalent	0.010 $9.5x10^{-3}$	0.018-0.181 $9.5x10^{-3}$-0.16	paraoxon malaoxon	3 months	67
MWCNTs–Au nanocomposites/GCE	amperometric	hydrophilic surface for biomolecule adhesion	$1.81x10^{-3}$	$3.0x10^{-3}$-3.027	malathion	30 days	69
AuNPs-PAN/Pt	amperometric	glutaraldehyde	$2.69x10^{-7}$	$(3.63\text{-}363)x10^{-7}$	Paraoxon	50 days	72
AuNPs-PPy/GCE	voltammetric	adsorption	$6.86x10^{-3}$	$(17.18\text{-}41.2)x10^{-3}$ 1.71-15.4	methyl parathion	30 days	73
AuNPs-$CaCO_3$/Au	amperometric	adsorption	$0.1x10^{-3}$ $0.1x10^{-3}$	$(0.1\text{-}100)x10^{-3}$ $(0.1\text{-}70)x10^{-3}$	malathion chlorpyrifos	60 days	74
HGNs/CHIT/GCE	voltammetric	covalent	$1.71x10^{-4}$	$2.85x10^{-4}$-0.43	Chlorpyrifos	60 days	75
NiO NPs-Carboxylic graphene-nafion/GCE	amperometric	entrapment	$5x10^{-8}$	$1.0x10^{-7}\text{-}1x10^{-4}$ $1x10^{-10}\text{-}1x10^{-8}$	methyl parathion chlorpyrifos	30 days	76
NiO NPs-Carboxylic graphene-nafion/GCE	amperometric	entrapment	$5x10^{-8}$	$1\times10^{-13}\text{-}1\times10^{-10}$ $1\times10^{-4}\text{-}1\times10^{-2}$	methyl parathion	30 days	77
SnO_2 NPs-Carboxylic graphene-nafion/GCE	amperometric	entrapment	5×10^{-8}	$1x10^{-7}\text{-}1x10^{-4}$ $1x10^{-4}\text{-}1x10^{-2}$	methyl parathion	30 days	78
Graphen oxide-AuNPs/GCE	amperometric	covalent	$2.85x10^{-4}$	$(1.42\text{-}28.5)x10^{-3}$ 0.028-0.285	chlorpyrifos	30 days	79
Graphen oxide-nafion/GCE	amperometric	adsorption	$9x10^{-3}$	$(22.6\text{-}452)x10^{-3}$ 4.52-90.5	dichlorvos	30 days	80
Graphite nanoplatelet–CHIT composite/GCE	voltammetric	covalent	$1.58x10^{-4}$	$1x10^{-4}$-1.0	chloropyrifos	10 days	81
AuNPs/PB/ZrO_2/CNTs/SPCE	DPV	adsorption	$2.44x10^{-6}$	$4.4x10^{-6}\text{-}4.4x10^{-2}$	dimethoate		83
Fe_3O_4NP/MWCNTs/Au	amperometric	covalent	$0.1x10^{-3}$	$(0.1\text{-}40)x10^{-3}$ $(0.1\text{–}50)x10^{-3}$ $(1\text{-}50)x10^{-3}$	malathion chlorpyrifos monocrotophos	60 days	84

Electrode/matrix	Technique	Immobilization method	LOD, μM	Linearity, μM	Analyte	Storage stability	Ref.
AuNPs–$CaCO_3$ bioconjugate/Au	amperometric	adsorption	0.1×10^{-3}	$(0.1\text{-}100) \times 10^{-3}$ $(0.1\text{-}70) \times 10^{-3}$	malathion chlorpyrifos	90 days	85
Fe_3O_4NP/MWCNTs/ITO	amperometric	covalent	0.1×10^{-3}	$(10\text{-}100) \times 10^{-3}$ $(0.1\text{-}70) \times 10^{-3}$ $(0.1\text{-}50) \times 10^{-3}$ $(0.1\text{-}70) \times 10^{-3}$	malathion chlorpyrifos monocrotophos	90 days	86
AuNPs/PB/GCE	amperometric	adsorption	3.5×10^{-9}	$(0.448\text{-}4.48) \times 10^{-2}$	monocrotophos	30 days	87
AuNPs/GCE	amperometric	adsorption	7×10^{-3}	$(28\text{-}170) \times 10^{-3}$	methamidophos	7 days	88
AuNPs–MWCNTs/GCE		adsorption	1×10^{-3}	$(0.1\text{-}7.0) \times 10^{-3}$			89
PB-CHIT/GCE	amperometric	glutaraldehyde	0.113×10^{-4}, 0.703×10^{-4} 0.194×10^{-4} 0.33×10^{-4}	$0.45 \times 10^{-4}\text{-}0.045$ $0.234 \times 10^{-3}\text{-}0.046$ $0.116 \times 10^{-3}\text{-}0.0194$ $0.167 \times 10^{-3}\text{-}0.0335$	dichlorvos methoate trichlorfon phoxim		90
MWCNTs/AuNPs–CHIT/GCE	Fourier transform	adsorption	0.01	0.1-10	monocrotophos	50 days	91

CHIT-chitosan; GCE-glassy carbon electrode; HGNs-hollow gold nanospheres; MWCNTs-multi-walled carbon nanotubes; PB-Prussian blue; PPy-polypyrrole; SPCE-screen printed carbon electrode.

CHIT-chitosan; GCE-glassy carbon electrode; HGNs-hollow gold nanospheres; MWCNTs-multi-walled carbon nanotubes; PB-Prussian blue; PPy-polypyrrole; SPCE-screen printed carbon electrode.

Table 1. Analytical performances of some nanostructured acetylcholinesterase based sensors for OPs determination.

http://dx.doi.org/10.5772/58310

5. Conclusion

This review addresses the recent trends in the development of nanomaterials based electrochemical biosensors for organophosphorus pesticides determination. The included examples demonstrate the great potential of the carbon nanotubes and the gold nanoparticles, as well as of the emerging graphene structures.

Current researches confirm that the adequate combination of nanomaterials, biological recognition events, and efficient electronic signal transduction result in biosensors with improved analytical performances, appropriate for the high sensitive determination of OPs, among other.

Author details

Margarita Stoytcheva and Roumen Zlatev

Autonomous University of Baja California, Engineering Institute, Mexicali, Mexico

References

[1] Codex pesticides residues in food online database, Codex Alimentarius Commission, 22nd Session, June 1997, http://www.codexalimentarius.net/mrls/pestdes/jsp/pest_q-e.jsp (accessed 14 October 2013)

[2] Council Directive 76/895/EEC (23 November 1976). Official Journal of the European Communities, 09.12.1976, L340

[3] Council Directive 86/363/EEC (24 July 1986). Official Journal of the European Communities, 03.07.1986, L164

[4] Council Directive 86/362/EEC (24 July 1986). Official Journal of the European Communities, 07.08.1986, L221

[5] Council Directive 90/642/EEC (27 November 1990). Official Journal of the European Communities, 14.12.1990, L350

[6] Council Directive 98/83/EC (3 November 1998). Official Journal of the European Communities, 5.12.1998, L330/32

[7] Balasubramanian K., Burghard M. Biosensors based on carbon nanotubes. Anal. Bioanal. Chem. 2006; 385 (3) 452-468.

[8] Eftekhari A. Nanostructured materials in electrochemistry. Weinheim: Wiley-VCH; 2008.

[9] Gorton L. Biosensors and modern biospecific analytical techniques. Elsevier; 2005.

[10] Guo S., Wang E. Synthesis and electrochemical applications of gold nanoparticels. Anal. Chim. Acta 2007; 598 (2) 181-192.

[11] Kerman K., Saito M., Yamamura S., Takamura Y., Tamiya E. Nanomaterial-based electrochemical biosensors for medical applications. Trends Anal. Chem. 2008; 27(7) 585-59.

[12] Kumar C. (2007). Nanomaterials for biosensors. Weinheim: Wiley-VCH; 2007.

[13] Luo X., Morrin A., Killard A., Smyth M. Application of nanoparticles in electrochemical sensors and biosensors. Electroanalysis 2006; 18(4) 319-326.

[14] Merkoçi A., Alegret S. Toward nanoanalytical chemistry: case of nanomaterial integration into (bio)sensong systems. Contributions to science 2005; 3(1) 57-66.

[15] Merkoçi A. (2009). Biosensing using nanomaterials. Hoboken, New Jersey: Wiley; 2009.

[16] Pumera M., Sánchez S., Ichinose I., Tang J. Electrochemical nanobiosensors. Sensors Actuators B 2007; 123(2) 1195-1205.

[17] Stoytcheva M., Zlatev R. Organophosphorus pesticides analysis. In: Stoytcheva M. (ed.) Pesticides in the Modern World-Trends in Pesticides Analysis. Rijeka: InTech; 2011. p. 143-164. Available from http://cdn.intechopen.com/pdfs/20989/InTech-Organophosphorus_pesticides_analysis.pdf (accessed 14 October 2013)

[18] Andreescu S., Marty J.-L. Twenty years research in cholinesterase biosensors: from basic research to practical applications. Biomol. Eng. 2006; 23(1) 1-15.

[19] Anzai J. Use of biosensors for detecting organophosphorus agents. Yakugaku Zasshi 2006; 126(12) 1301-1308.

[20] Jaffrezic-Renault N. New trends in biosensors for organophosphorus pesticides. Sensors 2001; 1(2) 60-64.

[21] Prieto-Simón B., Campàs M., Andreescu S., Marty J.-L. Trends in flow-based biosensing systems for pesticide assessment. Sensors 2006; 6(10) 1161-1186.

[22] Rodriguez-Mozaz S., Marco M.-P., Lopez de Alda M. J., Barceló D. Biosensors for environmental applications: future development trends. Pure Appl. Chem. 2004; 76(4) 723-752.

[23] Solé S., Merkoçi A., Alegret S. Determination of toxic substances based on enzyme inhibition. Part I. Electrochemical biosensors for the determination of pesticides using batch procedures. Crit. Rev. Anal. Chem. 2003; 33(2) 89-126.

[24] Solé S., Merkoçi A., Alegret S. Determination of toxic substances based on enzyme inhibition. Part I. Electrochemical biosensors for the determination of pesticides using flow procedures. Crit. Rev. Anal. Chem. 2003; 33(2) 127-143.

[25] Tran-Minh C. Immobilized enzyme probes for determining inhibitors. Ion-Selective Electrode Rev. 1985; 7 41-75.

[26] Turdean G., Popescu I.C., Oniciu, L. Biocapteurs ampérométriques a cholinestérases pour la détermination des pesticides organophosphorés. Can. J. Chem. 2002; 80 315-331.

[27] Martorell D., Céspedes F., Martínez-Fàbregas E., Alegret S. Amperometric determination of pesticides using a biosensor based on a polishable graphite-epoxy biocomposite. Anal. Chim. Acta 1994; 290(3) 343-348.

[28] Marty J.-L., Mionetto N., Rouillon R. Entrapped enzymes in photocrosslinkable gel for enzyme electrodes. Anal. Lett. 1992; 25(8) 1389-1398.

[29] Marty J.-L., Mionetto N., Noguer T., Ortega F., Roux C. Enzyme sensors for the detection of pesticides. Biosens. Bioelectron. 1993; 8(6) 273-280.

[30] Marty J.-L., Mionetto N., Lacorte S., Barceló D. Validation of an enzymatic biosensor with various liquid chromatographic techniques for determining organophosphorus pesticides and carbaryl in freeze-dried waters. Anal. Chim. Acta 1995; 311(3) 265-271.

[31] Sužnjević D.Ž., Veselinović D.S., Vukelić N.S., Pavlović D.Ž., Nikolić A.V. Investigation of the system butyrylthiocholineiodide-butyrocholinesterase by cyclovoltammetry and chronopotentiometry using inert working electrodes. J. Serb. Chem. Soc. 1985; 50(2) 83-88.

[32] Harlbert M., Baldwin R. Electrocatalytic and analytical response of cobalt phthalocyanine containing carbon paste electrodes toward sulfhydryl compounds. Anal. Chem. 1985; 57(3) 591-595.

[33] Hart J., Hartley I. Voltammetric and amperometric studies of thiocholine at a screen-printed carbon electrode chemically modified with cobalt phthalocyanine: studies towards a pesticide sensor. Analyst 1994; 119(2) 259-265.

[34] Skladal P. Determination of organophosphate and carbamate pesticides using a cobalt phthalocyanine-modified carbon paste electrode and a cholinesterase enzyme membrane. Anal. Chim. Acta 1991; 252(1-2) 11-15.

[35] Ricci F., Arduini F., Amine A., Moscone D., Palleschi G. Characterisation of Prussian blue modified screen-printed electrodes for thiol detection. J. Electroanal. Chem. 2004; 563(2) 229-237.

[36] Kulys J., D'Costa, E.J. Printed amperometric sensor based on TCNQ and cholinesterase. Biosens. Bioelectron. 1991; 6(2) 109-115.

[37] Martorell D., Céspedes F., Martínez-Fàbregas E., Alegret S. Determination of organophosphorus and carbamate pesticides using a biosensor based on a polishable, 7,7,8,8 –tetracyanoquino–dimethane-modified, graphite-epoxy biocomposite. Anal. Chim. Acta 1997; 337(3) 305-313.

[38] Evtugyn G., Budnikov H., Galyametdinov Yu., Suntsov E. Amperometric determination of thiocholine esters in the presence of butyrylcholinesterase. Zh. Anal. Khim. 1996; 51(4) 391-393.

[39] Neufeld T., Eshkenazi I., Cohen E., Rishpon J. A micro flow injection electrochemical biosensor for organophosphorus pesticides. Biosens. Bioelectr. 2000; 15(5-6) 323-329.

[40] Ovalle M., Stoytcheva M., Zlatev R., Valdez B. Electrochemical study of rat brain acetylcholinesterase inhibition by chlorofos: kinetic aspects and analytical applications. Electrochimica acta 2009; 55(2) 516-520.

[41] Luque de Castro M.D., Herrera M.C. Enzyme inhibition-based biosensors and biosensing systems: questionable analytical devices. Biosens. Bioelectron. 2003; 18(2-3) 279-94.

[42] Dumas D.P., Durst H.D., Landis W.G., Raushel F.M., Wild J.R. Inactivation of organophosphorus nerve agents by the phosphotriesterase from *Pseudomonas diminuta*. Arch. Biochem. Biophys. 1990; 227 155– 159.

[43] Munnecke D.M. (1980). Enzymatic detoxification of waste organophosphate pesticides. J. Agric. Food Chem. 1980; 28(1) 105-111.

[44] Efremenko E.N., Sergeeva V.S. Organophosphate hydrolase-an enzyme catalyzing degradation of phosphorus containing toxins and pesticides. Russian Chemical Bulletin, International Edition 2001; 50(10) 1826-1832.

[45] Lei C., Valenta M., Sapiralli K.P., Ackerman E.J. Biosensing paraoxon in simulated environmental samples by immobilized organophosphorus hydrolase in functionalized mesoporous silica. J. Environ. Qual. 2007; 36(1) 233-238.

[46] Mulchandani A., Chen W., Mulchandani P., Wang J., Rogers K.R. Biosensors for direct determination of organophosphate pesticides. Biosens. Bioelectron. 2001; 16(4-5) 225-230.

[47] Mulchandani P., Chen W., Mulchandani A. Flow injection amperometric enzyme biosensor for direct determination of organophosphate nerve agents. Environ. Sci. Technol. 2001; 35(12), 2562-2565.

[48] Wang J., Krause R., Block K., Musameh M., Mulchandani A., Schöning M.J. Flow injection amperometric detection of OP nerve agents based on an organophosphorus-hydrolase biosensor detector. Biosens. Bioelectron. 2003; 18(2-3) 255-260.

[49] Mulchandani P., Chen, Mulchandani, A. Microbial biosensor for direct determination of nitrophenyl-substituted organophosphate nerve agents using genetically engineered *Moraxella* sp. Anal. Chim. Acta 2006; 568(1-2) 217-221.

[50] Ureta-Zanartu M.S., Bustos P., Diez M.C., Mora M.L., Gutierrez C. Electro-oxidation of chlorophenols at a gold electrode. Electrochim. Acta 2001; 46 2545-2551.

[51] Borras C., Laredo T., Mostany J., Scharifker B.R. Study of the oxidation of solutions of p-chlorophenol and p-nitrophenol on Bi-doped PbO_2 electrodes by UV-Vis and FTIR in situ spectroscopy. Electrochim.Acta 2004; 49 641-648.

[52] Wang J., Deo R.P., Musameh M. Stable and sensitive electrochemical detection of phenolic compounds at carbon nanotube modified glassy carbon electrodes. Electroanalysis 2003; 15(23) 1830-1834.

[53] Ferreira M., Varela H., Toressi R., Tremiliosi-Filho G. Electrode passivation caused by polymerization of different phenolic compounds. Electrochim.Acta 2006; 52 434-442.

[54] Al Maznai H., Conway B.E. Efekti samo-inhibicije u reakciji anodne oksidacije fenola za elektrohemijsko prečišćavanje otpadnih voda. J. Serb.Chem. Soc. 2001; 66(11-12) 765-784.

[55] Lupu S., Ion I., Ion A.C. Voltammetric determination of phenol at platinum electrodes modified with polypyrrole doped with ferricyanide. Rev. Roumaine Chim., 2009; 54 351-357.

[56] Kim H.J., Chang S.C., Shim Y.B. α-Cyclodextrin modified screen printed graphite electrodes for detection of phenols. Bull. Korean Chem. Soc., 2002; 23 427-431.

[57] Arslan G., Yazici B., Erbil M. The effect of pH, temperature and concentration on electrooxidation of phenol. J. Hazardous Mater., 2005; B124 37-43.

[58] Obirai J., Bedioui F., Nyokong T. Electrooxidation of phenol and its derivatives on poly-Ni(OH)TPhPyPc modified vitreous carbon electrodes. J. Electroanal. Chem. 2005; 576(2) 323-332.

[59] Dejmkova H., Scampicchio M., Zima J., Barek J., Manni-no S. Determination of total phenols in foods by boron doped diamond electrode. Electroanalysis 2009; 21(9) 1014-1018.

[60] Shaoqin Liu, Lang Yuan, Xiuli Yue, Zhaozhu Zheng, Zhiyong Tang. Recent advances in nanosensors for organophosphate pesticide detection. Advanced Powder Technology 2008; 19 419–441.

[61] Periasamy A., Umasankar Y., Chen, S.-M. Nanomaterials-acetylcholinesterase enzyme matrices for organophosphorus pesticides electrochemical biosensors: a review. Sensors 2009; 9 4034-4055.

[62] Ajayan P.M. Nanotubes from carbon. Chem. Rev. 1999; 99 1787–1799.

[63] D. Tasis, N. Tagmatarchis, A. Bianco, M. Prato, Chemistry of carbon nanotubes, Chem. Rev. 106 (2006) 1105–1136.

[64] Britto P.J., Santhanam K.S.V., Rubio A., Alonso J.A., Ajayan P.M. Improved charge transfer at carbon nanotube electrodes. Advanced Materials 1999; 11(2) 154-157.

[65] Pedrosa V.A. Paliwal S., Balasubramanian S., Nepal D., Davis V., Wild J., Ramanculov E., Simonian A. Enhanced stability of enzyme organophosphate hydrolase interfaced on the carbon nanotubes. Colloids and Surfaces B: Biointerfaces 2010; 77 69–74.

[66] Deo R.P., Wang J., Block I., Mulchandani A., Joshic K.A., Trojanowicz M., Scholz F., Chen W., Lin Y. Determination of organophosphate pesticides at a carbon nanotube/organophosphorus hydrolase electrochemical biosensor. Analytica Chimica Acta 2005; 530 185–189.

[67] Ivanov A.N., Younusov R.R., Evtugyn G.A., Arduini F., Moscone D., Palleschi G. Acetylcholinesterase biosensor based on single-walled carbon nanotubes-Co phtalocyanine for organophosphorus pesticides detection. Talanta 2011; 85 216-221.

[68] Chen S., Huang J., Du D., Li J., Tu H., Liu D., Zhang A. Methyl parathion hydrolase based nanocomposite biosensors for highly sensitive and selective determination of methyl parathion. Biosens. Bioelectron. 2011; 26 4320– 4325.

[69] Du D., Wang M., Cai J., Qin Y., Zhang A. One-step synthesis of multiwalled carbon nanotubes-gold nanocomposites for fabricating amperometric acetylcholinesterase biosensor. Sensors and Actuators B 2010; 143 524–529.

[70] Li Y., Schluesener H.J., Xu S. Gold nanoparticle-based biosensors. Gold Bulletin 2010; 43(1) 29-41.

[71] Pingarrón J.M., Yáñez-Sedeño P., González-Cortés A. Gold nanoparticle-based electrochemical biosensors. Electrochimica Acta 2008; 53 5848–5866.

[72] Marinov I., Ivanov Y., Gabrovska K., Godjevargova T. Amperometric acetylthiocholine sensor based on acetylcholinesterase immobilized on nanostructured polymer membrane containing gold nanoparticles. Journal of Molecular Catalysis B: Enzymatic 2010; 62 67–75.

[73] Gong J., Wang L., Zhang L. Electrochemical biosensing of methyl parathion pesticide based on acetylcholinesterase immobilized onto Au-polypyrrole interlaced network-like nanocomposite. Biosens. Bioelectron. 2009; 24 2285–2288.

[74] Chauhan N., Narang J., Pundir C.S. Immobilization of rat brain acetylcholinesterase on porous gold-nanoparticle-$CaCO_3$ hybrid material modified Au electrode for detection of organophosphorous insecticides. Int. J. Biol. Macromol. 2011; 49 923– 929.

[75] Sun X., Zhai C., Wang X. A novel and highly sensitive acetyl-cholinesterase biosensor modified with hollow gold nanospheres. Bioprocess Biosyst. Eng. 2013; 36 273–283.

[76] Yang L., Wang G., Liu Y., Wang M. Development of a biosensor based on immobilization of acetylcholinesterase on NiO nanoparticles-carboxylic graphene-nafion modified electrode for detection of pesticides. Talanta 2013; 113 135–141.

[77] Yang L., Wang G., Liu Y. An acetylcholinesterase biosensor based on platinum nanoparticles-carboxylic graphene-nafion-modified electrode for detection of pesticides. Analytical Biochemistry 2013; 437 144–149.

[78] Zhou Q., Yang L., Wang G., Yang Y.. Acetylcholinesterase biosensor based on SnO_2 nanoparticles-carboxylic graphene-nafion modified electrode for detection of pesticides. Biosens. Bioelectron. 2013; 49 25–31.

[79] Liu T., Su H., Qu X., Ju P., Cui L., Ai S. Acetylcholinesterase biosensor based on 3-carboxyphenylboronic acid/reduced graphene oxide–gold nanocomposites modified electrode for amperometric detection of organophosphorus and carbamate pesticides. Sensors and Actuators B 2011; 160 1255– 1261.

[80] Wu S., Huang F., Lan X., Wang X., Wang J., Meng C. Electrochemically reduced graphene oxide and nafion nanocomposite for ultralow potential detection of organophosphate pesticide. Sensors and Actuators B 2013; 177 724– 729.

[81] Ion A.C., Ion I., Culetu A., Gherase D., Moldovan C.A., Iosub R., Dinescu A. Acetylcholinesterase voltammetric biosensors based on carbon nanostructure chitosan composite material for organophosphate pesticides. Materials Science and Engineering C 2010; 30 817-821.

[82] Lee J.H., Park J.Y., Min K., Cha H.J., Choi S.S., Yoo Y.J. A novel organophosphorus hydrolase-based biosensor using mesoporous carbons and carbon black for the detection of organophosphate nerve agents. Biosens. Bioelectron. 2010; 25 1566–1570.

[83] Gan N., Yang X., Xie D., Wu Y., Wen W. A disposable organophosphorus pesticides enzyme biosensor based on magnetic composite nano-particles modified screen printed carbon electrode. Sensors 2010; 10 625-638.

[84] Chauhan N., Pundir C.S. An amperometric biosensor based on acetylcholinesterase immobilized onto iron oxide nanoparticles/multi-walled carbon nanotubes modified gold electrode for measurement of organophosphorus insecticides. Analytica Chimica Acta 2011; 701 66-74.

[85] Chauhan N., Narang J., Pundir C.S. Immobilization of rat brain acetylcholinesterase on ZnS and poly(indole-5-carboxylic acid) modified Au electrode for detection of organophosphorus insecticides. Biosens. Bioelectron. 2011; 29 82–88.

[86] Chauhan N., Pundir C.S. An amperometric acetylcholinesterase sensor based on Fe_3O_4 nanoparticle/multi-walled carbon nanotube-modified ITO-coated glass plate for the detection of pesticides. Electrochim. Acta 2012; 67 79–86.

[87] Wu S., Lan X., Zhao W., Li Y., Zhang L., Wang H., Han M., Tao S., Controlled immobilization of acetylcholinesterase on improved hydrophobic gold nanoparticle/ Prussian blue modified surface for ultra-trace organophosphate pesticide detection. Biosens. Bioelectron. 2011; 27 82–87.

[88] Yan Rong L., Zhi Yong G., Yan Fen L., Qian L., Jian Chun B., Zhi Hui D., Min H. Immobilization of acetylcholinesterase on one-dimensional gold nanoparticles for detection of organophosphorous insecticides, Sci. China Chem. 2010; 53 820–825.

[89] Jha N., Ramaprabhu S., Development of Au nanoparticles dispersed carbon nanotube-based biosensor for the detection of paraoxon, Nanoscale 2010; 2 806–810.

[90] Sun X., Wang X. Acetylcholinesterase biosensor based on Prussian bluemodified electrode for detecting organophosphorous pesticides, Biosens. Bioelectron. 2010; 25 2611–2614.

[91] Norouzi P., Pirali-Hamedani M., Ganjali M.R., Faridbod F.A., A novel acetylcholinesterase biosensor based on chitosan–gold nanoparticles film for determination of monocrotophos using FFT continuous cyclic voltammetry. Int. J. Electrochem. Sci. 2010; 5 1434–1446.

Chapter 5

Conducting Polymer Nano Composite Epoxy Coatings for Anticorrosive Applications

Gazala Ruhi and S.K. Dhawan

Additional information is available at the end of the chapter

http://dx.doi.org/10.5772/58388

1. Introduction

Corrosion is a destructive phenomenon that causes severe damage to metals, alloys, composites etc [12]. The Electrochemical theory of corrosion involves two half-cell reactions, an oxidation reaction at the anode and a reduction reaction at the cathode. For metals corroding in water at neutral pH, the half cell reactions are presented as:

Anodic reaction: $M \Rightarrow M^{n+} + ne^-$

Cathodic reaction: $O_2 + 2H_2O + ne^- \Rightarrow 4OH^-$

These half cell reactions create microscopic anodes and cathodes covering the entire corroding metal surface, which turns to macroscopic anodes and cathodes with the progresses of time. Metals lose their useful properties like metallic lustre, strength, permeability to liquids and gases etc. with prolong exposure to corrosive atmosphere. Mild steel is frequently used as a structural material for marine applications, petroleum production and refining, pipelines, mining etc. because of its good machinability, superior strength and low cost of production. However, this steel is extremely vulnerable towards corrosion. Break down of mild steel occurs, when exposed to diffusive chloride ions, in the sea water. Chromate based conversion coatings were used for many years for corrosion protection of mild steel [3]. These coatings have chromate, an oxidising oxy anion of chromium with superb corrosion resistance. But chromate ion is considered as a serious environmental pollutant and is acutely toxic. As an alternative approach, organic paints and coatings are also applied to protect steel substrates against corrosion [4-8]. But, it is challenging to design corrosion resistant coatings having corrosion protection efficiency equivalent to traditional chromate coatings.

Conducting polymers have attracted immense importance because of their versatile processing applications. These polymers are reported as corrosion inhibitors for active

INTECH
open science | open minds

© 2014 The Author(s). Licensee InTech. This chapter is distributed under the terms of the Creative Commons Attribution License (http://creativecommons.org/licenses/by/3.0), which permits unrestricted use, distribution, and reproduction in any medium, provided the original work is properly cited.

metals and alloys [9-17]. The mechanism of corrosion protection by conducting polymers involves anodic protection of the underlying metal, by raising its potential to passive region [18]. The polymer exhibits barrier property in its electronically insulating state, where an important feature in its conducting state is the ability to store charge at the metal/coating interface. This charge can be effectively used to oxidize base metal to form a passive layer. Polyaniline and Polypyrrole are the most studied conducting polymers due to their environmental stability, biocompatibility, physical and electrical properties. Polypyrrole is reported as an effective material for corrosion protection purposes [19-23]. Polypyrrole based corrosion resistant coatings can be prepared either by chemical synthesis or it may be electrochemically deposited on the metal surface. Electrochemical deposition of polypyrrole coatings are reported to be carried out on steel substrates using aqueous media [24-30]. These coatings show good adherence to metal surface and thickness of the coatings can be controlled precisely [19-21]. Schaftinghen et.al. have reported that the surface per-treatment of steel surface prior to electrochemical deposition of polypyrrole coatings has strong influence on corrosion protection performance of the coatings [20]. However, mechanical integrity and thermal stability of coatings in harsh climatic conditions are still questionable. The valuable properties of the coatings lost with prolong exposure to diffusive anions (e.g. chlorides) [31]. To overcome this, conducting polymer based composite systems are interesting alternatives and have attracted substantial academic and industrial interest in recent years. These composites are conducting polymers with thoroughly dispersed small inorganic filler particles. The physical, chemical and mechanical properties of the polymer composites are superior as compared to polymer alone because of the synergistic interaction between the conducting polymers and filler particles [32-39]. Therefore, these composites can be explored as coating material having superior properties. Polyaniline/SiO_2 composite coatings have been reported to enhance the corrosion resistance of steel substrates [40, 41]. Incorporation of SiO_2 particles as filler in the polymer matrix has shown significant impact on thermal and mechanical properties of coatings [42]. Studies have shown that, silica has a large surface area and smooth non-porous surface [43]. This feature promotes contact between SiO_2 (as filler) and the polymer matrix. Inspite of this, the chemically synthesized polymer composites are difficult to apply as coatings because of their non solubility in common organic solvents. Conventional polymer coating like epoxy is known to have superior characteristics as compared to regular paints due to its good scratch resistance and adhesion to metal surface [5-6]. However, epoxy coating alone is not sufficient to protect the underlying metal and fails due to cathodic disbondment under harsh environmental conditions over prolong exposure [44]. Literature shows the use of SiO_2 particles as reinforcement for anti corrosive epoxy coatings [45]. The particles are found to be useful in enhancing properties of organic coatings in aggressive medium [46]. But this effort is not enough for long term corrosion protection of metals. Polymer composites having superior corrosion resistance can be incorporated in the existing epoxy paints to get the desired results. Very limited works are reported on epoxy paints containing polyaniline and polypyrrole as anticorrosive additives [47-49]. In a novel approach, Armelin et.al. [47] have used polyaniline and polypyrrole as additives in epoxy paint system. In this work, it is discussed that the added polymers enhance the corrosion protection ability of epoxy paint under accelerated conditions and also acts as adhesion promoter. Hema et.al. [99] have studied the corrosion preventive response of polyaniline composites where it has been observed that PANI-SiO_2 composites can be used as effective coatings for corrosion prevention in marine environment.

The crucial question faced by coating industries is to explore green alternatives for chromate conversion coatings with equally good corrosion protection ability. Chitosan is one of the potential candidates for coating purposes with excellent film forming ability, barrier property and affinity towards metals. It is a natural polysaccharide found in the shells of marine crustaceans, a common sea food waste. The proper disposal of the sea food wastes is a continuous problem as it creates major environmental concern associated with ground and drinking water pollution. Chitosan can be explored as an eco-friendly candidate for coating purpose because of its low cost, most plentiful abundance in nature, biodegradable and non-toxic nature [50, 51]. Previous reports have explored the potential use of chitosan coatings for contact lenses and wound dressing [52, 53]. Apart from this, chitosan forms complexes with metal ions and also adheres to negatively charged surfaces. This chemical uniqueness of chitosan can be exploited as corrosion resistant coatings for active metals. A few investigations are carried out to develop chitosan based corrosion resistant coatings for metals [54, 62]. However, chitosan is prone to absorb moisture from atmosphere leading to failure of the coating [63]. A few investigations are carried out to decrease the hydrophilicity and to improve bond strength of chitosan [64, 65]. However, these measures are not sufficient for long term corrosion protection by chitosan coatings. The interaction of chitosan and conjugated polymers like polyaniline, polypyrrole has been reported earlier [66-68]. The synergistic effect of chitosan and polypyrrole is reported to reduce the size and improve the solubility of polypyrrole [69]. The interactive advantage of chitosan and polypyrrole can be used as corrosion resistant coatings for metals like mild steel. Inspite of this, the potential of chitosan/polypyrrole composite as a corrosion resistant coating material is not studied well [70].

Present chapter explains the preparation of polypyrrole based nano composite coatings for mild steel substrate using powder coating technique. Powder coating technique is adopted because of the advantage of low cost, excellent finish and almost no emission of volatile organic compounds. The synthesis of polypyrrole/SiO_2 nano composites were carried out using in-situ chemical polymerization of pyrrole monomer on SiO_2 particles. The corrosion protection of the composites was evaluated by Tafel extrapolation, Electrochemical Impedance Spectroscopy (EIS) and salt spray tests in NaCl solution. In order to improve the corrosion resistance of mild steel further, chitosan is introduced in the polymer composite system. Studies are also focussed to highlight the role of chitosan in enhancing the mechanical, thermal and corrosion resistance properties of polypyrrole. SiO_2 particles are purposely added to further enhance the properties of the composite. In order to evaluate the role of chitosan in improving corrosion protection properties, polypyrrole/SiO_2 composites (without chitosan) are also synthesized. The crucial role of SiO_2 is justified by synthesizing chitosan/polypyrrole composites (without SiO_2).

Conducting polymer	Fillers/dopant	Substrates	References/Journal
Polyaniline	a) PO_4^{-3}	mild steel	[39]
	b) SiO_2	mild steel	[40]
	c) TiO_2	steel	[11]
	d) Al_2O_3	carbon steel	[15]
	(e) SiO_2/PFOA	mild steel	[99]
Polyethylene	Ni/Mo	steel	[32]
Polythiophene	------	steel	[17]
Polypyrrole	a) Diglycidyl ether of bisphenol A (DGEBA)	mild steel	[35]
	b) SO_4^{2-}, pTS^-, DBS	Ni/Ti alloy	[36]
	c) TiO_2	mild steel	[37]
	d) titanate nano tube	stainless steel	[33]
	e) chitosan	steel	[70]

Table 1. Corrosion resistant properties of conducting polymer composites.

2. Electrochemical studies

The electrochemical studies were carried out using Tafel extrapolation method and Electrochemical Impedance Spectroscopy (EIS) in 3.5% NaCl solution at room temperature. Experiments were carried in a conventional three electrode cell assembly using Autolab Potentiostat/Galvanostat, PGSTAT100 (Nova Software). In three electrode cell assembly, the steel strip of dimension 1 cm x 1 cm is taken as working electrode embedded in araldite epoxy, Pt as counter electrode and Ag/AgCl as reference electrode. The surface cleaning of working electrode was carried out by polishing metallographically polishing with 1/0, 2/0, 3/0 and 4/0 grade emery papers. Thereafter, electrodes were thoroughly cleaned with acetone to remove impurities present on the steel surface.

2.1. Tafel extrapolation method

Tafel extrapolation is a method to measure the rate of corrosion, where the working electrode is polarised in positive direction from cathodic to E_{corr} ending at anodic to E_{corr}. Figure 1 shows the potential vs. log current plot for an applied potential scan.

The anodic and cathodic curves are drawn by running the test results using electrochemical software. The anode is polarised upwards to more positive values by an amount η_a (anodic polarisation). Whereas, the cathode is polarised downwards to a more negative value by an amount η_c (cathodic polarisation). Here, η is over voltage (measure of polarisation with respect to equilibrium potential of an electrode). The measured electrode potential is called as corrosion (E_{corr}). The point of intersection between the two rectilinear polarisation curves in

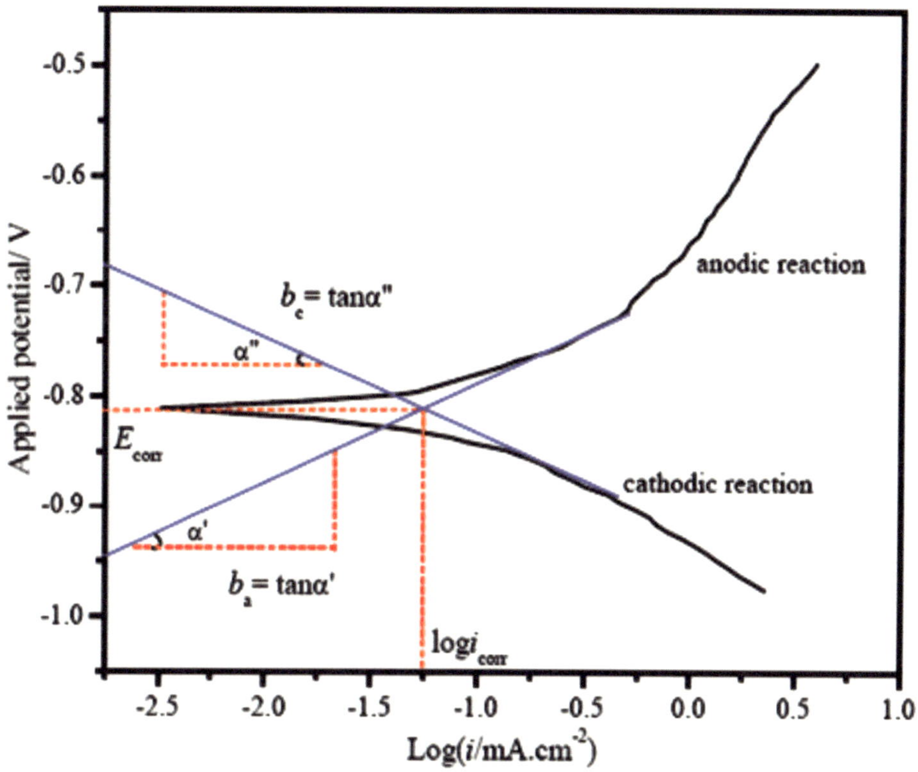

Figure 1. Polarisation curve of a corroding metal using Tafel extrapolation

the polarisation diagram (Figure 1) gives the value of corrosion current (i_{corr}). The values of anodic (βa) and cathodic (βc) Tafel constants are measured from the anodic and cathodic slopes respectively. It is possible to measure corrosion rate (C.R) in mm/year by knowing the value of i_{corr} using equation 1.

$$C.R=3.268x10^{3}x\frac{i_{corr}}{\rho}\frac{MW}{z} \tag{1}$$

Where MW is the molecular weight of the specimen (g/mole), ϱ is density of the specimen (g/m^3) and z is the number of electrons transferred in corrosion reactions.

2.2. Electrochemical Impedance Spectroscopy (EIS)

Electrochemical impedance spectroscopy (EIS) is a non-destructive technique which gives accurate kinetic and mechanistic information of a corroding system [71-79]. It is one of the powerful tools to study the corrosion rate. The impedance technique has been widely used to extract information concerning the interaction of metal/coating system with the corrosive environments, degradation of highly resistive organic coatings [80-83], corrosion rates at the defects and porosity in the coating. The impedance data is measured from Nyquist and Bode plots. Nyquist plot is also known as Cole-Cole plot (Figure 2a). In this plot, the real part of impedance $Z'=|Z|\cos\phi$ and imaginary part $Z''=|Z|\sin\phi$ of the impedance is plotted against each other at each excitation frequency. Usually the plot of Z′ vs Z″ is a semicircle and it cuts the Z′ axis at high frequency corresponding to R_s (solution resistance) and at low frequency corresponding to R_s+R_{ct}. Kinetic parameters like R_s, R_{ct} (charge transfer resistance) and C_{dl} (double layer capacitance) can be calculated from Nyquist plot. The disadvantages of the Nyquist plot are that the frequency does not appear explicitly and capacitance of the electrode can be calculated only after knowing the frequency. In Bode plots (Figure 2b), impedance (log Z) and phase angle (ϕ) are plotted against log of frequency. Bode plots give the information regarding the variation of impedance with respect to the frequency.

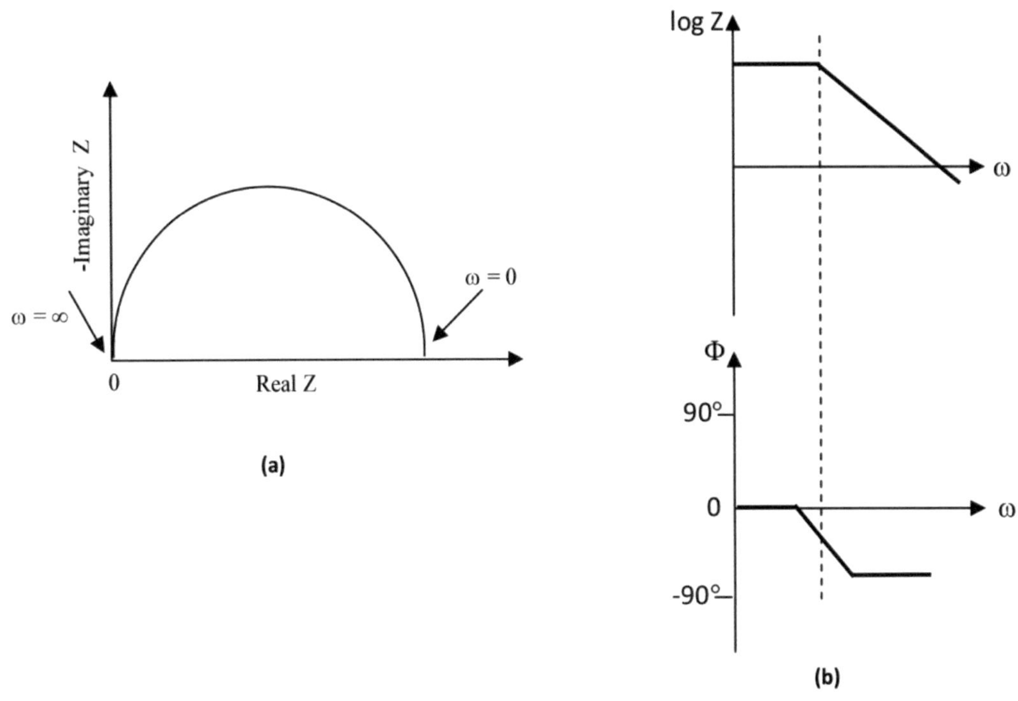

Figure 2. (a) Nyquist plot showing real vs imaginary part of impedance. (b) Bode plots showing the variation of impedance (log Z) and phase angle (Φ) with respect to change in frequency

2.3. Salt spray fog test

The salt spray fog test is a standardized method used to check corrosion resistance of coatings exposed to aggressive environments. Salt spray test is an accelerated corrosion test that produces a corrosive attack to the coated samples in order to predict their corrosion tolerance under high salt and humidity conditions. An intentional scribe mark is applied on the coated surface and the appearance of corrosion products (oxides) is evaluated after a period of time. Test duration depends on the corrosion resistance of the coatings. Salt spray test is popular because it is cheap, quick, well standardized and reasonably repeatable. Salt spray test is widely used in the industrial sector for the evaluation of corrosion resistance of coated surfaces or parts. The apparatus for testing consists of a closed testing chamber, where a salted solution (mainly, a solution of 5% NaCl) is atomized by means of a nozzle. This produces a corrosive environment of dense saline fog in the chamber so that parts exposed in it are subjected to severely corrosive conditions. Chamber construction, testing procedure and testing parameters are standardized under national and international standards, such as ASTM B 117 and ISO 9227. These standards contain the necessary information regarding testing parameters like temperature, air pressure of the sprayed solution, preparation of the spraying solution, concentration, pH, etc. Different coatings have different behaviour in salt spray test and consequently, test duration will differ from one type of coating to another.

3. Experimental work

3.1. Synthesis SiO_2 particles

The synthesis of mono disperse uniform sized SiO_2 particles was carried out by hydrolysis of tetra-ethyl orthosilicate (TEOS, Merck) in ethanol using ammonia as a catalyst. Aqueous ammonia (3.14ml, Merck) was added to a solution of ethanol in water. The solution was vigorously stirred for 1 hr. To this, 5 ml of TEOS was added slowly with constant stirring for the proper hydrolysis of TEOS. The appearance of turbidity indicated the formation of SiO_2 particles. The particles were retrieved by centrifuging at 15000 rpm for 5 minutes. The obtained powder was calcined at 600^0C in a furnace. The calcined silica was crushed by mortar pastel and kept in a desiccator for further use.

3.2. Synthesis of polymer composites (Polypyrrole/SiO_2)

Polypyrrole/SiO_2 composites (PCs) were synthesized by chemical oxidative polymerization of pyrrole (Acros Organics) using ferric chloride as an oxidant [84]. The schematic view of the synthesis is shown in Figure 3. Pyrrole and silica were taken to a weight ratio of 3:1 to form the composite. For this, synthesized SiO_2 particles were dispersed in deionised water using a mechanical stirrer. 0.01 M of sodium lauryl sulphate (SLS) was added to the above suspension with continuous stirring. Thereafter, 0.1 M of pyrrole was added slowly followed by drop wise addition of ferric chloride solution (0.1M). The appearance of black colour showed the start of polymerization of pyrrole. The polymerization was carried out at room temperature for a period of 4-5 hrs. The synthesized polymer composites (PCs) were retrieved from the suspension using G 4 filtration funnel. The obtained powder was washed with deionised water several times to remove oxidant and oligomers, followed by drying in a vacuum oven at 60^0C.

3.3. Preparation and deposition of powder coating

The polymer composite (PCs) was blended with epoxy powder coating formulation in various wt% loadings (1.0, 2.0, 3.0 and 4.0%) using a laboratory ball mill. For epoxy powder coating formulation has following composition: resin {epoxy (bisphenol A)+polyester} (70%), Flow agent (D-88) (2.3%), degassing agent (benzion) (0.7%), fillers (TiO2 and BaSO4) (27%). A homogenous mixture of well dispersed PCs in epoxy was obtained after ball milling. The blended epoxy-polymer composite formulation was applied on mild steel specimens using an electrostatic spray gun held at 67.4 KV potential with respect to the substrate. The powder coated mild steel specimens were cured in oven at 1800C for 30 minutes. The adhesion of the coatings was tested by tape test as per ASTM D3359-02 and found to pass the test.

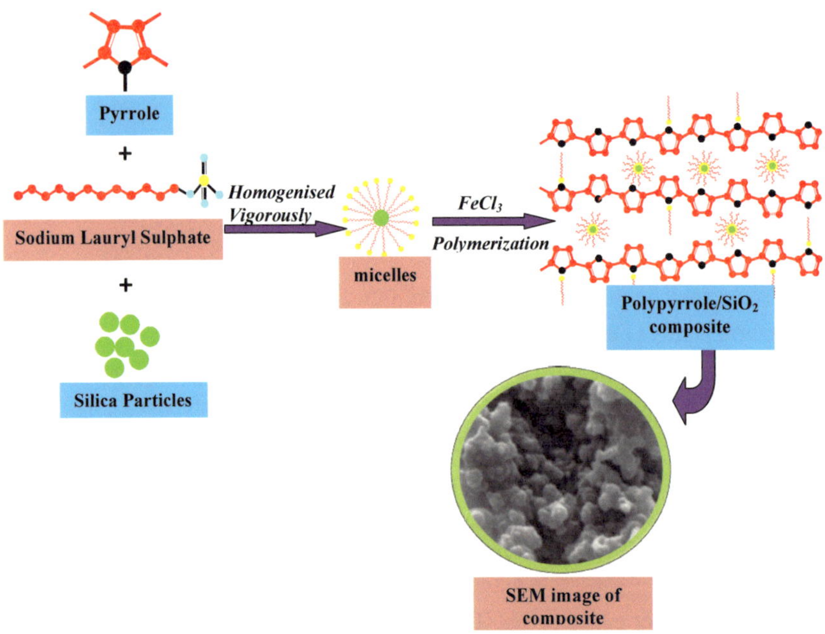

Figure 3. Schematic diagram of the synthesis of the polypyrrole/SiO_2 composite. (Progress in Organic Coatings DOI: 10.1016/j.porgcoat.2014.04.013).

4. Characterization of polypyrrole/SiO_2 composite

4.1. FTIR spectra

Figure 4 shows the FTIR spectra of SiO_2, polypyrrole alone and polypyrrole/SiO_2 composite (PCs). The FTIR spectrum of SiO_2 exhibits the characteristic bands of stretching and bending vibration of Si-O-Si at 1108 and 817 cm^{-1}, respectively. In FTIR spectrum of polymer composite, the observed peak at 1554 cm^{-1} corresponds to typical pyrrole ring vibration [85]. The low intensity peak at 1474 cm^{-1} is attributed to=CH in plane vibrations and the peak at 771 cm^{-1} is due to=CH out of plane vibrations [86-88]. A peak of considerable intensity at 1038 cm^{-1} is the N–H in plane deformation absorption of polypyrrole. The corresponding peak of C-N stretching [89] of polypyrrole at 1186cm^{-1} is masked by a band (at 1108 cm^{-1}) of SiO_2. For polypyrrole alone, the characteristic peaks of pyrrole rings, in and out of plane=CH vibrations are observed at 1559, 1480, 929 and 789 cm^{-1}, respectively [90]. On comparing the peaks of polypyrrole alone and polymer composites (PCs), it is observed that shifting of FTIR peak of polypyrrole occurred as a result of the incorporation of SiO_2 particles in the polymer matrix (Figure 4). The=CH out of plane vibration at 789 cm^{-1} of polypyrrole alone is shifted to a lower wave number (771 cm^{-1}) for the composite. The appearance of a broad band at 3440 cm^{-1} for the polymer composite (PCs) may be attributed to the interaction between SiO_2 and polypyrrole through hydrogen bonding between proton on N-H and oxygen atom on SiO_2. The above results show a strong interaction between polypyrrole and SiO_2 particles of the composite.

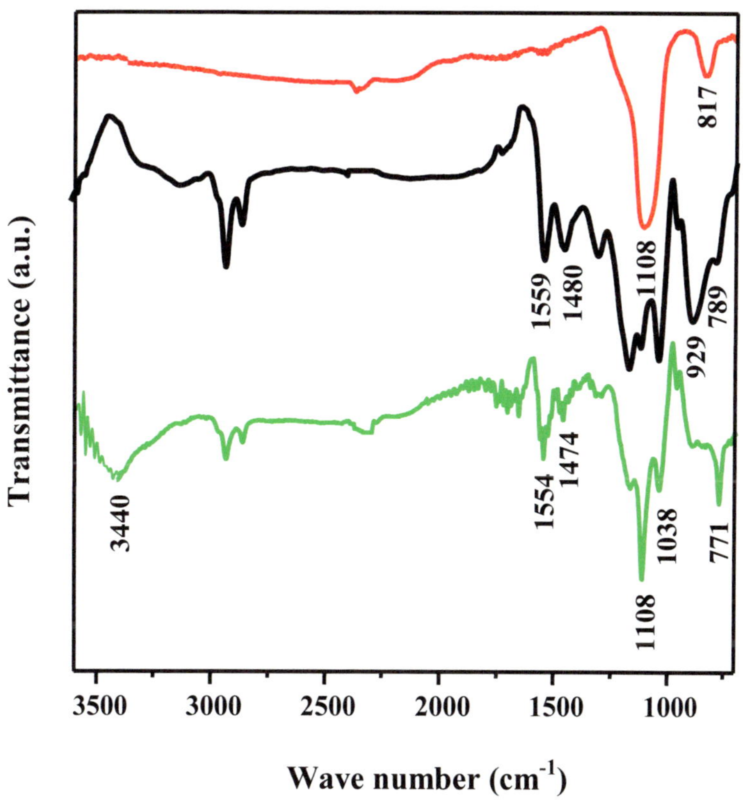

Figure 4. FTIR spectra of SiO_2 (▬▬), polypyrrole alone (▬▬), and polypyrrole/SiO_2 composite (▬▬). (Progress in Organic Coatings DOI: 10.1016/j.porgcoat.2014.04.013).

4.2. X-ray diffraction

Figure 5 shows the XRD powder diffraction pattern of SiO_2 and polymer composite. One can find a broad diffraction peak at 2θ of 24.3° corresponding to the semi crystalline nature of synthesized SiO_2 particles. For polypyrrole/SiO_2 composite, the broad characteristic peak has shifted to 22.2°. The shifting may be due to the SiO_2 particles embedded in the polymer matrix. A uniform covering of amorphous polypyrrole on the surface of SiO_2 is expected.

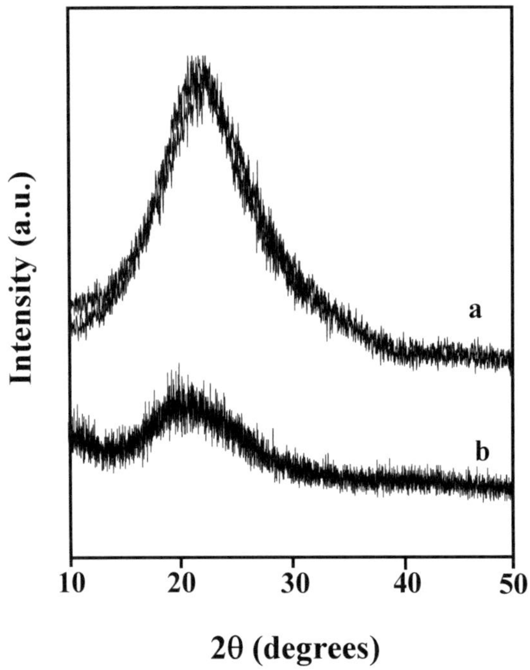

Figure 5. XRD graphs of (a) SiO_2 and (b) polypyrrole/SiO_2 composite. (Progress in Organic Coatings DOI: 10.1016/j.porgcoat.2014.04.013).

4.3. Thermo Gravimetric Analysis (TGA)

The thermogravimetric profiles of SiO_2, polypyrrole alone and polymer composite (PCs) are presented in Figure 6. The thermograms were recorded by heating the samples at a temperature range of 25 to 600°C under nitrogen atmosphere (60 ml/min). As shown in Figure 6, the SiO_2 particles are found to be thermally stable with almost no weight loss over the entire range of temperature. The thermogram of polypyrrole alone indicates three stage weight loss. In the first stage, 5-6% weight loss at temperature 100° C is attributed to the loss of water molecules from the polymer matrix [91].The weight loss at second stage that starts above 100°C until 220°C (about 7-8%) is basically due to the removal of dopants and low molecular weight oligomers. The gradual weight loss above 250°C represents the thermal decomposition of polypyrrole backbone [92]. The thermogram of polymer composite exhibits a three stage weight loss pattern similar to polypyrrole alone. It exhibits an initial weight loss of 6% (around 100°C) due to loss of residual water molecules followed by 7% weight loss (around 230° C) due to removal of dopants and oligomers. However, a considerable enhancement in the thermal stability is noticed for the polymer composite. With the addition of SiO_2, the thermal decomposition of polypyrrole (250°C) increases to 330°C for the polymer composite. The improvement in the thermal stability of the composite is certainly attributed to the presence of SiO_2 as filler. SiO_2 is stable in the range from room temperature to 600°C, when combined with polypyrrole, it restricts the thermal motion of the polypyrrole chains and shields the degradation of polymer [93]. The retarding effect of SiO_2 on the movement of polypyrrole chains also shows the relatively compact structure of polypyrrole/SiO_2 composite. The weight loss for polypyrrole alone and polymer composite at 600°C is found to be 63% and 57%, respectively. The decrease in weight loss indicates the interaction of polypyrrole with SiO_2 particles.

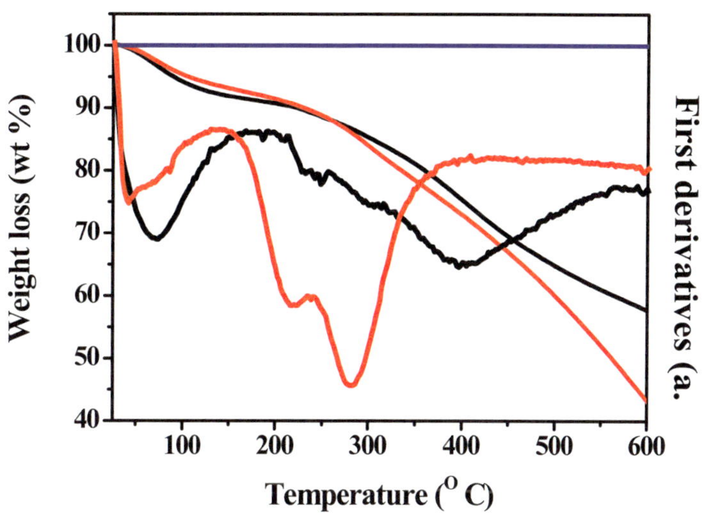

Figure 6. TGA graphs of SiO_2 (blue line), polypyrrole alone (red line) and polypyrrole/SiO_2 composite (black line). Their first derivatives are also shown here. (Progress in Organic Coatings DOI: 10.1016/j.porgcoat.2014.04.013).

4.4. SEM/TEM micrographs

Figure 7a depicts the scanning electron micrograph of synthesized SiO_2 particles. The particles exhibit spherical morphology with smooth texture. The average particle size of SiO_2 is 200 nm. The microstructural features of polymer composite (PCs), as shown in Figure 7b reveal a globular cluster of polypyrrole matrix incorporating SiO_2 particles. The entrapment of SiO_2 particles in the matrix can easily be noticed in Figure 7b. EDX analysis shows the presence of elements, Si (51.74%) and O (10.49%) from SiO_2, C (33.61%) and N (1.28%) from polypyrrole in the composite. The TEM image of polymer composite is shown in Figure 7c. It can be noticed that majority of SiO_2 particles are covered with polypyrrole. Some free separated SiO_2 particles can also be seen in the micrograph (Figure 7c). The SiO_2 particles are found to be well dispersed in polymer matrix. The large contact area of SiO_2 and polypyrrole in the polymer composite resulted in a more uniform composite structure. The microstructural analyses are in good agreement with FTIR and XRD results of polymer composite, confirming a strong interaction between polypyrrole and silica in the composite.

Figure 7. SEM micrographs showing morphology of (a) SiO_2 particles, (b) polypyrrole/SiO_2 composite. SiO_2 particles embedded inside the polymer matrix are visible in the micrograph. (c) TEM micrograph of polypyrrole/SiO_2 composite showing even distribution of SiO_2 particles in the composite. (Progress in Organic Coatings DOI: 10.1016/j.porgcoat.2014.04.013).

4.5. Corrosion studies of polymer composites in 3.5% NaCl solution

4.5.1. Tafel polarization

Figure 8 shows the Tafel polarisation behaviour of epoxy coated (EC) and epoxy with different wt% loading of polymer composites (PCs) coated steel specimens (PCs1, PCs2, PCs3 and PCs4) in 3.5% NaCl at room temperature (25± 3 °C).The plots were drawn to derive the values of different electrochemical parameters like corrosion potential (E_{corr}), corrosion current density (i_{corr}), anodic (β_a) and cathodic (β_c) Tafel constants. i_{corr}, β_a and β_c are measured by extrapolating the anodic and cathodic curves using Tafel extrapolation method (Table 1). The variability in the test results was found to be 2-3% for each measurement. Figure 8 exhibits a remarkable shift of E_{corr} towards positive potential for the epoxy with different wt% loading of PCs coated steel specimens as compared to the epoxy coated (EC) steel specimen. One can notice a slight reduction in E_{corr} for epoxy with 1 wt% loading of polymer composite coated steel specimen (PCs1) as compared epoxy coated steel (EC) specimen. However, the measured E_{corr} is found to be significantly increased for steel specimens coated with 2-4 wt% loading of polymer composite in epoxy. The E_{corr} occurred 177, 554 and 667 mV vs Ag/AgCl more positive for specimens PCs2, PCs3 and PCs4, respectively than the specimen EC (Table 1). The more positive values of E_{corr} for the epoxy with PCs coated specimens indicate the presence of a passive layer on the metal surface that control the anodic/cathodic reactions. The corrosion current density (i_{corr}) of epoxy coated steel specimen (Table 1) is observed to be eight times less than the bare mild steel (not shown here). Conventional polymeric coatings like epoxy are known to have good resistance towards diffusive ions and long exposure time is required to evidence degradation of polymeric chains [94]. The i_{corr} further reduced significantly with the addition of polymer composite in the epoxy resin. The trend of decrease of i_{corr} is in accordance with the increase of wt% loading of polymer composite in epoxy. The i_{corr} of PCs1 occurred four times less than the specimen EC. For specimens PCs2, PCs3 and PCs4, the measured i_{corr} is found to be two, more than four and four orders of magnitude less, respectively as compared to the specimen EC (Table 1). The significant reduction in the values of i_{corr} for the epoxy with polymer composite coated steel specimens indicate the effective corrosion protection performance of the polypyrrole/SiO_2 composite when added as additive in the epoxy resin. Further, the occurrence of notably higher value of anodic and cathodic Tafel constants for epoxy with polymer composite coated steel specimens as compared to epoxy coated specimens implies the effective role of polymer composite in controlling anodic and cathodic reactions. The combined effect of presence of conjugation in the polymer backbone and the presence of polar N-H group in the pyrrole ring acts as corrosion inhibitor [95]. Additionally, polypyrrole facilitates delocalisation of charge through it, which hinders the formation of localised anodic and cathodic currents. This ultimately makes the coated surface resistant towards corrosion that requires localization of charges [95]. Previous works carried out by Wessling [96], Sathiyanarayan et al. [97], Nguyen et al. [98] and Hema et. al. [99] explain the ability of conducting polymers to intercept electrons at metal surface and transport them. Similarly, works on PANI/SiO_2 composite in epoxy coatings also It is important to note that the epoxy with 4.0 wt% loading of PCs coated specimen (PCs4) exhibits more positive E_{corr} as compared to the epoxy with 3.0 wt% loading of PCs coated specimen (PCs3). But interestingly, the i_{corr} for specimen PCs3 is observed to be less than the specimen PCs4. This shows the better corrosion resistance of epoxy coating with 3.0 wt% loading of PCs. This could be the optimum limit of loading of PCs in the epoxy system.

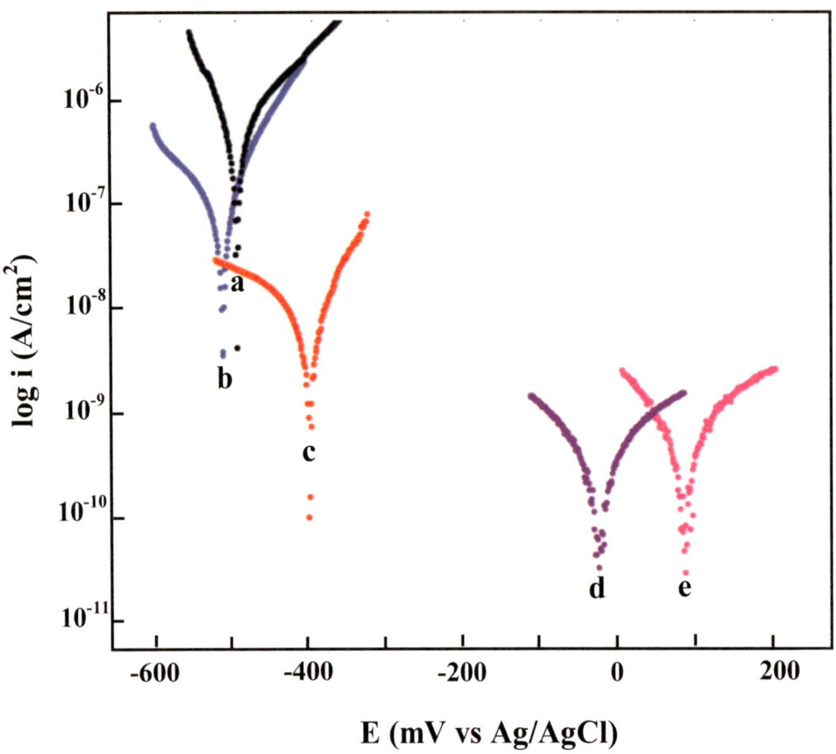

Figure 8. Tafel plots of (a) epoxy coated and epoxy with (b) 1.0 wt%, (c) 2.0 wt%, (d) 3.0 wt% and (e) 4.0 wt% loading of polypyrrole/SiO_2 composite (PCs) coated mild steel specimens in 3.5% NaCl solution at room temperature (25+3°C). (Progress in Organic Coatings DOI: 10.1016/j.porgcoat.2014.04.013).

Sample Name	Loading level of composite	E_{corr} (mV)	i_{corr} (A/cm^2)	β_a (mV/decade)	β_c (mV/decade)
EC	-----	-582.4	$8.4x10^{-7}$	92.1	189.8
PCs1	1.0 wt%	-523	$1.9x10^{-7}$	333.6	97.9
PCs2	2.0 wt%	-405.3	$1.8x10^{-8}$	659.4	126.7
PCs3	3.0 wt%	-28.3	$1.3x10^{-9}$	408.7	603.5
PCs4	4.0 wt%	83.5	$1.5x10^{-9}$	285.1	441.7

Table 2. Different electrochemical parameters obtained by Tafel extrapolation for epoxy and epoxy with different wt % loading of polymer composite in 3.5% NaCl solution.

The polarisation resistance (R_p) is inversely related to corrosion current density (i_{corr}) through Stern-Geary equation [100] as mentioned in equation 2.

http://dx.doi.org/10.5772/58388

$$R_p=\frac{\beta_a\beta_c}{2.3.3(\beta_a+\beta_c)}x\frac{1}{i_{corr}} \quad (2)$$

Since, the i_{corr} values reduced with the increase of percentage loading of PCs in the epoxy resin. Accordingly, the value polarisation resistance (R_p) increased with the increase of polymer composite in epoxy (Table 2). The measured R_p is found to be maximum for specimen PCs3 exhibiting its superior resistance towards diffusive chloride ions [101]. The corrosion protection efficiency (% P.E.) calculated from R_p values are shown in Table 2. % P.E. was determined from the measured R_p {polarisation resistance of epoxy coated mild steel (R_{ps}) and polarisation resistance of epoxy with different wt% loading of PCs coated mild steel (R_{pc})} values by using equation 3.

$$\%P>E>=\frac{R_{pc}-R_{ps}}{R_{pc}}x100 \quad (3)$$

The P.E. is observed to be 81.54%, 98.73% and 99.92% for specimens PCs1, PCs2 and PCs4, respectively. Up to 99.95 % P.E. was achieved for sample PCs3 (epoxy coating with 3 wt % loading of polymer composite).

Sample Name (%P.E)	R_p (Ω)	Protection Efficiency
EC	31.8 x 10^3	-----
PCs1	172.6 x 10^3	81.5
PCs2	2523.4 x 10^3	98.7
PCs3	77280 x 10^3	99.9
PCs4	45310 x 10^3	99.9

Table 3. Protection efficiency (%P.E.) calculated using polarisation resistance (R_p)

Another parameter that allows the evaluation of barrier nature of a surface coating is the cross linking density. It is reported that higher the cross linking density of the coating, lower is the diffusion of electrolyte through it [48]. The values of glass transition temperature (T_g) shows the extent of cross linking and is directly proportional to the cross linking density. Figure 9 shows the DSC curves of epoxy and epoxy with different wt% loading of polymer composites (PCs). The observed values of T_g are mentioned in Table 3. The T_g for epoxy is measured to be less than the epoxy with polymer composites. It shows the higher cross linking density of epoxy with PCs. Further, the T_g increases gradually with the increase of wt% loading of polymer composite in the epoxy system. This clearly indicates a better cross linking and hence low extent of electrolyte diffusion with the increase wt% of polymer composite. The results are in good agreement with the Tafel extrapolation test results.

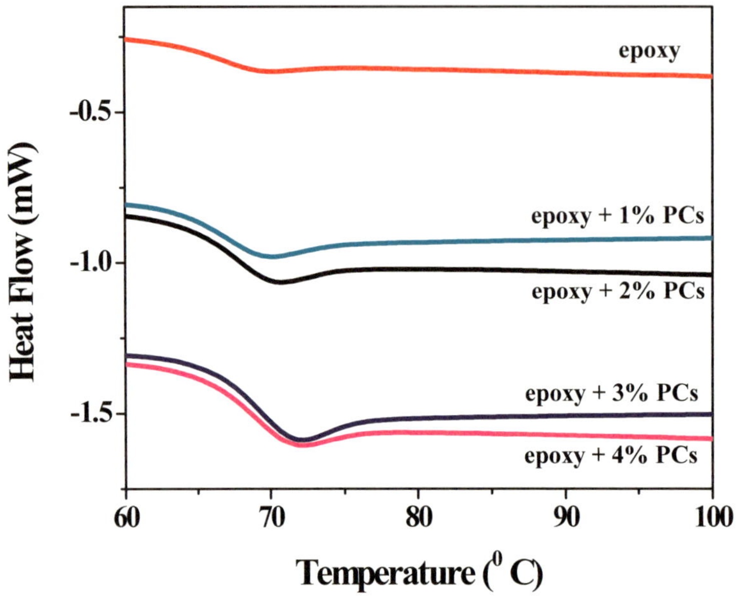

Figure 9. DSC thermograms of epoxy alone and epoxy with different wt% loading of polypyrrole/SiO_2 composites (PCs) recorded after second heating, at a heating rate of 10° C/min in nitrogen atmosphere. (Progress in Organic Coatings DOI: 10.1016/j.porgcoat.2014.04.013).

Loading level of Composite	T_g (° C)
0%	65.2
1.0%	66.6
2.0%	67.3
3.0%	69.2
4.0%	69.8

Table 4. Glass transition temperature (T_g) obtained after second heating.

4.5.2. EIS study

Electrochemical Impedance Spectroscopy (EIS) is a non-destructive electrochemical method used to estimate the performance of the coatings against corrosion. Impedance measurements were carried out at open circuit conditions in 3.5% NaCl solution at room temperature. The impedance graphs obtained for epoxy coated (EC) and epoxy with different wt% loading of polymer composites (PCs1, PCs2, PCs3 and PCs4) coated steel specimens are displayed in Nyquist (Figure. 10) and Bode plots (Figure. 11). The Nyquist plots for epoxy with polymer composite coated mild steel specimens show much higher impedance value (almost a straight line, Figure. 10c and d) as compared to epoxy coated specimen. However no apparent time constant can be resolved in the plots except for epoxy coating with 2 wt% loading of polymer composite (Figure 10b). So, an attempt to fit the equivalent circuit for the extraction of impedance parameters failed for coated specimens. The Nyquist plot for specimen PCs2 can be differentiated into a high frequency capacitive behaviour followed by low frequency diffusion controlled behaviour of the coating. The high frequency part of the plot represents the property of the coating, whereas, the low frequency part is associated with the Faradiac processes occurring at coating/metal interface through pores and cracks present in the coating [102]. The high frequency capacitive behaviour of the specimen PCs2 demonstrates the corrosion resistance property, while low frequency diffusion behaviour indicated occurrence of diffusion process at the coating/metal interface. Warburg impedance Z_w represents the porosities present in the coating, is expressed in equation 4.

$$Z_W = \sigma\omega^{-1/2} \tag{4}$$

Here, σ is the Warburg coefficient and ω is the angular frequency ($2\pi f$) at which the Warburg diffusion starts. The value of Warburg coefficient (σ) is calculated to be 17.7 x 10^5 Ωcm^{-1}(by putting the value of Z_w at frequency 12.1 Hz, where diffusion process starts). Further, the Nyquist plots for epoxy coating with 3 and 4 wt% loading of polymer composite (Figure 10 c and d) show capacitive property similar to an undamaged coating with significantly high impedance in the high frequency region. However, scattering of data points is noticed in the lower frequency region, is basically due to the surface heterogeneities.

The corresponding Bode plots are more informative, as it gives simultaneous measurement of modulus of impedance $|Z|$ with respect to frequency. The Bode plots, as shown in Figure 11 exhibit high frequency region due to the coating capacitance (C_c) and low frequency region due to the charge transfer processes occurring at solution/coating interface [103-105]. The Bode plots of specimens PCs3 and PCs4 show a straight line (slope-1) with high modulus of impedance ($|Z|$) in the high frequency region. This signifies the excellent barrier property of the surface film [106]. However, the slope of the plot becomes-1/2 in the lower frequency region (Figure 11). This is could be due to the adsorbed water on the surface of the coated sample. The value of $|Z|$ in the low frequency region signifies the good barrier property of the surface film [106]. The measured value of $|Z|$ (Table 4) is found to be significantly high (64940 x 10^3 Ωcm^2) for the specimen PCs3 (epoxy coating with 3.0 wt% loading of PCs). The magnitude of $|Z|$ at low frequency dropped gradually following the trend from specimens PCs3, PCs4, PCs2, PCs1 to EC (Table 4). The capacitance of the coating (C_c) is an important parameter to evaluate the failure of the coating, as it determines the extent of water uptake [106, 107]. The capacitance of a coating is related to $|Z|$ by equation (4).

$$|Z| = \frac{1}{2\pi f C_c} \tag{5}$$

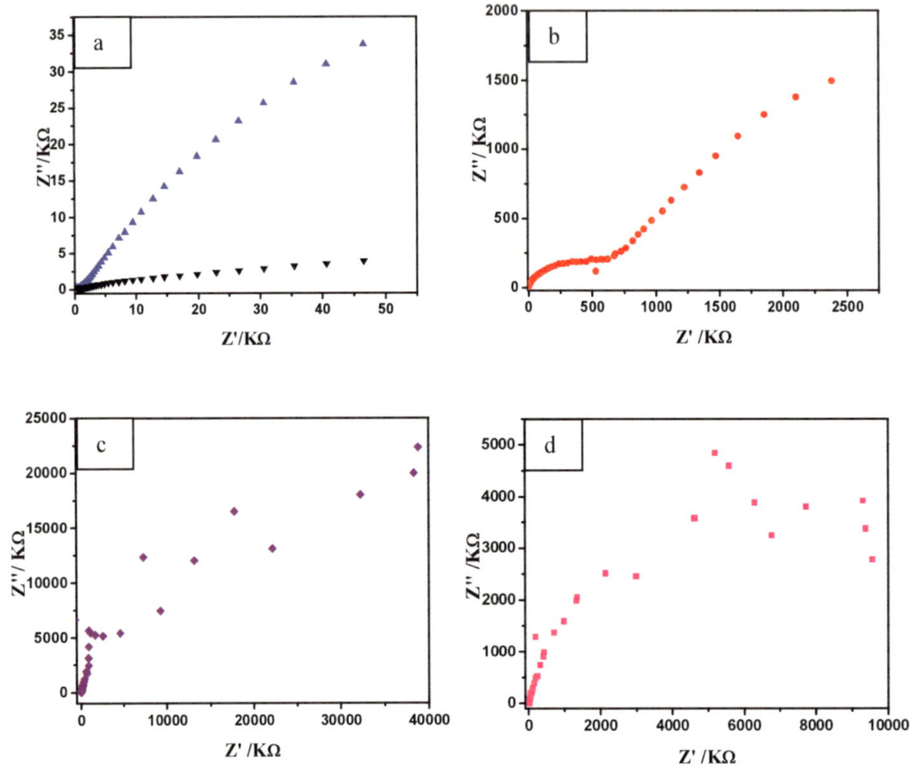

Figure 10. Nyquist plots of epoxy coated (▼) and epoxy with 1.0 wt% (▲),2.0 wt% (●), 3.0 wt% (♦) and 4.0 wt% (■) loading of polypyrrole/SiO_2 composite (PCs) coated mild steel specimens in 3.5% NaCl solution at room temperature (25+3°C). (Progress in Organic Coatings DOI: 10.1016/j.porgcoat.2014.04.013).

Where, f is the frequency of the applied AC voltage. The values of coating capacitance (C_c) are mentioned in Table 4. The measured C_c is found to be minimum for specimen PCs3 (2.44 x 10^{-8} F/cm^2) exhibiting the undamaged coating formed with 3.0 wt% loading of polymer composite in epoxy. The EIS data clearly shows the superior corrosion resistance property of epoxy coating with 3.0 wt% loading of PCs. The EIS data are in accordance with the Tafel polarisation test results.

From the above electrochemical measurements, it is clear that the polymer composite has made a significant improvement in the corrosion resistance properties of the conventional epoxy coatings in 3.5% NaCl solution. However, it is necessary to evaluate the corrosion resistance of the coating for relatively longer period of exposure to corrosive electrolyte. Since, epoxy with 3 wt% loading of polymer composite coated steel specimen (PCs3) has shown the superior corrosion resistance amongst the coated specimens, therefore, it has been considered for periodic impedance measurement. Figure 12 shows the variation of modulus of impedance (|Z|) and coating capacitance (C_c) with the immersion time. The magnitude of impedance reduced significantly for the first 10 days of immersion. This is basically due to the weakening

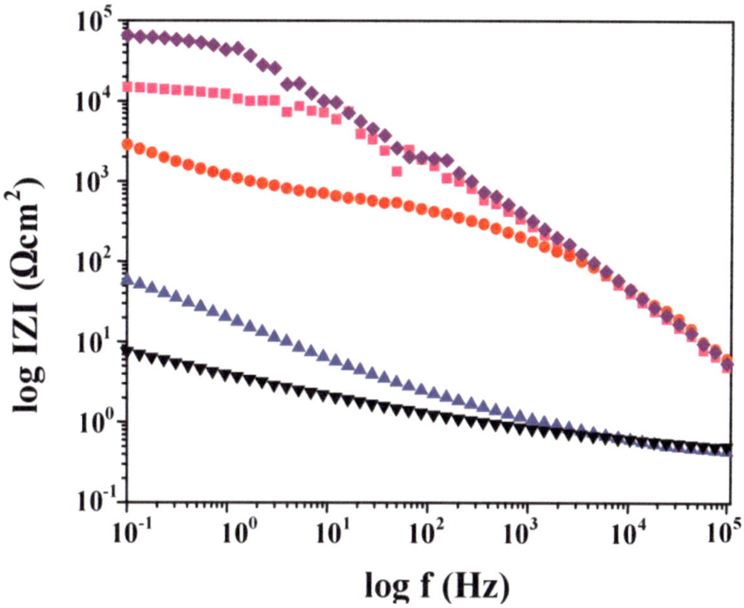

Figure 11. Bode plots of epoxy coated (▼) and epoxy with 1.0 wt% (▲), 2.0 wt% (●), 3.0 wt% (♦) and 4.0 wt% (■) loading of polypyrrole/SiO_2 composite (PCs) coated mild steel specimens in 3.5% NaCl solution at room temperature (25 ± 3°C). (Progress in Organic Coatings DOI: 10.1016/j.porgcoat.2014.04.013).

Sample Name	Loading level of composite	|Z| (KΩ cm²)	Cc (F/cm²)
EC	0 %	7.5	2.1×10^{-4}
PCs1	1.0 wt%	57.4	2.7×10^{-5}
PCs2	2.0 wt%	2810	5.6×10^{-7}
PCs3	3.0 wt%	64940	2.4×10^{-8}
PCs4	4.0 wt%	14750	1.0×10^{-7}

Table 5. Different electrochemical parameters extracted by EIS measurements for epoxy and epoxy with different wt % loading of polymer composite in 3.5% NaCl solution.

of the barrier property of the surface film due to diffusive chloride ions. Accordingly, an increase in coating capacitance (C_c) is noticed for this period. The increase in the value of coating capacitance is due to the electrolyte uptake with the passage of immersion time [108, 109]. The high dielectric constant of water (ε=80) increases the dielectric constant of the organic coatings when exposed to it leading to increase of coating capacitance [110]. The above observations reveal the weakening of the barrier property of the coating with immersion time. However, the coating capacitance becomes almost constant after 15 days of immersion. This could be due to the clogging of the pores of the coatings by the corrosion products. Interestingly, the measured |Z| for specimen PCs3 after 20 days of immersion (16.6 KΩ cm²) is still observed to be higher than the freshly immersed epoxy coated specimen (7.5 KΩ cm^2). Similarly, the C_c for specimen PCs3 after 20 days of immersion (9.4 x 10^{-5} F/cm^2) is lower than the freshly immersed epoxy coated specimen (2.1 x 10^{-4} F/cm^2). This clearly shows the superior barrier property shown by epoxy coatings with polymer composite.

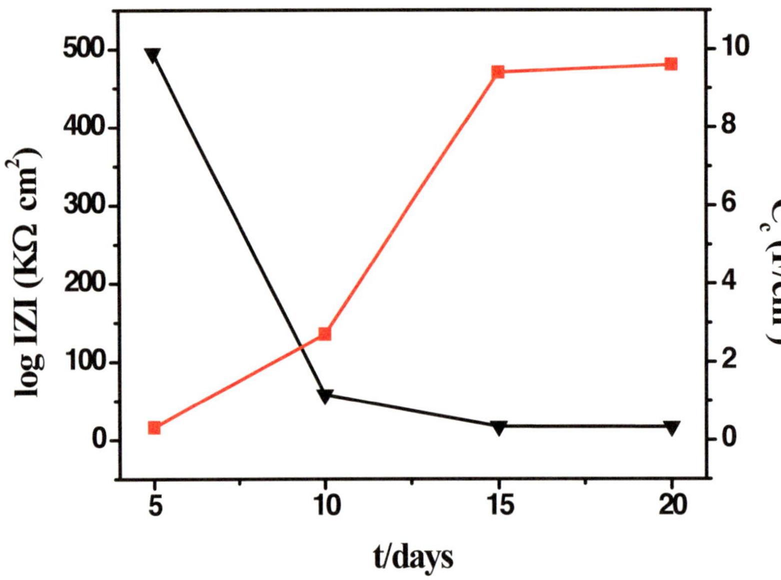

Figure 12. Variation of magnitude of impedance (|Z|) and coating capacitance (Cc) with time for epoxy with 3 wt% loading of polymer composite coated steel specimen exposed to 3.5% NaCl solution. (Progress in Organic Coatings DOI: 10.1016/j.porgcoat.2014.04.013).

4.5.3. Salt spray test

Figure 13 shows the photographs of epoxy coated (EC) and epoxy with different wt% loading of polymer composites (PCs1, PCs2, PCs3 and PCs4) coated steel panels after exposure to salt spray fog for 65 days. Figure 13a clearly shows the occurrence of severe rusting throughout the scribe mark for epoxy coated steel panel. The under film blisters were also noticed visually. The presence of rust and blisters indicates loss of adherence of the epoxy coating during prolong exposure to salt fog. The presence of polymer composites in epoxy improves the corrosion resistance properties of the steel, which can be noticed from the photographs (Figure 13b-e). The epoxy with 1.0 and 2.0 wt% loading of PCs coated steel panels exhibit less extended corrosion along the scribe mark (Figure 13b and c). Further, the extended corroded area along the scribe mark almost disappeared for steel panels coated with epoxy with 3.0 and 4.0 wt% loading of PCs (Figure 13d and e). The results clearly show that the epoxy powder coating formulations incorporating polypyrrole/SiO_2 composites are very effective in preventing corrosion and blistering near the scribe mark when exposed to environmental conditions of high humidity and high salt content. The remarkably high corrosion protection of epoxy coatings with polymer composites as compared to epoxy coating alone is due to the dual protection mechanism by the synthesized composite in the coating system. Polypyrrole forms a passive layer and simultaneously act as a physical barrier between metal and electrolyte. On the other hand, SiO_2 particles as filler in polypyrrole matrix provide mechanical integrity and reduce the degradation of polymer composite coating under corrosive conditions.

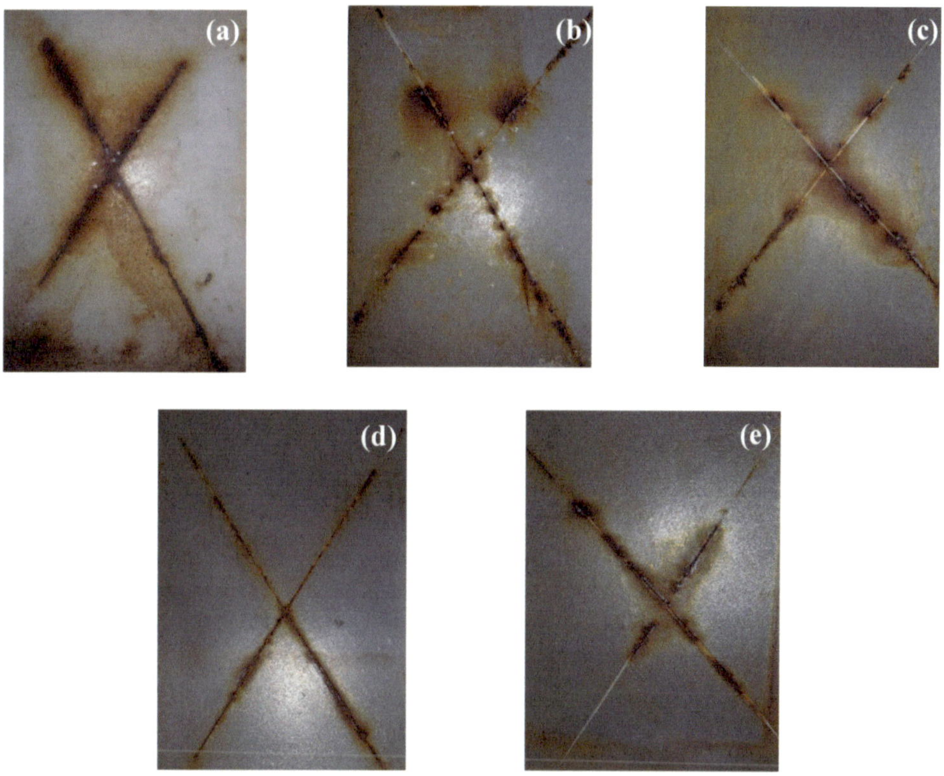

Figure 13. Photographs of (a) epoxy coated and epoxy with (b) 1.0 wt%, (c) 2.0 wt%, (d) 3.0 wt% and (e) 4.0 wt% loading of polypyrrole/SiO_2 composite (PCs) coated mild steel specimens exposed to salt spray fog after 65 days. (Progress in Organic Coatings DOI: 10.1016/j.porgcoat.2014.04.013).

5. Anti corrosive properties of chitosan/polymer composite in 3.5% NaCl solution

5.1. Synthesis of chitosan/polypyrrole/SiO_2 composite (CsPC)

Chitosan with a degree of deacetylation 90% was a fine pale yellow powder (100 mesh) ground from shrimp shell was used for the synthesis of the composite. The synthesis of mono dispersed uniform-sized SiO_2 particles were carried out by hydrolysis of tetra-ethyl orthosilicate (TEOS) in ethanol using ammonia as a catalyst as mentioned in section 3.1. Chitosan/polymer composites (CsPC) were synthesized by chemical oxidative polymerization of pyrrole monomer in presence of 1wt% chitosan solution (in 1% acetic acid) [111]. For this, a mixture of 1 gm chitosan (Cs), 1 gm acetic acid and 98 gms of deionized water was stirred till a clear pale solution (pH ~ 3) of chitosan acetate appears [112]. The synthesized silica particles were dispersed in suspension of sodium lauryl sulphate (SLS) in distilled water. The suspension was added to chitosan solution and was allowed to stir for 20 minutes. It is reported that chitosan interacts with anionic surfactants by electrostatic attraction to form soluble complexes [113, 114]. Thereafter, pyrrole was introduced in it, followed by drop by drop addition of freshly prepared $FeCl_3$ to initiate the polymerization of pyrrole. The molar ratio of $FeCl_3$: pyrrole : SLS was taken to be 1:1:0.5. The resultant composite was filtered and washed with water and ethanol to remove oligomers and oxidant. Thereafter, it was dried in a vacuum oven at 60^0 C. The schematic of the synthesis of chitosan/polymer composite is shown in Figure 14.

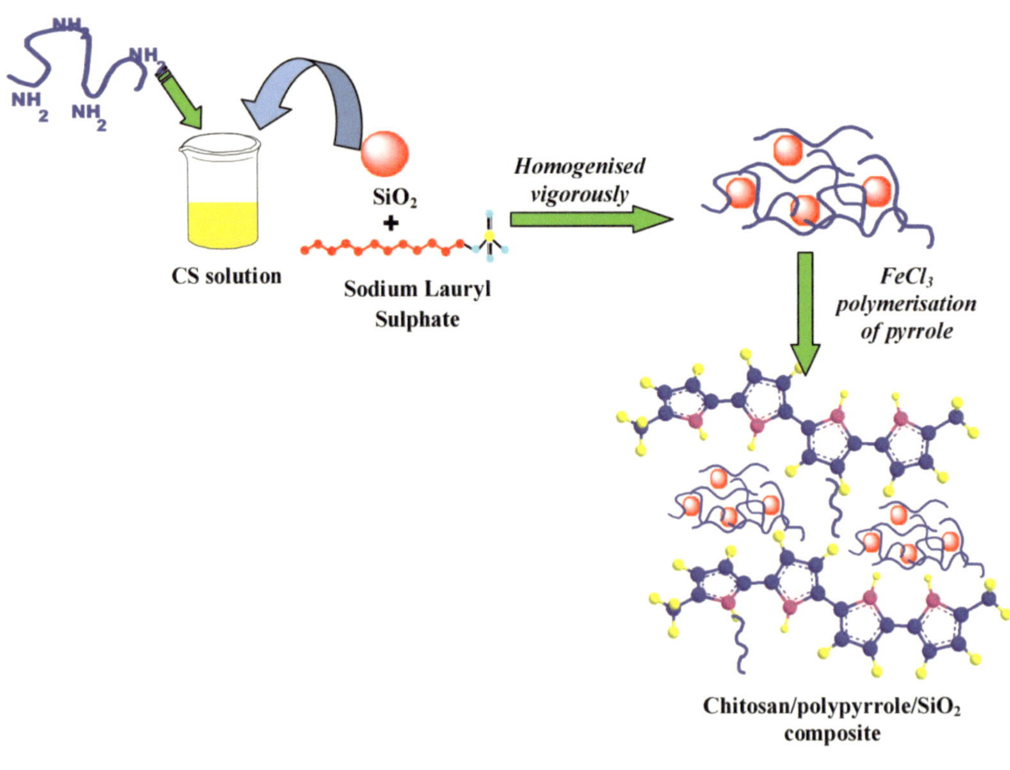

Figure 14. Schematic of the synthesis of chitosan/polypyrrole/SiO_2 composites. (Communicated to Progress in Organic Coatings, Ref. 110, Elsevier)

5.2. Powder coating on mild steel surface

Mild steel specimens were polished metallographically, prior to the development of coating. The powder coating formulations were prepared by blending composites in epoxy resin. A homogenous mixture of well dispersed composites in epoxy was obtained after ball milling treatment. The coatings were applied on mild steel specimens using an electrostatic spray gun held at 67.4 KV potential with respect to the substrate (grounded). The powder coated mild steel specimens were cured at $140^{0}C$ for 30 minutes. The adhesion of the coating was tested by tape test as per ASTM D3359-02 and found to pass the test.

6. Characterization of chitosan/polymer composites

6.1. FTIR spectra

The FTIR spectrum of chitosan, as shown in Figure 15 exhibits an intense broad band in the spectral range of 3200-3650 cm^{-1}. It is attributed to the axial stretching of O-H and N-H bonds. A small band near 2870 cm^{-1} is due to the C-H stretching vibration. A peak of considerable intensity at 1650 cm^{-1} is assigned to amide I vibration and peaks at 1426 and 1382 cm^{-1} is the result of coupling of C-N axial stretching and N-H angular deformation. The peaks at 1155, 1074 and 1024 cm^{-1} corresponds to the stretching vibration of C-O-C linkage in the glucosamine rings. The FTIR spectrum of chitosan-polypyrrole composite (CP) exhibits a broad band at 3421 cm^{-1} showing the N-H stretching of pyrrole and O-H stretching of chitosan. The characteristic peaks of polypyrrole at 1546 cm^{-1} (C=C benzoic form), 1458 cm^{-1} (C-N stretching) was observed with considerable intensity [115-117]. The peak at 1172 cm^{-1} (S-O stretching) and 1035 cm^{-1} (S-C stretching) confirm the formation of chitosan/polypyrrole composite doped with SLS. It is important to note that the reported peak for S-O stretching for polypyrrole occurs at 1168 cm^{-1}, which has been shifted to 1172 cm^{-1} for the composite. It clearly shows interaction of chitosan and polypyrrole with SLS. The corresponding peak for amide vibration at 1650 cm^{-1} of chitosan was not observed for the composite. This shows the presence of mainly protonated amino groups as a result of dissolution of chitosan in acetic acid. The other peaks at 902, 768 and 670 cm^{-1} are due to the C-H out of plane deformation vibration of the ring. An additional peak of considerable intensity appears at 1108 cm^{-1}, solely due to the presence of SiO_2 in chitosan/polypyrrole/SiO_2 composite (CsPC). However, this peak is slightly shifted, as the peak for Si-O-Si stretching in SiO_2 occurs at 1080 cm^{-1}. The shifting of the peak could be due to the adsorption of chitosan/polypyrrole on the surface of silica particles. It is reported that chitosan provides the formation of active sites for grafting of polypyrrole chains on silica particles and acts as a stabilizer of silica-polypyrrole particles [118]. The FTIR results exhibit strong interaction between chitosan, polypyrrole and silica in the composite.

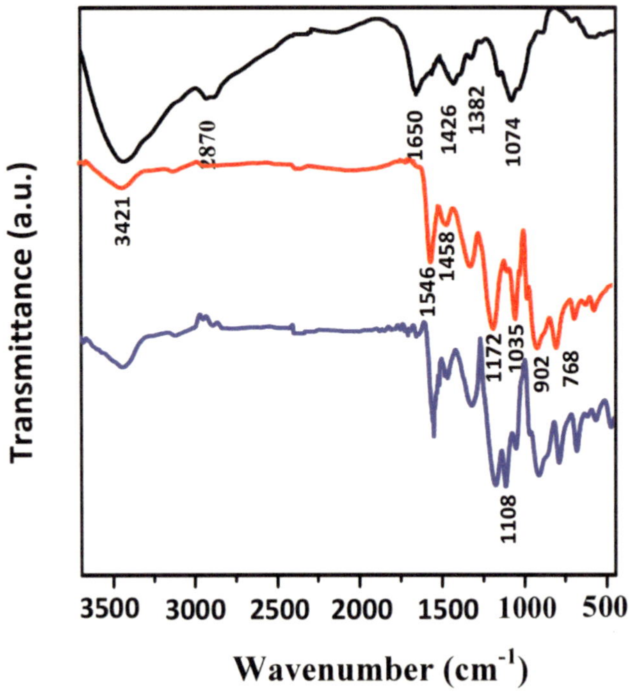

Figure 15. FTIR spectra of chitosan (——), chitosan/polypyrrole (——) and chitosan/polypyrrole/SiO_2 (——) composite. (Communicated to Progress in Organic Coatings, Ref. 110, Elsevier)

6.2. XRD diffraction

The XRD diffraction patterns of chitosan (a) and chitosan/polypyrrole/SiO_2 composite (CsPC) are presented in Figure 16. The presence of a scattering peak at 2θ value 19.88° corresponding to (020) plane is basically due to the semi crystalline nature of chitosan [119, 120]. The diffractogram of CsPC showed a broad scattering peak at approximately 26°-27°, which indicates a highly amorphous structure of the composite. This shows the interaction between chitosan and polypyrrole. The absence of corresponding peaks of silica is due to the uniform covering of polypyrrole chains on the silica particles [118].

6.3. Thermogravimetric analysis of composites

TGA curves for chitosan, chitosan/polypyrrole (CP) and chitosan/polypyrrole/SiO_2 (CsPC) composites are presented in Figure 17. The curves were recorded by heating the samples from 25° to 800° C under constant heating rate of 15°C/min under nitrogen atmosphere (60 ml/min). The first stage of all the three curves shows 10% weight loss between 25° to 145° C. The weight loss is attributed to the loss of moisture and other volatile compounds. The second weight loss of 34 % for chitosan occurs at 240° C, is mainly due to thermal decomposition of chitosan [121, 122]. The second weight loss due to thermal decomposition of polymer composites is delayed to 260° and 265° C for CP and CsPC, respectively. Aditionally, the chitosan and CP showed a

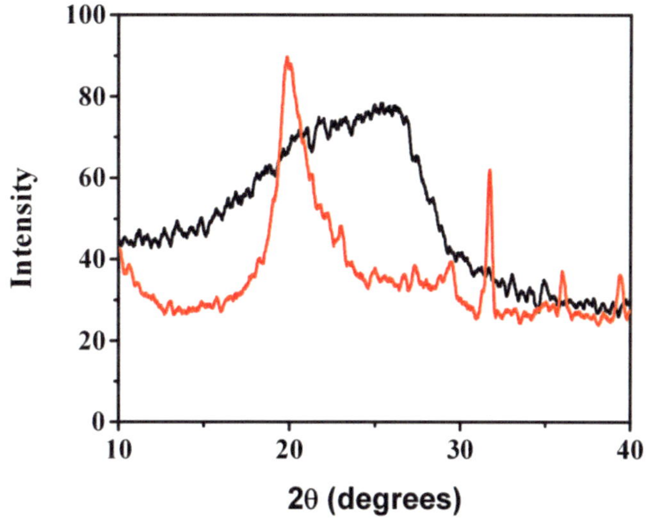

Figure 16. XRD graphs of (a) chitosan and (b) chitosan/polypyrrole/SiO_2 composite. (Communicated to Progress in Organic Coatings, Ref. 110, Elsevier)

weight loss of 61% and 54%, respectively at 800° C. The comparatively less weight loss of chitosan/polypyrrole composite (CP) as compared to chitosan shows the better thermal stability of composite. The incorporation of SiO_2 in the polymer matrix further improves the thermal stability of the composite as the weight loss at 800° C was only 49%. The increase in thermal stability of CsPC can be attributed to the better thermal stability of silica particles and to the interactions between silica and chitosan/polypyrrole.

6.4. SEM/TEM images of chitosan/polymer composites

Figure 18 (a-c) depict the FE-SEM and TEM images of composites. The FE-SEM image of the chitosan/polypyrrole/SiO_2 composite (CsPC) reveals a clustered structure composed of spherical nano particles (diameters ~ 40-50 nm) of polypyrrole and chitosan (Figure 18 a). The presence of SiO_2 particles embedded in chitosan/polypyrrole matrix can easily be noticed in the micrograph. The TEM image of CsPC (Figure 18 b) exhibits the presence of distinct spherical nano particles of chitosan and polypyrrole arranged in a regular pattern. It is reported that the polymerisation of pyrrole in absence of chitosan, forms polypyrrole particles having diameters 100-150 nm with aggregated cauliflower morphology [69]. On the other hand, the in situ chemical polymerisation of pyrrole in presence of chitosan causes the formation of much smaller polypyrrole nano spheres (40-60 nm). Here, chitosan acts as a steric stabiliser and hinders the formation of large size polypyrrole particles [67]. The TEM micrograph of polypyrrole/SiO_2 composite (PS) (Figure 18c) shows the presence of large size polypyrrole particles with diffused morphology. The EDAX spectrum of CsPC confirms the presence of carbon, nitrogen, sulphur, oxygen and silicon in the composite. The microstructural features of the composites are in good agreement with the FTIR results showing strong interaction between polypyrrole and chitosan.

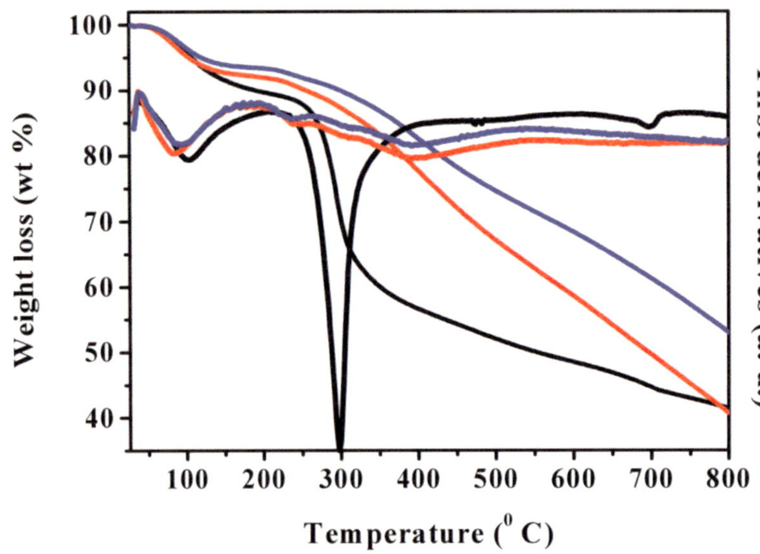

Figure 17. TGA curves of chitosan (▬▬), chitosan/polypyrrole (▬▬) and chitosan/polypyrrole/SiO_2 (▬▬) composite. (Communicated to Progress in Organic Coatings, Ref. 110, Elsevier)

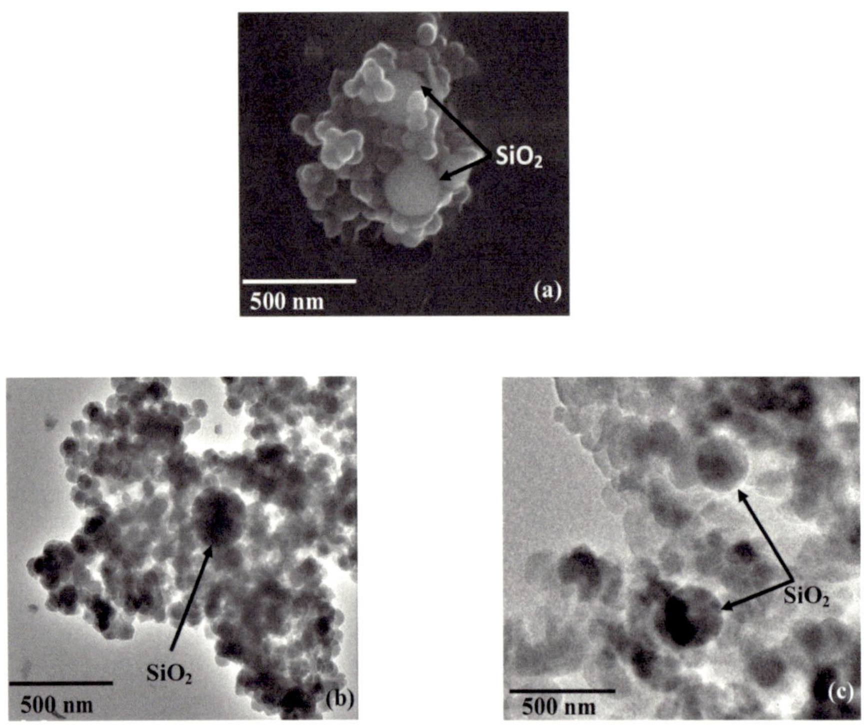

Figure 18. (a) FESEM micrograph of chitosan/polypyrrole/SiO2 composite. TEM micrographs of (b) Chitosan/polypyrrole/SiO_2 and (c) polypyrrole/SiO_2 composites. (Communicated to Progress in Organic Coatings, Ref. 110, Elsevier)

6.5. Corrosion studies of chitosan/polymer composites in 3.5% NaCl solution

6.5.1. Tafel polarisation

Figure 19 shows the Tafel polarisation curves for epoxy coated (EC) and epoxy with different wt% loadings of chitosan/polymer composite coated steel specimens (CsPC1 for 1.0%, CsPC2 for 2.0% and CsPC3 for 3.0%) exposed to 3.5% NaCl solution. For comparative studies Tafel plots are drawn for polypyrrole/SiO_2 composite with 4 wt% loading in epoxy,PCs4 (without chitosan) and chitosan/polypyrrole composite (CP) (without SiO_2) coated steel specimens. Different electrochemical parameters, such as corrosion potential (E_{corr}) and corrosion current density (i_{corr}) are derived from curve fitting method and are summarized in Table 5.

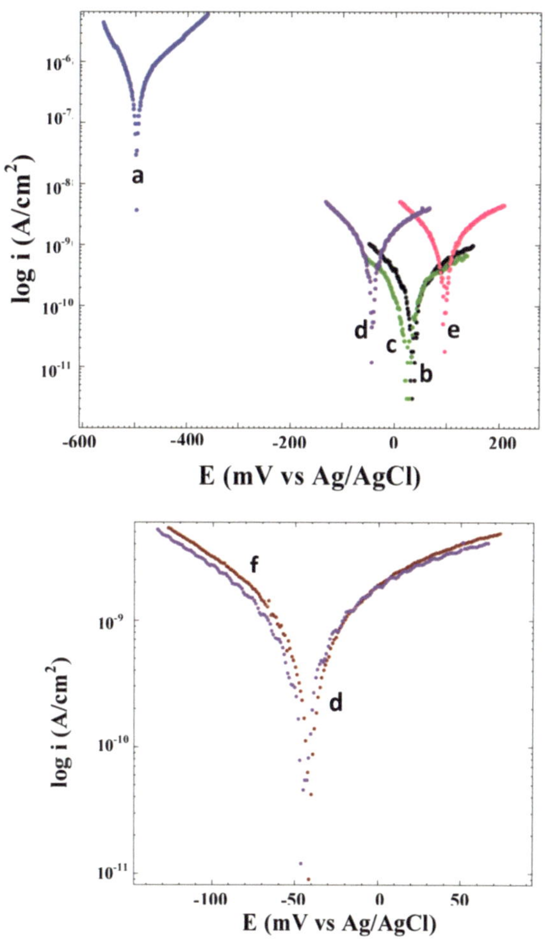

Figure 19. (a) Tafel plots of (a) epoxy coated and epoxy with (b)1.0 wt%, (c) 2.0 wt%, (d) 3.0 wt% chitosan/polypyrrole/SiO_2 composite, (e) polypyrrole/SiO_2, (f) chitosan/polypyrrole composite coated mild steel specimens in 3.5% NaCl solution at room temperature (25+3°C). (Communicated to Progress in Organic Coatings, Ref. 100, Elsevier)

The E_{corr} for epoxy coated mild steel is observed to be-582.09 mV. One can notice a remarkable shift of E_{corr} in the noble direction for all the composite coated steel specimens exhibiting the presence of corrosion resistant passive layer on the metal surface. The E_{corr} of specimen PCs4 occurs at 95.43 mV. It is important to mention that the polypyrrole/SiO_2 composite coating (PCs4) that was taken for the comparative studies is developed by 4.0 wt % loading in epoxy powder formulation. Accordingly, the i_{corr} value of PCs4 is observed to be three orders of magnitude less than the epoxy coated steel, showing its superior corrosion resistance property. On the other hand, the chitosan/polypyrrole/SiO_2 composite coated steel specimens, CsPC1 and CsPC2 show the E_{corr} values 36.46 mV and 20.75 mV, respectively, (Table 5) which was slightly less than the E_{corr} observed for specimen PCs4. However, the corrosion current density (i_{corr}) values of CsPC1 and CsPC2 were found to be almost 7.73 and 16 times respectively less as compared to the specimen PCs4. The significantly less i_{corr} values for the chitosan/polypyrrole/SiO_2 composite coated specimens clearly exhibit the efficient role of chitosan in enhancing the corrosion protection tendency of the composite. It is be due to the synergistic effect of the chitosan and polypyrole that resulted in the formation of coatings, which is less susceptible to moisture and prevent the penetration of corrosive ions. Further, the i_{corr} value increased slightly for specimen CsPC3 as compared to CsPC1 and CsPC2. This reveals the deterioration of corrosion protection tendency of the composite coating with further increase of the percentage loading in the epoxy. The 2.0 wt% loading of CsPC in epoxy could be the best possible combination of coating formulation. The role of SiO_2 particles cannot be ignored in the composite system, as the E_{corr} and i_{corr} values of specimen CP is found to be almost equal to the specimen CsPC3 (the chitosan/polymer composite coated specimen with least corrosion resistance). The better thermal stability of chitosan/polypyrrole/SiO_2 composite is discussed in TGA analysis. The remarkably high corrosion resistance of chitosan/polypyrrole/SiO_2 composite coatings could be due to the combined effect of chitosan and polypyrrole as corrosion inhibitors. The polarisation resistance (R_p) values are related to corrosion current density (i_{corr}), which is directly proportional to corrosion rate through Stern-Geary equation as mentioned in equation (2) are also shown in Table 5. The maximum R_p value measured for sample CsPC2, further confirms the superior corrosion resistance property of the epoxy with 2 wt% loading of chitosan/polypyrrole/SiO_2 composite. The maximum of 99.97% corrosion protection efficiency (P.E.) is achieved for specimen CsPC2. So, chitosan/polypyrrole/SiO_2 composite coating with 2.0 wt% loading in epoxy powder formulation exhibits superior corrosion resistance among all test specimens in 3.5% NaCl solution.

Figure 20 shows the DSC thermograms of epoxy and epoxy with chitosan/polypyrrole/SiO_2 (CsPC2), chitosan/polypyrrole (CP) composite. In order to evaluate the role of chitosan in the composite, DSC thermograms of epoxy with polypyrrole/SiO_2 composites (PCs4) are also shown here. The glass transition temperature (T_g) is related to cross linking density [85], that relates to the barrier nature of the surface film. The observed T_g is found to increase with the incorporation of polymer composite in the epoxy resin. The T_g for epoxy with chitosan/polypyrrole/SiO_2 composite (75.1°C) is found to be almost 10°C more than the epoxy resin, showing the higher cross linking density with superior barrier property of the chitosan/polymer composite. On the other hand, the T_g for epoxy with polypyrrole/SiO_2 composite

Sample Name	Loading level of polymer	E_{corr} (mV)	i_{corr} (A/cm^2)	R_p (KΩ)	Protection Efficiency (% P.E.)
EC	-----	-582.49	$8.45x10^{-7}$	31.87	----
CsPC1	1.0 wt%	36.46	$0.50x10^{-9}$	105580	99.96
CsPC2	2.0 wt%	20.75	$0.24x10^{-9}$	156200	99.97
CsPC3	3.0 wt%	-46.68	$3.79x10^{-9}$	22632	99.85
PCs4	4.0 wt%	83.522	$1.51x10^{-9}$	45310	99.92
CP	1.0 wt%	-41.62	$3.27x10^{-9}$	20580	99.84

Table 6. Different electrochemical parameters obtained by Tafel extrapolation in 3.5% NaCl solution.

occurs 6°C less than the epoxy with chitosan/polypyrrole/SiO_2 composite. The excellent film forming tendency of chitosan and the synergistic interaction of chitosan and polypyrrole are the reasons for the occurrence of higher cross linking density of chitosan/polypyrrole/SiO_2. It is reported that higher the cross linking density of the coating, lower is the diffusion of electrolyte through it [85]. Therefore, epoxy with chitosan/polypyrrole/SiO_2 composite has the superior barrier property. The results are in accordance with the Tafel extrapolation test results.

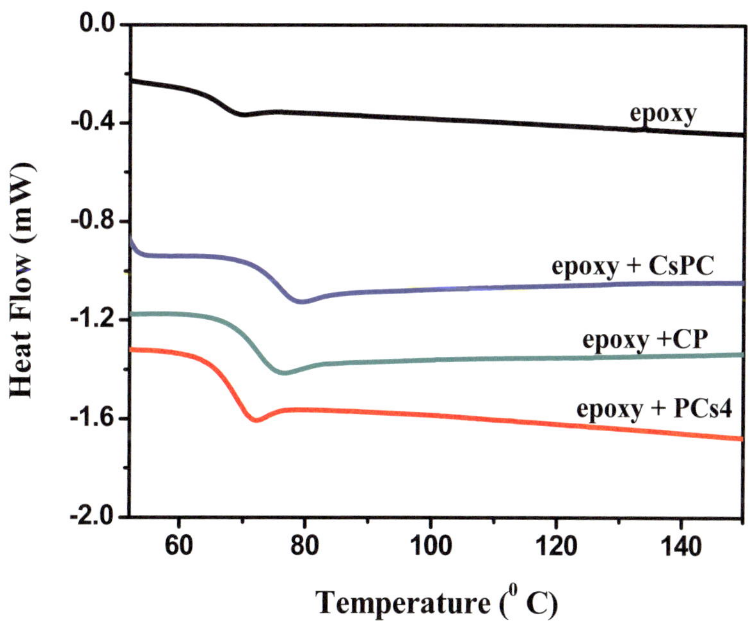

Figure 20. DSC thermograms of epoxy and epoxy with 2% loading of chitosan/polypyrrole/SiO_2 (CsPC), chitosan/polypyrrole composites (CP) and polypyrrole/SiO_2 (PCs4) composite recorded after second heating, at a heating rate of 10° C/min in nitrogen atmosphere. (Communicated to Progress in Organic Coatings, Ref. 110, Elsevier)

Sample name	T_g (° C)
epoxy	65.2
epoxy+CsPC	75.1
epoxy+CP	71.3
epoxy + PCs4	69.2

Table 7. Glass transition temperature (T_g) obtained after second heating.

6.5.2. Electrochemical impedance spectroscopy (EIS)

Impedance measurements were carried out in terms of Bode plots at open circuit conditions in 3.5% NaCl solution at room temperature (25 ± 3° C). The Bode plots obtained for epoxy coated (EC), chitosan/polypyrrole/SiO_2 composite coated with different wt% loading in epoxy (CsPC1, CsPC2 and CsPC3), chitosan/polypyrrole composite (CP) coated and polypyrrole/SiO_2 composite (PS) coated steel specimens are displayed in Figure 21. Bode plot is informative as it gives simultaneous measurement of modulus of impedance |Z| with respect to frequency. As shown in Figure 21, a high magnitude of impedance (|Z|) with a slope of-1 at frequency ~ 10 Hz, for the composite coated steel specimens specimens signify an excellent barrier property of the surface film and correspond to the less corrosive ions being in contact with the metal surface [123]. The respective region for the epoxy coated steel specimen shows a significantly less |Z| (Table 7) revealing its weak barrier property against corrosive ions. Further, the Bode plots show a resistive platue (horizontal line) in the low frequency region (< 1Hz). The total impedance (|Z|) in this region is mainly due to the impedance of the coating defects and can be considered as a measure of coating protectiveness [81, 124]. The magnitude of total impedance for coated specimens, CsPC1 and CsPC2 in this region is found to be four orders of magnitude higher as compared to the epoxy coated steel specimen (Table 7). Although, the |Z| values for specimen CP and PCs4 is almost three orders of magnitude higher as compared to epoxy coated steel (EC). But, |Z| values occur an order of magnitude less than the specimens CsPC1 and CsPC2. A much high magnitude of impedance for chitosan/polypyrrole/SiO_2 composite coated specimens shows the presence of surface film with superior corrosion resistance. However, the low value of impedance for epoxy coated steel is associated with the high defectiveness of the coating. Further, the capacitance of the coating (C_c) is an important parameter to evaluate the failure of the coating as it determines the extent of water uptake [106]. The capacitance of a coating is related to the magnitude of the impedance (|Z|) by the equation (3) is given in Table 7. The value of C_c is found to be least for specimen CsPC2 exhibiting the undamaged coating formed with 2.0 wt% loading of chitosan/polypyrrole/SiO_2 composite in epoxy coating formulation. As expected, the highest value of C_c for epoxy coated specimens (EC) shows its greater electrolyte uptake tendency. The observed results are in accordance with the Tafel polarisation results.

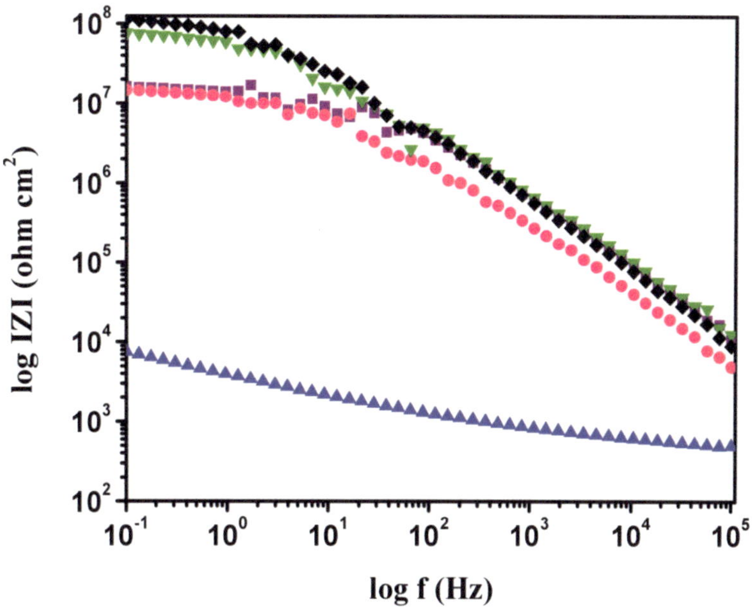

Figure 21. Bode plots of epoxy coated (▲) and epoxy with 1.0 wt% (▼), 2.0 wt% (◆), 3.0 wt% (■) and polypyrrole/ SiO_2 composite (*) coated mild steel specimens in 3.5% NaCl solution at room temperature (25+3°C). (Communicated to Progress in Organic Coatings, Ref. 110, Elsevier)

Sample Name	Loading level of polymer	|Z| (KΩ cm²)	C_c (F/cm²)
EC	0 %	7.55	2.1 x 10^{-4}
CsPC1	1.0 wt%	76669	2.0 x 10^{-8}
CsPC2	2.0 wt%	104200	1.5 x 10^{-8}
CsPC3	3.0 wt%	16220	9.9 x 10^{-8}
PCs4	4.0 wt%	14750	1.0 x 10^{-7}
CP	1.0 wt%	16281	2.7 x 10^{-7}

Table 8. Different electrochemical parameters extracted by EIS measurements in 3.5% NaCl solution.

6.6. Salt spray tests

Salt spray tests results are displayed in Figure 22. Figure shows the photographs of epoxy coated (a) and epoxy with different wt% loading of polymer composites, CsPC1 (b), CsPC2 (c), CsPC3 (d), CP (e) and PCs4 (f) coated steel panels after exposure to salt spray fog for 65 days. Figure 22a clearly shows rusting throughout the scribe mark for epoxy coated steel panel. The presence of rust and blisters indicates the loss of adherence of the epoxy coating during prolong exposure to salt fog. The incorporation of chitosan/polypyrrole/SiO_2 composites (CsPC) in epoxy improve the corrosion resistance properties of the steel that can be noticed from the photographs (Figure 22b-d). The epoxy with 1.0 and 2.0 wt% loading of CsPC coated steel panels exhibit a less corrosion along the scribe mark (Figure 22b and c). However, the corrosion near scribe mark is visible for steel panels coated with epoxy with 3.0 % loading of CsPC (Figure 22d). The results clearly show that the epoxy powder coating formulations incorporating chitosan/polypyrrole/SiO_2 composites are very effective in preventing corrosion and blistering near the scribe mark when exposed to environmental conditions of high humidity and high salt content. Extended corrosion is noticed for steel panel coated with epoxy with chitosan/polypyrrole composite exhibiting the less protective nature of surface film (Figure 22e). The appreciable high E_{corr} noticed for steel coated with epoxy with 4.0 % loading of polypyrrole/SiO_2 composite (PS) is supported by less extended corroded area along the scribe mark (Figure 22f). Amongst all, the remarkably high corrosion protection of chitosan/polypyrrole/SiO_2 composite coatings is due to the protection offered by chitosan and polypyrrole present in the coating system. The film forming tendency of chitosan and passive layer formation of polypyrrole acts as a physical barrier between metal and electrolyte. On the other hand, SiO_2 particles as filler in chitosan/polypyrrole matrix provide mechanical integrity and reduce the degradation of polymer composite coating under corrosive conditions.

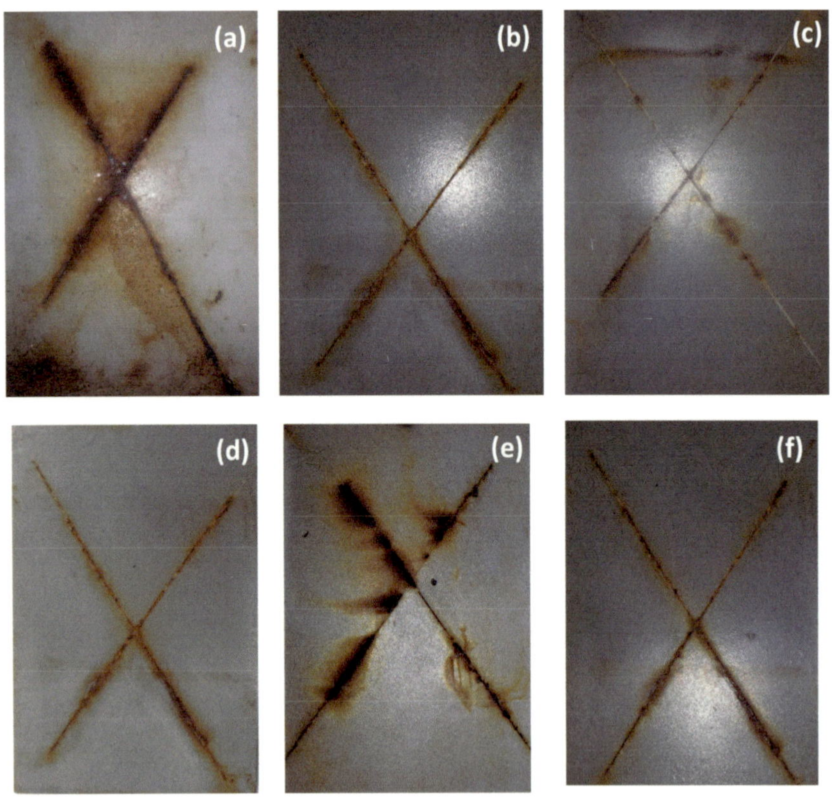

Figure 22. Photographs of (a) epoxy coated and epoxy with (b) 1.0 wt%, (c) 2.0 wt%, (d) 3.0 wt% loading of chitosan/polypyrrole/SiO_2, (e) chitosan/polypyrrole and (f) polypyrrole/SiO_2 composite coated mild steel specimens exposed to salt spray fog after 65 days. (Communicated to Progress in Organic Coatings, Ref. 110, Elsevier)

http://dx.doi.org/10.5772/58388

7. Conclusions

In present study, corrosion resistance of mild steel specimens coated with conventional epoxy coating and epoxy with different wt% loading of polypyrrole/SiO_2 and chitosan/polypyrrole/SiO_2 composites were evaluated by means of Tafel extrapolation, electrochemical impedance spectroscopy and salt spray tests in highly corrosive conditions. The electrochemical measurements and salt spray tests suggest that mild steel specimens coated with epoxy coatings with polypyrrole/SiO_2 composites (PCs) makes a remarkable improvement in the corrosion resistance properties as compared to conventional epoxy coatings. The corrosion protection efficiency (% P.E.) increases with the increase of percentage loading of PCs in epoxy resin. A maximum P.E. of 99.95 % is achieved for coating with 3.0 wt% loading of PCs, thereafter, it reduced slightly for coating with 4.0 wt% loading of PCs. The incorporation of chitosan (boi-polymer) in the polypyrrole/SiO_2 system further improves the corrosion resistance of the steel specimens. Corrosion protection efficiency (%P.E.) of 99.97 is achieved for the 2 wt% loading of chitosan/polypyrrole/SiO_2 composites in the epoxy resin. The salt spray test results show a highly extended corroded area along the scribe mark for the epoxy coated steel panel. Whereas, presence of polymer composites (polypyrrole/SiO_2 and chitosan/polypyrrole/SiO_2) in the epoxy system appears to inhibit the extended corrosion as almost no corrosion or blistering is noticed for the mild steel specimen coated with epoxy with polymer composites. The above results reveal that the synthesized polymer composites interacts with the epoxy resin to form well adherent, high quality, superior anti corrosive coatings on mild steel surface using powder coating technique.

Acknowledgements

Authors thank the Director, National Physical Laboratory, CSIR, New Delhi for providing the laboratory facilities for conducting experiments and characterizations. Their thanks are extended to Mr. Brij Bisht for salt spray test results. One of the authors Gazala Ruhi is thankful to CSIR for her RA fellowship.

Author details

Gazala Ruhi and S.K. Dhawan

Polymeric & Soft Materials Section, National Physical Laboratory, (CSIR), Dr. K.S. Krishnan Marg, New Delhi, India

References

[1] Wan Nik W B, Zulkifli F, Rahman M M and Rosliza R. International Journal of Basic & Applied Sciences 2011; 11, 75.

[2] Jason L, Richard R, Edward L, Alexander F and Brenda L. Biofouling 2004; 20, 237.

[3] Hernandez L A, Hernandez L S and Rodriguez-Reyna S L. International Journal of Corrosion 2012; 2102, 1.

[4] Hanga T X, Truca T A, Olivierb M G, Vandermiersb C, Guéritc N and Pébèred N. Progress in Organic Coatings 2010; 69 (4) 410.

[5] Shokry H. Chemistry of Metals and Alloys (2009); 2, 202.

[6] Rani B E A, and Basu B B J. International Journal of Corrosion 2012; Review Article ID 380217, 15 pages.

[7] Lee H and Neville K. Handbook of Epoxy Resins: Mc Graw Hill, New York; 1952.

[8] Barletta M, Lusvarghi L, Mantini F P and Rubino G. Surface and Coatings Technology 2007; 20, 7479.

[9] Yan M, Vetter C A and Gelling V J. Corrosion Science 2013; 70, 37.

[10] Sharifirad M, Omrani A, Rostami A A and Khoshroo M. Journal of Electroanalytical Chemistry 2010; 645, 149.

[11] Khanmohammadi M, Mizani F, Khaleghi M B, Garmarudi A B. Protection of Metals and Physical Chemistry of Surfaces 2013; 49 (6) 662.

[12] Gergely A, Pfeifer E, Bertóti I, Török. T and Kálmán E. Corrosion Science 2011; 53 (11) 3486.

[13] El-Shazly A H and Al-Turaif H A. International Journal of Electrochemical Science 2012; 7, 211.

[14] Ohtsuka T. International Journal of Corrosion 2012; 1.

[15] Al-Dulaimi A A, Hashim S and Khan M I. Pertanika Journal of Science and Technology 2011;19(2) 329.

[16] Kousik G, Pitchumani S and Renganathan N G. Progress in Organic Coatings 2001; 43 (4) 286.

[17] Sabouri M, Shahrabi T, Faridi H R and Salasi M. Corrosion Engineering Science and Technology 2009; 44 (1) 51.

[18] Spinks G M, Dominis A J, Wallace G G and Tallman D E. Journal of Solid State Electrochemistry 2002; 6, 85.

[19] Xu L, Chen W and Mulchandani A. Angewandte Chemie International 2005; 44, 6009.

[20] Schaftinghen T V, Deslouis C, Hubin A and Terryn H. Electrochimica Acta 2001; 51, 1695.

[21] Hien N T L, Garcia B, Pailleret A and Deslouis C. Electrochimica Acta 2005; 50, 1747.

[22] Ferreira C A, Aeiyach S, Coulaud A and Lacaze P C. Journal of Applied Electrochemistry 1999; 29, 259.

[23] Troch-Nagels G, Winand R, Weymeersch A and Renard L. Journal of Applied Electrochemistry 1992; 22, 756.

[24] Jassen W and Beck F. Polymer 1989; 30, 353.

[25] Hulser P and Beck F. Journal of Applied Electrochemistry 1990; 20, 596.

[26] Ferreira C A, Aeiyach S, Aaron J J and Lacaze P C. Electrochimica Acta 1996; 41, 1801.

[27] Pirvu C, Mindroiu M and Demetrescu I. Key Engineering Materials 2009; 415, 65.

[28] Subathira A and Meyyappan R. Recent Research in Science and Technology 2010; 2(1) 124.

[29] Liu A S and Oliveira M A S. Materials Science-Poland 2009; 27 (1) 265.

[30] Raotole P, Patil P P and Raotole M. International Journal of Emerging Technology and Advanced Engineering 2013; 3 (11) 62.

[31] Sathiyanarayanan S, Azim S S and Venkatachari G. Electrochimica Acta 2007; 52, 2068.

[32] Niedbala J. Bulletin of Material Science 2011; 34 (4) 993.

[33] Herrasti P, Kulak A N, Bavykin D V, Ponce de Léon C, Zekonyte J, Walsh F C. Electrochimica Acta 2011; 56 (3) 1323.

[34] Feng X M, Yang G, Xu Q, Hou W H and Zhu J J. Macromolecular Rapid Communications 2006; 27, 31

[35] Riaz U, Ashraf S M and Ahmad S. Progress in Organic Coatings 2007; 59, 138.

[36] Qi K, Chen Z, Zhang G and Guo X. The Open Corrosion Journal 2011; 4, 18.

[37] Lenz D M, Delamar M and Ferreiraa C A. Journal of Electroanalytical Chemistry 2003; 540 35.

[38] Hosseini M G, Sabouri M and Shahrabi T. Progress in Organic Coatings 2007; 60 (3) 178.

[39] Hosseini M G, Sabouri M and Shahrabi T. Materials and Corrosion 2006; 57 (5) 407.

[40] Kumar S A, Bhandari H, Sharma C, Khatoon F and Dhawan S K. Polymer International 2012; 4066.

[41] Al-Dulaimi A A, Shahrir H and Khan M I. Sains Malaysiana 2011; 40 (7) 757.

[42] Marjanovic G C, DragicQevic L, Milojevic M, Mojovic M, Mentus S, DojcQinovic B, Marjanovic B and Stejskal J. Journal of Physics and Chemistry B 2009; 113, 7116.

[43] Kin S H, Ahn S H and Hirai T. Polymer 2003; 44, 5625.

[44] Dominis A, Spinks G M and Wallace G.G. 56th Annual Technical Conference on Society of Plastic Engineers 1998; 1229.

[45] Shi H, Liu F,Yang L and Han E. Progress in Organic Coatings 2008; 62, 359.

[46] Li H, Zhang Z, Ma X. Hu M, Wang X and Fan P. Surface and Coatings Technology 2007; 201, 5269.

[47] Armelin E, Pla R, Liesa F, Ramis X, Iribarren J.I. and Aleman C. Corrosion Science 2008; 50, 721.

[48] [48]Radhakrishnan S, Sonawane N and Siju C R. Progress in Organic Coatings 2009; 64, 383.

[49] Saravanan K, Sathiyanarayan S, Muralidharan S, Azim S and Venkatachari G. Progress in Organic Coatings 2007; 59, 160.

[50] Srinivasaa P C and Tharanathana R N. Food Reviews International 2007; 23 (1) 53.

[51] Portes E, Gardrat C, Castellan A and Coma V. Carbohydrate Polymers 2009; 76, 578.

[52] Gupta B, Saxena S, Arora A and Alam M S. Indian Journal of Fibre and Textile Research 2011; 36, 272.

[53] Liu B S, Huang T B, Yao C H, Fang S S and Chang C J. Journal of Medical and Biological Engineering 2009; 29, 60.

[54] Sugama T and Cook M. Progress in Organic Coatings 2000; 38, 79.

[55] Lu X, Wiren A and Albertson A. Proceedings of ACS Division of Polymeric Material: Science and Engineering 1998; 79, 242.

[56] Vathsala K, Venkatesha T V, Praveen B M and Nayana K O. Engineering 2010; 2, 580.

[57] Carneiro J, Tedim J, Fernandes S C M, Freire C S R, Gandini A, Ferreira M G S and Zheludkevich M L. Surface and Coatings Technology 2013; 226, 51.

[58] Rath P C, Singh B P, Besra L and Bhattacharjee S. Journal of American Ceramic Society 2102; 95 (9) 2725.

[59] Cordero-Arias L, Cabanas-Polo S, Gao H, Gilabert J, Sanchez E, Roether J A, Schubert D W, Virtanene S and Boccaccinia A R. RSC Advances; 2013; 3, 11247.

[60] Ahmed R A, Farghali R A and Fekry A M. International Journal of Electrochemical Science 2012; 7, 7270.

[61] Redepenning J, Venkataraman G, Chen J and Stafford N. Journal of Biomedical Material Research 2003; 66A, 412.

[62] Yeh J M, Chin C P and Chang S. Journal of Applied Polymer Science 2003; 88, 3264.

[63] Sugama T and Jimenez S M. Journal of Material Science 1999; 34, 2003.

[64] Bumgardner J D, Wiser R and Gerard P D. Journal of Biomedical Science, Polymer Edition 2003; 14, 423.

[65] Hahn B D, Park D S, Choi J J, Ryu J, Yoon W H, Choi J H, Kin H E and Kim S G. Surface and Coatings Technology 2011; 205 (8-9) 3112.

[66] Yang S, Tirmizi S A, Burns A, Barney A A and Risen W M. Synthetic Metals 1991; 32, 191.

[67] Khor E and Whey J L H. Carbohydrate Polymers 1995; 183.

[68] Yang X and Lu Y. Polymer 2005; 46, 5324.

[69] Yingmei L, Guicun L, Hongrui P and Kezheng C, Polymer International 2011; 60, 647.

[70] Yalcinkaya S, Demetguil M, Timur V and Colk N. Carbohydrate Polymers 2010; 79 (4) 908.

[71] Gacrielli G and Keddam V. Corrosion 1992; 48, 794.

[72] Walter G W. Corrosion Science 1991; 32, 1041.

[73] Wang J H, Duh J G and Shih H C. Surface and Coatings Technology 1996; 78, 248.

[74] Singh I B. Corrosion Science 2003; 45, 2285.

[75] Moutarlier V, Gigandet M P, Normand B and Pagetti J. Corrosion Science 2005; 47, 937.

[76] Souto R M, Laz M M and Ries R L. Biomaterials 2003; 24, 4213.

[77] Patil R N, Sharma B V and Mahanwar P A. Der Chemica Sinica 2012; 3 (2) 458.

[78] Liu X P, Zheng T L and Xiong J P. International Journal of Electrochemical Science 2013; 8, 11588.

[79] Manohar A K, Bretschger O, Nealson K H and Mansfeld F. Bioelectrochemistry 2008; 72, 149.

[80] Tsai C H and Mansfeld F. Corrosion Science 1993; 49, 726.

[81] Nocodemo L. Corrosion 1993; 49, 235.

[82] Grundmeier G, Schmidt W and Stratmann M. Electrochimica Acta 2000; 45, 2515.

[83] Kending M and Scully J. Corrosion 1990; 46, 22.

[84] Ruhi G, Bhandari H and Dhawan S K. (Communicated to) Surface and Coatings Technology, 2013.

[85] Radhakrishnan S, Sonawane N and Siju C R. Progress in Organic Coatings 2009; 64, 383.

[86] Saravanan K, Sathiyanarayan S, Muralidharan S, Azim S S and Venkatachari G. Progress in Organic Coatings 2007; 59, 160.

[87] Fu Y, Su Y S and Manthiram A. Journal of Electrochemical Society 2012; 159 (9) 1420.

[88] Lei J, Liang W and Martin C R. Synthetic Metals 1992; 48, 301.

[89] Cheah K, Forsyth M and Truong M V. Synthetic Metals 1998; 94, 215.

[90] Dai T, Yang X and Lu Y. Material Letters 2007; 61, 3142.

[91] Chandrakanthi R L N and Careem M A. Thin Solid Films 2002; 417, 51.

[92] Bose S, Kuila T, Uddin M E, Kim N H, Lau A K T and Lee J H. Polymer 2010; 48, 5921.

[93] Zhang W H, Fan X D, Tian W and Fan W W. eXPRESS Polymer Letters 2012; 6 (7) 532.

[94] Alam A, Sherif S M and Al-Zahrani S M, International Journal of Electrochemical Science 2013; 8, 3121.

[95] Iroh J O and Su W. Electrochimica Acta 2000; 46, 15.

[96] Wesseling B. Synthetic Metals 1997; 85, 1313.

[97] Sathiyanarayanan S, Muthukrishnan S and Venkatachari G. Progress in Organic Coatings 2006; 55, 5.

[98] Nguyen T D, Nguyen T A, Pham M C, Piro B, Normand B and Takenouti H. Journal of Electroanalytical Chemistry 2004; 572, 225.

[99] Bhandari H, Kumar S A and Dhawan S K. (2012), Conducting Polymer Nanocomposites for Anticorrosive and Antistatic Applications, Nano Composites-New Trends and Developments, In Tech Publishers, Croatia.

[100] Creus j, Mazille H and Idrissi H. Surface and Coatings Technology 2000; 130, 224.

[101] Shinde V, Sainkar A R and Patil P P. Corrosion Science 2005; 47,1352.

[102] Pebre N, Picaud T, Duprat M and Dabosi F. Corrosion Science 1989; 29, 1073.

[103] Ruhi G, Modi O P, Sinha A S K and Singh I B. Corrosion Science 2008; 50, 639.

[104] Ruhi G, Modi O P and Singh I B. Surface and Coatings Technology 2009; 204, 359.

[105] Ruhi G, Modi O P and Singh I B. Corrosion Science 2009; 51, 3057.

[106] Loveday D, Peterson P and Rodgers B. Journal of Coatings Technology 2004; 1 (10) 88.

[107] Metikos-Hukovic M, Tkalcec E, Kwokal A and Piljac J. Surface and Coatings Technology 2003; 165 40.

[108] Belluci F and Nicodemo L. Corrosion 1993; 49 235.

[109] Zheludkevich M I, Shchukin D G, Yasakau K A, Mohwald H and Ferreira M G S. Chemistry of Materials 2007; 19, 402.

[110] Rosero-Navarro M C, Pellice S A, Duran A and Apricio M. Corrosion Science 2008; 50, 1283.

[111] Ruhi G, Dhawan S K. (Communicated to) Progress in Organic Coatings, 2013.

[112] Modrzejewska Z, Stawczyk J, Matyka K,. Matyka M, Mróz I and Ciszewski A. Polish Journal of Environmental Studies 2006;15 (4A) 84.

[113] Thogngam M and Mc Clements D J. Langmuir 2005; 21, 79.

[114] Vikhoreva G A, Babak V G, Galich E F and Galbraikh G S. Polymer Science A 1997; 39, 617.

[115] Zhang H, Zhong X, Xu J. J and Chen H.Y. Langmuir 2008; 24, 13748.

[116] Li X, Wan M, Wei Y, Shen J and Chen Z. Journal of Physics and Chemistry B 2006; 110, 14623.

[117] Yoon H and Jang J. Advanced Functional Materials 2009; 19, 1567.

[118] Yang X, Dai T and Lu Y. Polymer 2006; 47 (1) 441.

[119] Marguerite R. Progress in Polymer Science 2006; 31, 603.

[120] Aimoli C G, Torres M A and Beppu M. Material Science and Engineering C 2006; 26, 78.

[121] Sakurai K, Maegawa T and Takahashi T. Polymer 2000; 41, 7051.

[122] Zeng M, Fang Z and Xu C. Journal of Membrane Science 2004; 230, 175.

[123] Zheludkevich M L, Serra R, Montemor M F, Yasakau K A, Salvado I M and Ferreira M G S. Electrochimica Acta 2005; 51 (2) 208.

[124] Scully J R. Journal of Electrochemical Society 1989; 136 (4) 979.

Chapter 6

Monitoring of the Surface Modification of Nanoparticlesby Electrochemical Measurements Using Scanning Electrochemical Microscope

David Hynek, Michal Zurek, Petr Babula,
Vojtech Adam and Rene Kizek

Additional information is available at the end of the chapter

http://dx.doi.org/10.5772/57203

1. Introduction

1.1. SECM as a tool for multi-purpose characterisation of materials

The first papers related to the scanning electrochemical microscopy (SECM) were published in the nineties of the 20th century [1, 2]. This technique is based on the scanning of the studied surface by ultramicroelectrode (UME) and electrochemical analysis of surface [1]. Such technique gives us an electrochemical picture of the surface. SECM employs an UME probe (tip) to induce chemical changes and collect electrochemical information while approaching or scanning the surface of interest (substrate). The substrate may also be biased and serves as the second working electrode. Many different types of UMEs have been fabricated, e.g., microband electrodes, cylindrical electrodes, microrings, disk-shaped, and hemispherical electrodes [3]. The disk geometry of the UME is the most preferred design from many reasons discussed below. UME is the most important part of the electrochemical microscope, which determines the results of scanning. Electrodes in micro or nano dimensions offer important advantages for electroanalytical applications including greatly diminished ohmic potential drop in solution and double-layer charging current, the ability to reach a steady state in seconds or milliseconds, and a small size allowing make experiments in microscopic dimensions [4].

SECM requires minimal sample preparation than other spectroscopy techniques (e.g. fluorescence or electron microscopy). Another advantage is that the evaluation of image outputs can be numerically modelled, allowing to obtain information and analysis the reaction kinetics.

INTECH
open science | open minds

© 2014 The Author(s). Licensee InTech. This chapter is distributed under the terms of the Creative Commons Attribution License (http://creativecommons.org/licenses/by/3.0), which permits unrestricted use, distribution, and reproduction in any medium, provided the original work is properly cited.

Therefore, it is possible to study effectively both material characteristic and biological properties [4]. Due to the non-destructive measurement method, experiments can be successfully repeated several times. The SECM surface scanning is connected with one major risk of damage to the scanning electrodes or the studied surface due to distance between the tip and surface interference. This distance should be optimized and the probe or a tip should be held at a constant height above the surface during measurement. Analysed surface can act as a second working electrode and therefore four electrodes configuration is used in SECM [5].

Application of SECM is very wide and allows studying various structures and processes in micro and submicrometer-sized systems [5-7]. The target of detection could be electron, ion, and molecule transfers, and other reactions at solid-liquid, liquid–liquid, and liquid-air interfaces. In recent years the study of nanoparticles and its application rapidly has been increasing [8]. The SECM study of nanoparticles increases rapidly too because the surface modification by this nanoparticles brings many positives and allows to open the whole new research area [8]. The wide range of SECM applications allows to investigate a wide variety of processes, corrosion of metals, characterization of materials, membrane and enzymatic reactions, photosynthesis, DNA hybridization and metabolism in single living cells [4]. This direction has had the great influence for the increase of published papers related to the SECM technique since the year 1995 (approximately) [2]. The bio-applications of SECM are one of the most developed areas in the recent years and the usage of nanoparticles in this area is very widespread [4, 8, 9].

1.2. Measuring modes and principles of SECM

Although the theory of SECM measurements has been described in many books and reviews, we still consider beneficial to present here the short overview of this. At first, Figure 1 shows a schematic diagram of the basic SECM instrument. An UME tip is attached to a 3D piezo positioner controlled by a computer, which is also used for data acquisition. A bi-potentiostat controls the potentials of the tip and/or the substrate versus the reference electrode and measured tip and substrate currents are in pA and sub-pA orders. 10 pA is the lowest detectable current, the maximum is at 250 mA. Measured current resolution is 0.0015% of current range (minimum 0.3 fA). These informations are related to the model ChInstrument 920D which is shown on Figure 2. Usually detected currents are in pico or nano amper order. The most SECM measurements take place in feedback (FB) or generation/collection (GC) modes [4, 10, 11].

The feedback mode usually uses the ultramicroelectrode (UME), which serves as a working electrode in a three or four-electrode system. The four electrode system is completed with a sample (substrate) that serves as a second working electrode. The electrodes are immersed in a solution containing redox mediator. Specifically, one redox form of a quasi-reversible redox couple. In the following discussion it is assumed that the reduced form R is added, although the same explanations hold for the addition of the oxidized form O of the mediator if the reaction directions are reversed.

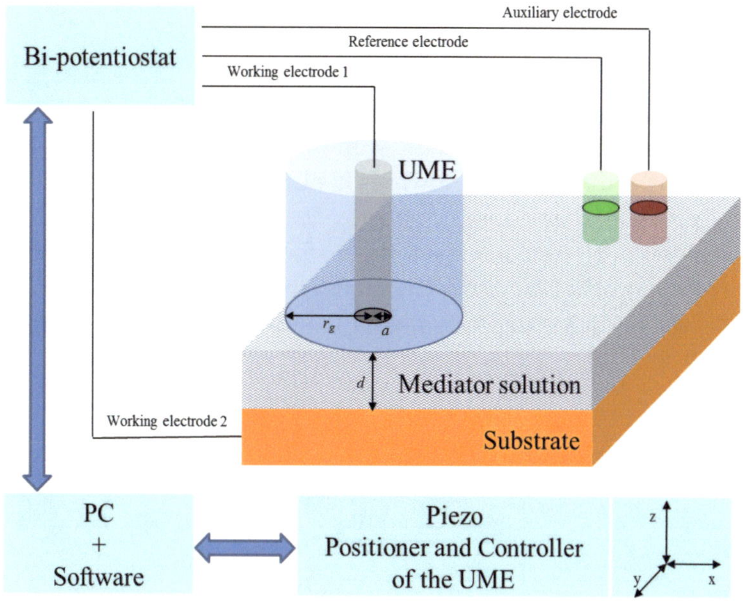

Figure 1. Scheme of SECM instrumentation. The basic parts are as follows: a piezo positioner and a controller of the UME, a bi-potentiostat and PC (with measuring software). The UME scans the substrate surface due to moving the UME by the piezo positioner. The system can work in a three- or four-electrode setup (with connected second working electrode).

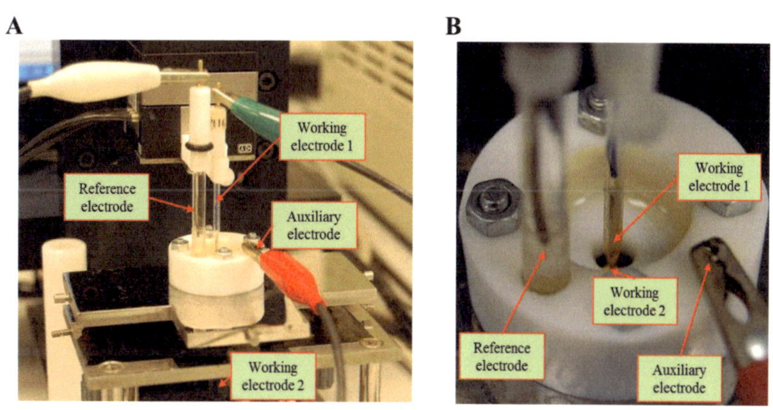

Figure 2. Real picture of measurement apparatus ChInstruments 920D. (A) The overview of the whole electrochemical cell with support. (B) The detail image of the electrochemical cell with connected electrodes.

In the simplest case, the UME is inserted only in mediator solution, a potential is applied to the tip, and diffusion-controlled conversion of the mediator occurs according to equation as it follows:

$$R \rightarrow O + ne^- \quad (1)$$

and thus a steady-state faradaic current $i_{T\infty}$ could be detected. This situation is necessary to understand as the limiting case where the distance d of the tip from the substrate electrode is infinite (Fig 3A). If the UME is in proximity with substrate electrode and this is an insulator, the surface hinders the diffusion of R towards the UME (Fig 3B) and the faradaic current i_T decreases. Such situation is named as a negative feedback. The opposite situation is the positive feedback, which differs by conducting substrate where heterogeneous reaction takes place. This reaction regenerates the R by electrochemical conversion and thus the new reagent source becomes available (Fig 3C). From this reason i_T increases, especially with respect to the dependence $i_T(d)$ [2, 5, 12].

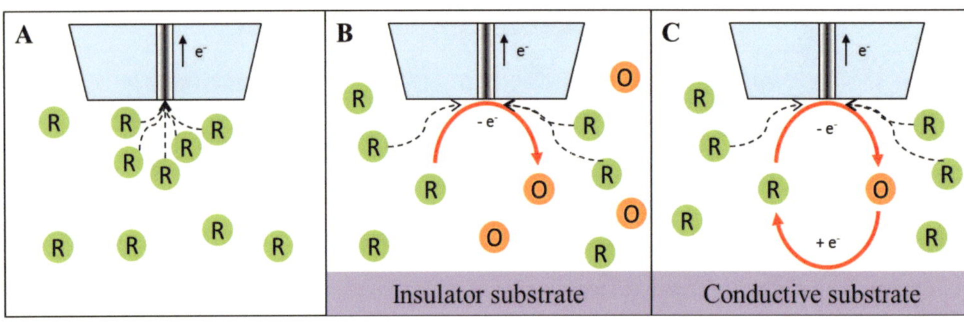

Figure 3. Scheme of feedback mode of the SECM operation. (A) The UME tip is far from the substrate and the tip reaction is caused just form species R from the solution. (B) Negative feedback: the insulating substrate averts the regeneration of species O and diffusion of R to the tip. (C) Positive feedback: species R is regenerated at the substrate and the electrochemical loop is closed.

Generation/collection mode differs from feedback mode in one important thing, the presence of the mediator is solution. This mode works in a solution that does not initially contain any substance that can be converted at the UME at a potential E_T. In the generation/collection (G/C) mode, both tip and substrate can be used as working electrodes. One working electrode generates some species, which are collected at the other electrode. The G/C mode is more sensitive because the background signal is very weak [2, 12].

If an oxidizable or reducible substance is formed on the surface of substrate, this compound can be detected at the UME if it is located close to the active region (Fig 4A). From such description it is obvious the name of the mode, sample-generation/tip-collection (SG/TC) mode. When the tip is moved through the thick diffusion layer produced by the substrate, the changes in i_T reflect local variations of concentrations of redox species. The collection efficiency in SG/TC mode is much lower than that in the TG/SC mode, and the true steady state can be achieved only by using a micrometer-sized substrate. Other disadvantages of this mode are the high sensitivity to noise and the difficulty in controlling the tip/substrate separation distance [2, 12].

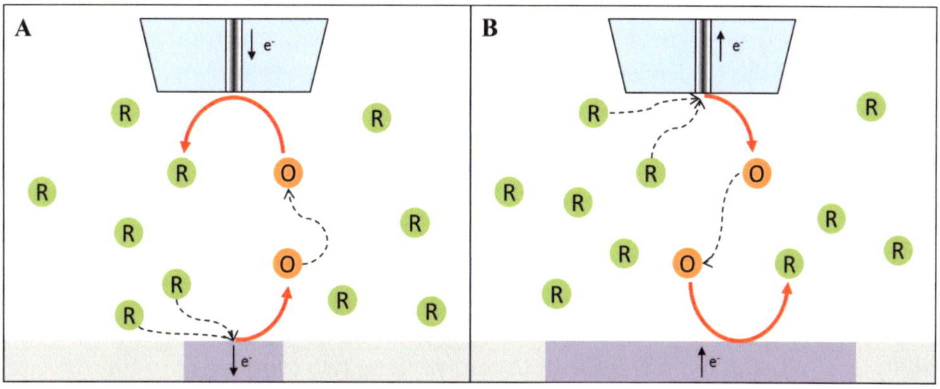

Figure 4. Scheme of SG/TC mode (A) and TG/SC mode (B). (A) O is electrogenerated on the substrate surface and collected at the tip. The tip and substrate currents are recorded in both cases. (B) The tip generates species O by oxidation of R in solution; O diffuses toward the substrate and is reduced back to R.

The application of the contrary processes leads to the creation of the tip-generation/sample-collection (TG/SC) (Fig 4B). In the TG/SC mode experiments, the tip generates an electro-active species that diffuses across the tip/substrate gap to react on the surface of substrate. A TG/SC experiment includes simultaneous measurements of both tip and substrate currents (i_T and i_S). For a one-step heterogeneous electron transfer (ET) reaction at steady state, these quantities are almost identical if d is not very large. Under these conditions, the tip-generated species R predominantly diffuse to the large substrate, rather than escape from the tip-substrate gap. For a process with a coupled homogeneous chemical reaction, there may be large differences between i_S and i_T. If the reaction is slow, the process is diffusion-controlled, and $i_s/i_T \rightarrow 1$ at short separation distances. For the other hand if homogeneous reaction is very fast, most species R get converted to the electro-inactive product before reaching the substrate and hence, the very low substrate current is detected, $i_s/i_T \rightarrow 0$. Between these two extremes, the homogeneous kinetics can be determined by measuring the collection efficiency as a function of d.

As it is obvious in analytical chemistry, connection of various analytical methods to the one sample detection is very popular. Combination of methods allows the new view on the samples. SECM does not stand out of this direction. Studying of surface structures and dynamics is possible due to combination of SECM with other analytical methods. A number of analytical techniques have been combined with SECM [2, 7, 9, 11-17] including near-field scanning optical microscopy (NSOM), surface plasmon resonance (SPR), electrochemical quartz crystal microbalance (EQCM), fluorescence spectroscopy (FS), electrogenerated chemiluminescence (ECL), electrochemical scanning tunnelling microscopy (ESTM) and atomic force microscopy (AFM) [2]. Among above mentioned methods there are two last which would be useful to describe more. Scanning tunnelling microscopy (STM) is based on the concept of quantum tunnelling. When a conducting tip is brought very near to the surface to be examined, a bias (voltage difference) applied between the two can allow electrons to tunnel through the vacuum between them. The resulting tunnelling current is a function of tip position, applied voltage, and the local density of states of the sample [18, 19]. The extension of STM with electrochemical detection leads to the electrochemical scanning tunnelling microscopy (ESTM) as a powerful new technique for detailed structural and topographical characterization of electrode/electrolyte interfaces. Knowledge of surface structures could be crucial to the understanding of electrochemical processes that are taking place at the electrode surface [18, 19].

Opposite this, the most widely used hybrid method is AFM-SECM. The atomic force microscope maps the topography of a substrate with nanometer vertical resolution by monitoring the interaction force between the sample and a sharp tip, which is attached to the end of a cantilever. AFM methodology is often used to provide a feedback mechanism for height control in scanning near-field optical microscopy. For samples that show variations in both reactivity and topography, it is difficult to resolve these two components with conventional SECM fixed height measurements. The strength of the AFM-SECM combination is in its positioning capabilities and simultaneous possibility of electrochemical detection [20]. Using the AFM mode, one can position a submicrometer-sized tip near the substrate surface and keep a constant tip-substrate distance during the surface raster. Maintaining a constant distance between the tip and the substrate is essential for avoiding tip crashes and the damage to sample during the imaging of a substrate with non-uniform surface reactivity and/or complex topography.

2. Modification of gold electrode

2.1. Self-Assembled Monolayer (SAM)

Self-assembled monolayers (SAMs) are highly ordered molecular assemblies formed by the adsorption of an active surfactant on a solid surface. Understanding and controlling long-range electron transfer across such molecular assemblies are important for the development of technological systems such as molecular electronics and sensors. The most common SAMs are n-alkanethiols adsorbed on gold by Au–S linkage. The manipulation of individual alkanethiolate by SECM [15] is mainly based on regeneration of a gold surface either by electrochemically initiated removal of the chemisorbed thiolate from the gold surface (Eq. 2) or by oxidative decomposition (Eq. 3).

$$AuSR + e^- \longrightarrow Au0 + RS^- \tag{2}$$

$$AuSR + 2H_2O \longrightarrow Au0 + RSO_2^- + 3e^- + 4H^+ \tag{3}$$

The way of preparing the SAMs on metal surface is an immersion of freshly prepared or clean metal electrodes into a dilute solution of alkanethiols for several hours at room temperature. Many authors deal with the optimization of this procedure related to the best reproducibility of the produced SAMs [21]. The most frequently studied alkanethiol compounds with different features presented in literature are as it follows: n-alkanethiol; α,ω-alkanedithiol; ω-mercaptoalkanol; ω-mercaptoalkane carboxylic acid; ω-mercaptoalkane amino acid [22]. Application of other types of thiols such as aromatic compounds is possible too [22]. Because the metal–sulphur bond is stable between thiols and metal surface, the way for the modification of metal surfaces with organic monolayers could be easily modified with any particular molecule that is firstly thiol-derived and then attached to the metal surface [23].

The formation of a SAM is often depicted in schemes as uniformly aligned molecules on a flat surface. However, it has been recognized that a wide variety of defects are possible such as pinholes, collapsed sites, islands, and domains [24]. Typically, the first assumption is that the metallic surface, upon which the SAM is formed, is uniformly flat. Scanning techniques show a variety of features such as terraces, steps, and crystalline boundaries. The choice of electrode metal is an important factor in determining the SAM integrity. A second assumption is that the organic molecules of the SAM align uniformly. The tilt angle of alkane thiols can cause significant variation in the monolayer packing [25].

SECM has almost been used for studying the structure of alkane thiols, because SECM signals respond very sensitively to defects in the monolayer. A defect-free monolayer passivates the gold electrode so strongly that it behaves like an insulator in SECM. No passivation occurs at defects, and results as high currents. Application of nanometer-sized electrodes showed that direct electron transfer through the layer, rather than electron transfer at the defects, was dominant and the size of the defects was 1–100 nm [26].

Model for electron transfer at self-assembled monolayers measured by scanning electrochemical microscopy has been suggested by *Liu and Bard, et al.* [27]. The developed models were used to measure independently the electron transfer rates mediated by monolayer-attached redox moieties and a direct electron transfer through the film as well as the rate of a bimolecular electron transfer reaction between the attached and dissolved redox species.

When the monolayer absorbed on the substrate contains an electroactive group, the regeneration of the mediator occurs via a bimolecular reaction with these surface-bound redox centres with a rate constant k_{BI}. The rate of the tunnelling ET is governed by the relative contributions from the forward and backward reactions, k_f and k_b, which are both dependent on the value of substrate potential E_s. If bimolecular ET reaction is irreversible and the electron tunnelling through the monolayer is quasi-reversible, the effective rate constant obtained from fitting an experimental approach curve to theory is k_{eff}. After some approximations and limited conditions presented in [17] the one important result is formed as follows: the concentration and potential will not affect k_{eff} and the overall rate of ET is limited by the bimolecular ET rate constant, k_{BI}.

Forming of SAMs on the surface of substrate electrode is one of the most applied procedures in SECM techniques. The great advance of this lies in the immobilisation of target molecules and possibility of creating the surface electrochemical image. Application of SAMs systems in SECM measurements have many utilizations as the study of electron transfer through the creating monolayer [27-32], investigation of various functionalizations of the SAMs parts [33,

34], monitoring the adsorption/desorption processes [35, 36] and interactions between monolayer and target molecules [37].

2.2. Paramagnetic particles

Paramagnetic particles are usually used as an isolation tool in biological applications [38]. Its specificity is related with the paramagnetic properties of the metal core, which reacts on external magnetic field and the other specificity is caused by the surface modification. Connection of SECM detection with nanoparticles, in general, leads to the generation of the one basic problem. What is the measured signal? It is the signal of individual particles, or the average signal of the layer? The most application leads to the second variant. However, some authors focused on the detection of single particle [8]. *Quinn et al.* [39] reported one of the first uses of SECM to investigate the properties of metal NPs. They followed the compression of a monolayer of monodisperse thiol-capped AgNPs as the investigating of NP-film conductivity changes with particle interdistance. *Tel-Vered and Bard* [40] used SECM for the first time to synthesize and to characterize a single particle on a conductive carbon support. Using this approach, they avoided the problems arising from multi-NP synthesis (e.g., narrow size distribution, surface agglomeration, occurrence of lattice defects, and poisoning of the metal by impurities).

Usage of paramagnetic particles in SECM techniques is focused on the improvement of the binding of bio-compound(s) (enzymes, antibodies, etc.) onto the electrode surface [8]. Direct connection of bio-compounds with electrode surface is very often disrupted by many phenomena as follows: loss of biochemical activity, nonspecific binding, blocking of the electrode surface, and regeneration of the sensing surface [41]. Application of paramagnetic particles onto the electrode surface with subsequent SECM detection has been presented by some authors as it follows.

Wijayawardhana, C. A. and Wittstock, G. et al. [41, 42] present miniaturized sandwich enzyme immunoassay for the model analyte mouse-IgG. After capturing mouse IgG by the primary anti-IgG antibody (Ab) coupled with magnetic particles, a second anti-IgG Ab conjugated to an enzyme-label (alkaline phosphatase – ALP) is bound to the analyte. ALP catalyses the hydrolysis of 4-aminophenyl phosphate (PAAP). Platinum microelectrode scans surface over the beads domains and detect product of hydrolysis, 4-aminophenol, by oxidation. The scheme is presented in Fig. 5.

The other Wittstock's cooperation, this time with *Zhao C.* [43, 44], was focused on the system based on the activity of glucose dehydrogenase (GDH). This enzyme was chosen because of its high activity and independence of dissolved oxygen in the sample catalysing the transfer of electrons from glucose to an electron mediator. Biotinylated glucose dehydrogenase was bound to streptavidin-coated paramagnetic beads, which were located on a hydrophobic surface. The catalytic activity of immobilized GDH was mapped in SECM feedback mode and generation-collection mode using ferrocenemethanol, ferrocenecarboxylic acid, *p*-aminophenol, and ferricyanide as electron mediators, respectively [44].

http://dx.doi.org/10.5772/57203

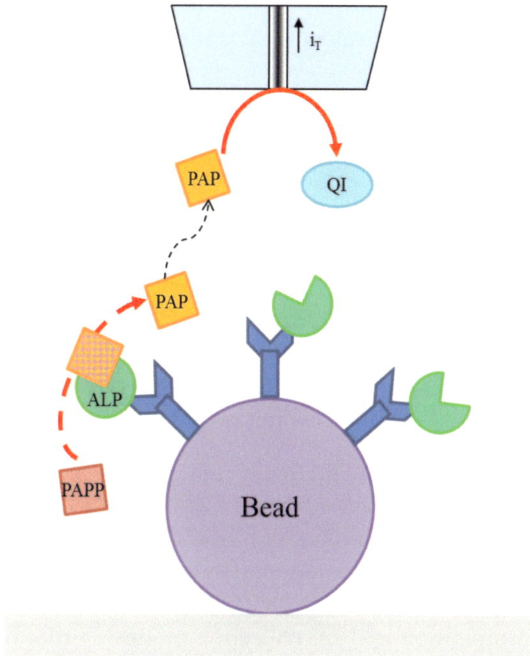

Figure 5. The principle of detection with magnetic beads combined with enzyme reaction. The surface of the bead is coated with primary antibody that has accessible Fab groups responsible for binding the analyte. The label alkaline phosphatase (ALP) is attached to the captured analyte as a conjugate with the second Ab. The ALP catalyses the conversion of redox inactive 4-aminophenyl phosphate (PAPP) in solution to the redox active *p*-aminophenol (PAP) not presented in the bulk of the solution. The faradaic current at the tip, i_T, is caused by the oxidation of PAP to the quinone imine (QI). A quasi-stationary and hemispherical diffusion of PAP can be expected over the beads.

2.3. Quantum dots

Other kinds of NPs widely used as labels for determination of proteins and DNA are semiconductor nanocrystals, quantum dots (QDs). Efficient methods for the preparation of quantum dots and their functionalization with biomaterials have been recently developed [45]. These nanoparticles are applied for labelling of biomaterials in biorecognition processes (e.g., DNA sensing) [46]. In generally, the QDs are employed as an electro sensitive label and detected by various voltammetry techniques, almost by differential pulse voltammetry [8]. Connection of QDs with SECM method is very outstanding. However, several works are presented in this area of research as follows. Morphological analysis of PbSe was shown by *Zimin, S., et al.* [47]. They investigated the morphology of the surface of PbSe/CaF2/Si(111) epitaxial structures after anodic electrochemical etching. Two possible morphological types of PbSe nanostructured surficial porous layers were distinguished. Agglomerated nanoislands with 80-400 nm lateral sizes and 70 nm average separating gaps were observed for the anodized films with initial flat terraced surface. *Liu, G. J.* group [48] presents the investigation of semiconductor photocatalyst - quaternary-alloyed Zn*x*Cd1-*x*S*y*Se1-*y*, by SECM. The photoelectrochemical properties were evaluated by scanning electrochemical microscopy in the photoelectrochemical mode using an optical fibre tip attached to a Xe lamp as an excitation source. The spot with a precursor composition Zn 0.3 Cd 0.7 S 0.8 Se 0.2 (elemental ratio, 1:2.12:1.75:0.81) showed the highest photocurrent under 150 W Xe lamp irradiation.

It is obvious from the above mentioned application that connection of QDs with SECM is very unusual and disparity in application point of view. This research area forms the big unexplored and perspective potential for the further investigation.

2.4. Gold nanoparticles

Creation of the small specific modified area on the gold surface is possible due to modification of the gold nanoparticles instead of the modification of the whole surface of the substrate electrode. This way can offer specific modification in the small dimensions. The other important reason for the application of nanoparticles lies in the fact that, whilst the bulk material is incredibly inert, nanostructured gold can exhibit exceptional catalytic activity [49].

Application of gold nanoparticles is obviously affected by many conditions. The most important are the size and the shape of the particles. The size effect has been clearly presented by *Wain, A. J.* [50]. In this work the influence of particle size on the electrocatalytic behaviour of commercially available spherical gold nanoparticles (AuNPs) is demonstrated. The experiment focused on the oxygen reduction reaction (ORR) and the electrooxidation of hydrogen peroxide using redox competition and tip collection modes. For both reactions studied the electrocatalytic activity, normalized to surface area, was observed the increase with decreasing particle diameter in the range 5-50 nm. It is necessary to connect this result with the role of mass transport on the observed particle size effects. A 50 nm AuNP has a surface area 100 times greater than a 5 nm AuNP. Therefore, for a fixed surface area, a spot consisting of 50 nm AuNP has a factor of 100 fewer particles than a spot consisting of 5 nm AuNPs. However, under conditions of convergent diffusion, the mass transport limited flux to a 50 nm particle is ten times greater than to a 5 nm particle. Taking these two factors into account, the total flux to a spot of 5 nm particles is expected to be ten times greater than to a spot of 50 nm particles which, under mass transport limited conditions, could lead to an enhancement in the apparent electrocatalytic activity.

Influence of the shape of the gold particles was investigated by *Sanchez-Sanchez, C. M. et al.* [51]. The authors used three types of nanoparticles: semi-spherical particles with a particle size of 3.1±0.46 nm, cubic shaped (≈ 85%) particles in dimension 43.8±1.1 nm and short nanorods with an average width of 10.9±1.2 nm and length of 42.9±5.3 nm. The influence of the nanoparticle shape was investigated through the electrocatalytic activity for the ORR. The results demonstrate that cubic gold NPs are the most active towards ORR in 0.1M NaOH, followed by the spherical gold NPs and, finally, by the short gold nanorods.

The same authors investigated the influence of platinum nanoparticles on the ORR [52]. They employed TG/SC mode to study ORR on shape-controlled Pt nanoparticles. The catalytic activities of four types of Pt nanoparticles, spherical, cubic, hexagonal, and tetrahedral/octahedral, were compared. Hexagonal particles exhibited the highest activity towards ORR in acid electrolytes.

As it is presented above, the influence of the size and shape was investigated through the electrocatalytic measurements - the aqueous phase oxygen reduction reaction (ORR). The ORR

http://dx.doi.org/10.5772/57203

on gold can take place via a 2-electron pathway yielding hydrogen peroxide or a 4-electron route generating water. The reaction mechanism and kinetics are known to be structure sensitive. On Au(1 0 0) single crystal surfaces the 4-electron mechanism dominates, on Au(1 1 1) and Au(1 1 0) the 2-electron mechanism is more significant. The activity of these facets is known to decrease in the order Au(1 0 0) > Au(1 1 0) > Au(1 1 1) [53, 54]. Therefore by changing the relative proportions of these low index planes present on the surface of AuNPs one can essentially tune their ORR electrocatalytic activity [50, 55]. This is in agreement with above-mentioned experiment comparing various types of nanoparticle shape. It has been concluded that the highest activity is for the cubic structure where the faces of the cube is oriented along the (1 0 0) direction, in all cases, which is in agreement with the theoretical symmetry expected for a cube.

The very good example of the SECM study of gold nanoparticles (and theirs modification) was presented by *Huang et al.* [56]. They show the combination of AFM and SECM methods for the detection of individual ~20 nm sized gold nanoparticles. The scheme is shown in Fig. 6. The physical and electrochemical properties of functionalized particles by redox-label PEG chains were studied. The redox PEGylated nanoparticles were assembled onto a gold electrode surface, forming a random nanoarray, and interrogated *in situ* by a combined AFM-SECM nanoelectrode probe. In so-called mediatortethered (Mt) mode, AFM-SECM affords the nanometer resolution required for resolving the position of individual nanoparticles and measuring their size, while simultaneously electrochemically directly contacting the redox-PEG chains they bear.

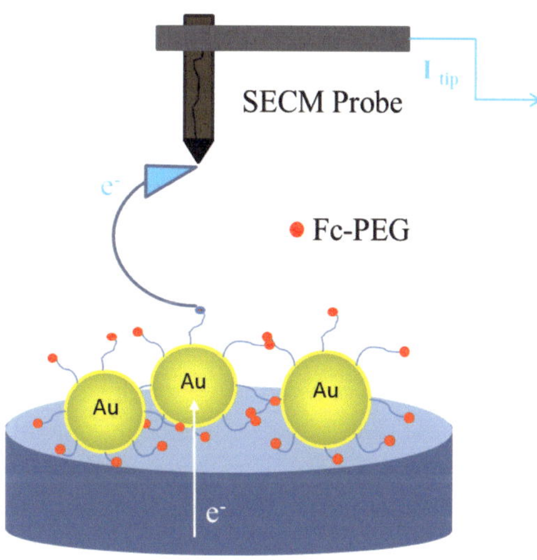

Figure 6. Combination of atomic force and electrochemical scanning microscopy (AFM-SECM) is presented as an analytical tool for simultaneous study of the physical and electrochemical properties of individual ≈ 20 nm sized gold nanoparticles functionalized by redox-labelled PEG chains. The measurements were done in a 0.1 M pH 6 citrate buffer solution. The tip was biased at a potential of E_{tip} = +0.3 V/SCE, while a potential of E_{sub} = -0.1 V/SCE was applied to the substrate.

The nanoparticle acts both as a scaffold and as a nano-electrode, mediating electron transfer between the underlying gold substrate and the redox labels of the PEG. The AFM-SECM probe is brought in "molecular" contact with the macromolecular PEG layer covering the nanoparticle. The tip and substrate are biased so that the redox label is alternatively oxidized at the tip and re-reduced at the substrate. The tip was biased at a potential of E_{tip} = +0.3 V/SCE, while a potential of E_{sub} = -0.1 V/SCE was applied to the substrate in 0.1 M citrate buffer. This redox cycling generates a specific tip electrochemical current whose intensity depends on the (time-averaged) tip-substrate distance, the Brownian dynamics of the PEG, and the local surface concentration in PEG [56]. The use of tapping mode allows the tip-substrate distance to be fixed so that the tip current specifically probes the presence of the labelled molecule and its surface concentration (coverage) on the nanoparticle. The dual measurement of the size and current response of single nanoparticles uniquely allows the statistical distribution in grafting density of PEG on the nanoparticles that was determined and correlated to the nanoparticle diameter.

3. Electrochemical imaging

3.1. Characterizing of molecules and their interactions

Scanning electrochemical microscopy is the useful tool for the investigation of the electro-behaviour of micro and nano objects. The wide range of applications is possible; for the simplicity, it is divided into several fields of research as follows: characterization of entities, measurement of charge and mass transport, biochemical processes, reaction kinetics, energy measurement, and degradation processes [2, 4, 12]. Each presented direction of the research can be further divided into the subunits. The first mentioned area, characterization of entities, could be divided in point of view of the material background, size of detected objects, physicochemical properties, reactivity etc. Interaction phenomenon is closely connected with characterization of entities, on which participates almost at least one biocompound (oligonucleotide, protein, enzyme, antibodies, lipids or the whole cell).

3.1.1. DNA

The first SECM images of DNA were reported by *Guckerberger et al.* in 1994. Since this point much progress has been made and the advance to the development of the fabrication techniques of DNA microarrays has been achieved. The detection of various base pairs is critical for the study, diagnosis and treatment of genetic diseases [6].

The simplest form of nucleic acid detection was described by *Wang et al.* who exploited the intrinsic electrochemical properties of the guanine residue in the DNA duplex [57]. Surface-immobilized DNA molecules were detected by the generation/collection mode of SECM through the oxidation of guanine residues by tip generated Ru complex. The second approach to imaging DNA and the hybridisation event is to use electrochemically active labels. *Wang et al.* [58] reported application of the silver nanoparticles in combination with SECM to image a DNA microarray. Silver nanoparticles were deposited at sites where hybridisation had occurred. The result was an area over which positive feedback could occur as a consequence of mediator recycling (site of silver deposition).

The development of reagentless procedures, by which DNA hybridisation could be detected, was presented by *Turcu et al.* [59, 60]. They detected the hybridisation process through coulomb interactions between the negatively charged sugar phosphate back bone of the immobilised oligonucleotide and the $[Fe(CN)_6]^{3-}$ as a free diffusing redox mediator.

Hybridisation process of DNA was investigated by *Roberts et al.* too [6]. They studied adsorption of genomic DNA and subsequent interactions between adsorbed and solvated DNA. They prepared microarray of polyethylenimine (PEI) film deposited on a screen-printed carbon substrate using the SECM. There was single stranded DNA electrostatically adsorbed on the surface of the polyethylenimine. The further adsorption of complementary single stranded DNA on the surface was observed [6].

3.1.2. Proteins

The SECM protein detection was widely investigated especially by *Girault* group [61-64]. They used a detection system based on the tagging adsorbed proteins with silver nanoparticles (Fig. 7). A positive feedback was obtained when a platinum tip approached the silver-tagged proteins in a 1mM $[Os(bpy)_3]^{2+}$ aqueous solution. The feedback loop was created by $[Os(bpy)_3]^{3+}$ generated by the tip, diffused toward the silver-tagged proteins and next reduced back to $[Os(bpy)_3]^{2+}$. Compound $[Os(bpy)_3]^{2+}$ was chosen as a mediator because the standard redox potential for the couple $[Os(bpy)_3]^{3+/2+}$ (0.631 V vs SCE) is positive enough so that its oxidized form is able to accept an electron from Ag atoms attached to the protein on the substrate (Ag^+/Ag standard potential is 0.558 V vs SCE). Furthermore, this potential is not too close to the positive limit of the potential window in water.

Another *Giraults* work [64] presents the protein detection based on the tagging of free cysteines and other nucleophiles in proteins and peptides by benzoquinone. The tagged proteins are detected by the mediated reduction of benzoquinone with a redox species ($K_3[Fe(CN)_6]$ or $K_3[IrCl_6]$) produced electrochemically at the SECM tip. In addition, the SECM investigation of adsorption and binding of protein to the polymer surface was done by *Glidle et al.* [65]. This phenomenon was studied according to the application of proteins to the biosensor platform.

3.1.3. Enzymes

The recent SECM studies of immobilized enzymes focus on two areas: measuring the catalytic enzyme activity and micropatterning the surface with enzymes [2]. Two SECM-based approaches, generation/collection and feedback mode, are applied to the redox enzymes detection. The GC mode is more sensitive and can be employed when enzyme kinetics are too slow for feedback measurements. Opposite this, the feedback mode is more suitable for enzymes with high activity [12].

Both directions of the enzyme research by SECM are covered by the work of *Wittstocks* group [43, 66-70]. The first case is represented by the experiment in which the direct mode of SECM was used for the local deposition of oligonucleotide patterns on thin gold films and the GC

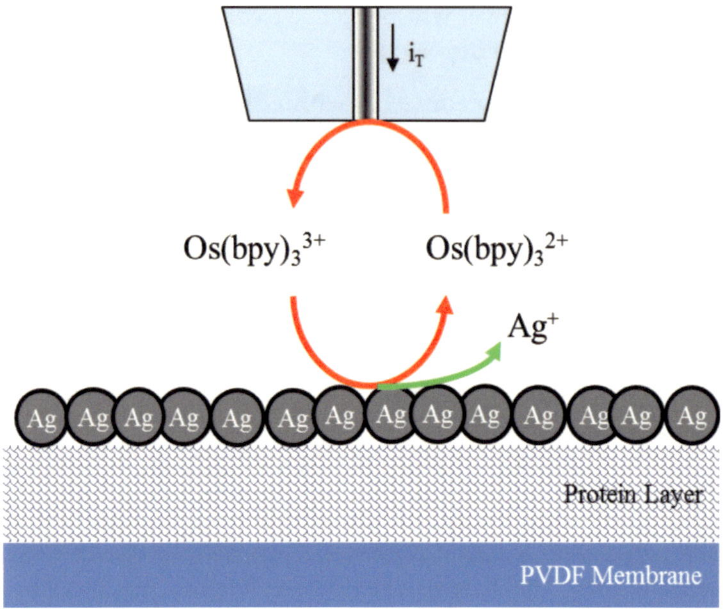

Figure 7. Schematic representation of the electrochemical principle used for protein detection through silver nanoparticles. The working electrode is held at +0.7 V vs Ag/AgCl; at this value the oxidation of the redox mediator is observed. When the tip is close enough to the sample, the oxidized form of the mediator reacts with the silver nanoparticles leading to the production of Ag^+. 1 mM $[Os(bpy)_3]^{2+}$ aqueous solution as a mediator and 0.1 M KNO_3 as a supporting electrolyte were used.

mode was applied for the determination of the amount of surface-accessible oligonucleotides [66]. This process was monitored through the specific system created by biotin-labelled complementary strand connected with streptavidin and biotin-labelled β-galactosidase. The activity of the linked β-galactosidase was mapped with SECM in the GC mode by monitoring the oxidation of *p*-aminophenol (PAP) created in the enzyme-catalysed hydrolysis of *p*-aminophenyl-β-d-galactopyranoside. Micropattering of the surface with enzymes is presented in another *Wittstock*s work [67] where the monitoring of patterned enzyme layers containing glucose oxidase and horseradish peroxidase was described. These layers were created by a combination of microcontact printing (MCP) and local electrochemical desorption followed by chemisorption of functionalized thiols or disulphides. The reactions of two enzymes, horseradish peroxidase (HRP) and glucose oxidase (GOx), were studied.

In the last year one interesting work related to enzyme systems was published [71]. This work is focused on the study of multi enzyme systems by SECM. There is presented study of the kinetics of a two-enzyme coupled reaction, specifically acetylcholine esterase and choline oxidase. This type of experiment push the boundaries further because of approach to the real systems. The biological metabolism process is obviously given via a group of enzymes working together in sequential pathways. The exploration of the metabolism mechanism requires the understanding of the multi-enzyme coupled catalysis systems.

3.1.4. *Antibodies*

The most of SECM interaction experiments is limited through the fact that one of reaction compound is bound with (onto) the surface (glass, gold electrode, carbon electrode) [72]. The specificity of the antibodies binding allows performing label-free experiments. This trend is obvious in the last three years, where some works related to this phenomenon have appeared. *Le et al.* [73] presents label-free immunosensor based on a biotinylated single-chain variable fragment antibody immobilized on co-polypyrrole film. Immunosensor device formed by immobilization of a biotinylated single-chain antibody on an electropolymerized copolymer film of polypyrrole using biotin/streptavidin system was demonstrated for the first time. The response of the biosensor toward antigen detection was monitored by electrochemical analysis of the polypyrrole response by DPV, not by SECM. The SECM system monitoring the interrogation of antibody–antigen interactions was presented by *Holmes et al.* [74, 75]. Screen-printed carbon electrodes were used as a substrate for the deposition of a dotted array, where the dots consisted of biotinylated polyethyleneimine. These were then further derived, first with neutravidin and then with a biotinylated antibody to the protein neuron specific enolase (NSE). In the next experiments biotinylated antibody towards a relevant antigen of interest (PSA, NTx, ciprofloxacin) were used. These antigens were chosen for their clinical relevance but also since, they display a broad range of molecular weights to determine the influence of the size of the antigen. SECM using a ferrocene carboxylic acid mediator showed clear differences between the array and the surrounding unmodified carbon. Binding of the antigen was accompanied by a measured increase in current response, which may be explained in terms of protein electrostatic interactions and hydrophobic interactions to the mediator. Control experiments with other proteins showed only a non-specific binding across the whole of the substrate, thereby confirming that specific binding occurs between the antibody and antigen on the surface of the dots.

3.1.5. *Lipids*

The application of lipids by SECM is focused on the characterization of lipid mono and bilayers, especially on the charge transfer through lipid membranes [9, 17, 76]. The research in this area could be divided in three main streams.

The first case is about applying of SECM technique for investigation of membrane transport, mainly through the pores [77, 78]. *Wilburn et al.* [79] presents the utilisation of SECM for the creation of voltage-gated biological ion channels. These were simulated by insertion of the peptaibol antibiotic alamethicin into reconstituted phosphatidylcholine bilayer lipid membranes (BLMs). SECM was utilized to probe initial BLM resistivity, the insertion of alamethicin pores, and mass transport across the membrane.

Membrane transport of redox species through the membrane, this time without pores or channels, is the second direction. *Zhan and Bard* [80] present experiment based on the immobilisation of liposome structure (15-50 μm in diameter) filled with the $Ru(2,2'\text{-bipyridine})_3^{2+}$ as a redox mediator on the glass substrates (Fig. 8). Conical carbon fibre tips of submicrometer size were used to approach, image, and puncture. This way the leakage of redox mediator through the lipid membrane was probed. The last direction of research is focused to the charge/

electron transfer through the lipid layers. Out of the study of the basic charge transfer processes [81-83], there were studied the absorption procedures as metal ion absorption [84] so as drug absorption [85].

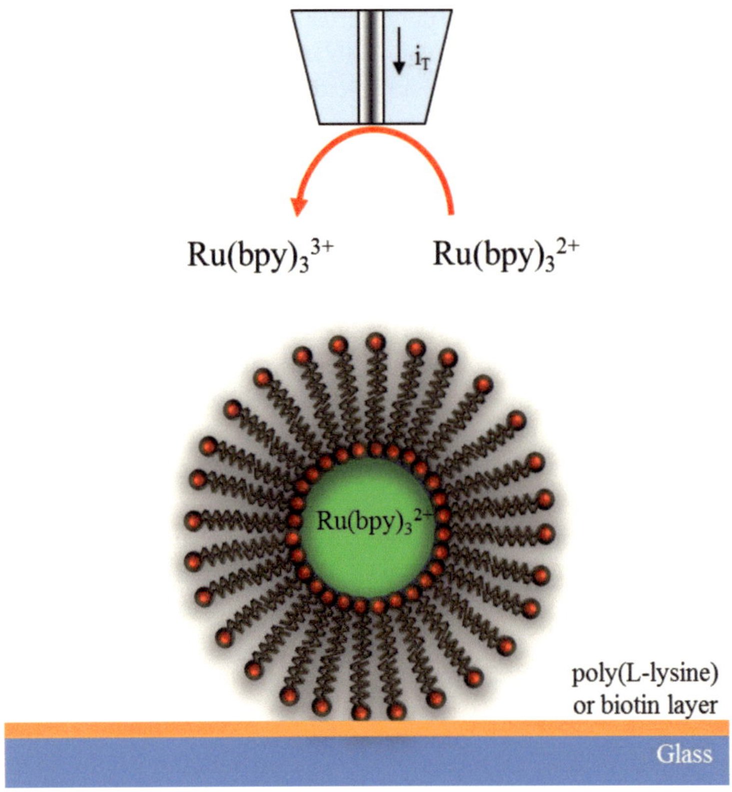

Figure 8. Liposomes of 15-50 nm were immobilized on glass substrates via either poly(L-lysine) or biotin-avidin-biotin sandwich structure; these liposomes were probed by a submicrometer-sized carbon fibre tip controlled by the SECM. The immobilized liposomes were imaged by SECM in a solution of 1.0 mM ruthenium^{3+} hexamine in 100 mM phosphate buffer, pH 7.0. The leakage of $Ru(bpy)_3^{2+}$ from individual liposomes was detected by placing a tip near the liposome and monitoring the current for $Ru(bpy)_3^{2+}$ oxidation.

3.1.6. Cells

The issue of cell studying by SECM is so wide and varied that the short overview is not possible. There are various imaging ways of an individual cell based not only on its surface topography, but also on cellular activities such as photosynthesis, respiration, electron transfer, single vesicular exocytosis, and membrane transport. Therefore the two best reviews in this field are cited [9, 86].

3.2. Various current levels — Images of modification

Modification of surfaces is the introduction of well-defined, directly local changes in the chemical or physical properties of such surfaces. This includes etching processes, deposition of materials, and chemical modification of the surface material. As a scanning probe technique, SECM is predestined to generate chemical changes locally. Instead of detection, the SECM can serve as the preparation tool for microstructures creation. In general, two main ways of application are reported, the direct mode and the feedback mode [2]. The usage of direct mode is based on the fact that the tip is held in a close proximity to the substrate, and the voltage is applied between them to cause the desired reaction at the substrate. In the feedback mode surface patterning, a redox mediator is present in the solution. The tip-generated redox species must be able to induce the desired reaction (i.e., deposition or etching) at the substrate and be regenerated on its surface. The benefit of the feedback mode deposition is that the substrate needs not be a conductor [2]. This phenomenon was described very well in a review of *Sun et al.* [2], where the authors presented works up to year 2007, let us focus the next text on the applications presented just since the year 2007 till now.

The surface modifications are still focused on the enzyme applications. *Li et al.* prepared nanometer-sized carbon fibre disk electrode and applied them to micropattern active horse-radish peroxidase (HRP) coupled with a carbon fibre disk electrode as the SECM tip. Active HRP substrates were fabricated on the substrates coated with streptavidin by binding biotinylated HRP. In one work [87] for the deactivation of other regions on the HRP micro-patterns the authors used the reactive species generated on the electrode as the tip of SECM held at 1.7 V through oxidation of Br^- in 0.20 M phosphate buffer (PB) containing 2.5 x 10^{-2} M KBr and 2.0 x 10^{-3} M BQ (pH 7.0). The micropatterns of active HRP were characterized using the feedback mode of SECM in PB containing 2.0 x 10^{-3} M BQ and 2.0 x 10^{-3} M when the tip potential was held at -0.2 V. Another work from *Li et al.* [87] presents deactivation of other region of the HRP micropatterns by OH- generated at the tip held at -1.7 V in 1.0 M KCl containing 2.0 x 10^{-3} M benzoquinone (BQ) (pH 8.0). The feedback mode of SECM with a tip potential of - 0.2 V was used for characterizing the active HRP micropatterns in 1.0 M KCl containing 2.0 x 10^{-3} M BQ and 2.0 x 10^{-3} M H_2O_2.

Another application of HRP in patterning experiments has been described by *Roberts et al.* [88]. In this case the immobilisation in a patterned fashion onto glass slides is presented. Microarray of HRP islands was deposited on amino-modified glass slides using glutaraldehyde cross-linking combined with the SECM and this was used as a micro-deposition device [88].

Application of nitrogen compounds were shown as a good tool for the patterning technique by SECM. *Cougnon et al.* [89] showed production of diazonium salts from nitro compounds. The nitro precursor is reduced at the tip to the amine, which is diazotized in the interelectrode space. The tip acts as a source of diazonium salts, allowing sample derivatization just under the tip. Extension of this work has been published in another Cougnons work [90] where the authors present the second mechanism of microstructuring. This time the process starts with the reduction of a nitro-containing compound at the scanning electrochemical microscope tip. After diffusion in the interelectrode gap, the amine is oxidized on a gold surface, thus resulting in a local derivatization. *Janin et al.* [91] describes another way of pattering based on the electrochemical grafting of nitrophenyl groups onto platinum ultramicroelectrode (UME). The grafting was made using the electrochemical reduction of nitrophenyldiazonium and the possibility to reduction the diazonium onto Pt UME was observed. Extension of the above-mentioned procedure by creating of a *p*-nitrothiophenol SAM was described by *Schwamborn et al.* [92]. The reduction of the terminating nitro groups of a *p*-nitrothiophenol SAM under formation of hydroxylamine or amino groups is invoked using the direct mode of SECM. The local modification of the redox states was visualized by using the feedback mode of SECM. The current at the Pt tip electrode is determined by the electron-transfer rate for re-oxidation of the redox mediator on the sample surface. Selective post-functionalization with an avidin-alkaline phosphatase conjugate allows visualization of the microstructure using the generator-collector mode of SECM too. Simplification of the surface modification by grafting procedure on the gold substrate by 4-azidobenzenediazonium with SECM was presented by *Coates et al.* [93]. The electrografted spots of diazonium were in this case performed by positioning a SECM platinum tip at a given distance above the gold substrate and the SECM was used in a three-electrode configuration (the Pt tip serving as a microanode). As the mediator acetonitrile containing 5 mM 4-azidobenzenediazonium and 0.1 M Bu_4NBF_4 was used.

Usage of various nanoparticles or nano structured materials expands through the wide range of applications and techniques. SECM is no exception. Preparation of some nanoparticles was described in literature in recent few years. Preparation of gold nanoparticles is described by many authors [94], and possibility of SECM application for preparation of gold nanostructures exists. Process of local electrodeposition of gold nanoobjects, namely gold nanorods, is described in work of *Fedorov* and *Mandler* [95]. In this process a gold microelectrode was the source of the gold ions whereby double pulse chronoamperometry was employed to generate initially Au seeds, which were further grown under controlled conditions.

The silver nanoparticles were prepared either at liquid interface [96] or on the SAM monolayer [97]. The creation at the liquid interface is based on the electron transfer reaction between aqueous Ag^+ ions generated by anodic dissolution of an Ag disk UME, and bis(pentamethyl-cyclopentadienyl) iron (decamethylferrocene, DMFc) in a 1,2-dichloroethane (DCE) phase [96]. The second way, usage of SAM monolayer, is based on the local deposition of Ag nanoparticles (NPs) on ω-mercaptoalkanoic acid SAMs. SECM was used for generating a controlled flux of silver ions by anodic dissolution of silver microelectrode close to the SAMs modified Au(1 1 1) surface [97].

3.3. Quantification of electrochemical images

SECM is a powerful tool for studying structures and processes in micro and submicro meter-sized systems. It can probe electron, ion, and molecule transfers as well as other reactions at solid-liquid, liquid–liquid, and liquid-air interfaces. With the same setup, several SECM modes of operation can be realized, see section 1.2. The most widespread is the feedback mode, tip generation/substrate collection (TG/SC) mode and the substrate generation/tip collection (SG/TC) mode.

The quantitative SECM theory has been developed for various heterogeneous and homogeneous processes and different tip and substrate geometries. In general, theoretical SECM dependencies can be generated by numerically solving partial differential equations. In some cases, analytical approximations allow for easier generation of theoretical dependencies and analysis of experimental data. The most quantitative studies were carried out under steady-state conditions, because the non-steady-state SECM response depends on too many parameters that allow any possible presentation set of working data. The steady-state theory is simpler and often can be expressed in the form of dimensionless working curves or analytical approximations.

For the understanding of mathematical description of the SECM experiments, it is necessary to present some general characteristic and notations. The first assumption, which is generally accounted, is the fact that experiments are made with a redox mediator in a solution where only its diffusion is important (migration and convection are neglected). Its bulk concentration is written as c and D is used for its diffusion coefficient (same value assumed for red or ox form in most of the cases). The tip electrode is generally a disk shaped microelectrode where is characterised the radius of the conductive core, a; and the total tip radius, r_g. The dimensionless parameter RG is defined as follows: $RG = r_g/a$ and is normally ≤ 10. The tip placed at a distance d from the substrate, the dimensionless parameter L is defined as d/a). The potential of the tip is almost always set to a value where the redox reaction of the mediator is rapid and reversible: only mass transport defines the current value. The absolute value of the current going through the tip electrode (respectively through the substrate) is usually written as i_T (respectively i_S). Normalized current is commonly used in SECM experimental reports, defined by the comparison with the tip current when the tip is very far from the studied substrate: $i_T/i_{T\infty}$. Finally the positive feedback situation is referred with the "con" subscript (tip above a conductive substrate) and the negative one with "ins" subscript (tip above an insulating substrate).

The simplest situation in SEMC measurements is the steady state feedback mode, when the tip is immersed in a solution containing a redox mediator (an oxidizable species R) [2]. By applying the positive potential to the tip, the oxidation of R occurs

$$R - ne^- \longrightarrow O \tag{4}$$

at a rate governed by diffusion of R to the UME surface. If the tip is far (greater than several tip diameters) from the substrate the steady-state current, $i_{T\infty}$ is given as follows

$$i_{T\infty} = 4nFDca \tag{5}$$

where F is the Faraday constant, n is the number of electrons transferred in reaction, D and c are the diffusion coefficient and the bulk concentration of R and a is the tip radius.

When the tip is brought to within a few tip radii of a conductive substrate surface, the O species formed in the reaction (Eq. 4) diffuses to the substrate where it can be reduced back to R

$$O + ne^- \longrightarrow R \tag{6}$$

This process produces an additional flux of R to the tip and hence "positive feedback", the increase in tip current ($i_T > i_{T\infty}$). With shorter the tip-substrate separation distance (d), the tip current is larger. When reaction (Eq. 6) is rapid, than $i_T \to \infty$ as $d \to 0$.

If the substrate is an inert electrical insulator, the tip generated species, O, cannot react at its surface. At small d, $i_T < i_{T\infty}$, because the insulator blocks the diffusion of species R to the tip surface ("negative feedback"). The closer the tip to the insulator substrate, the smaller the i_T, with $i_T \to 0$ as $d \to 0$.

The specification that has to be underlined in the context of SECM is that this exact analytical expression is valid only for an infinite RG value: Application of microelectrode with finite RG value led to the modification of equation 1 as follows:

$$i_{T\infty} = 4nFDca\beta(RG) \quad (7)$$

Various approximate expressions of β(RG) can be found in the literature and the better expressions are as follows:

$$\beta = 1 + \frac{0.1380}{(RG - 0.6723)^{0.8686}} \quad (8)$$

$$\beta = 1 + 0.639\left(1 - \frac{2}{\pi} ArcCos \frac{1}{RG}\right) - 0.186\left[1 - \left(\frac{2}{\pi} ArcCos \frac{1}{RG}\right)^2\right] \quad (9)$$

3.3.1. Homogeneous reaction in solution

The first theoretical study of SECM feedback for homogeneous reactions in solution published in 1994 [98] showed that there is a linear relationship between the current measured through the tip, i_T, and the current going through the substrate, i_S. This relationship can be rewritten as follows:

$$i_T = i_S\left(1 - \frac{i_T^{ins}}{i_T^{con}}\right) + i_T^{ins} \quad (10)$$

i_T^{ins} and i_T^{con} being the negative and positive feedback currents that would been obtained in the same geometric configuration (same L). It is necessary to kept in mind that this equation describes the reaction in an electrolyte. As soon as no homogeneous reaction occurs on the substrate or tip (with connection vice versa), the relationship between the tip and substrate steady-state current is very easy [99]:

$$i_T = i_S = i_T^{con} \quad (11)$$

3.3.2. *Heterogeneous reaction*

Study of heterogeneous reaction, exactly heterogeneous electron transfer, is widely used application in SECM determinations. The kinetics of heterogeneous electron transfer can be determined with high lateral resolution while scanning a tip parallel to the surface. Distance-dependent measurements provide quantitative information on sample properties.

One step heterogeneous electron transfer reaction is the simplest case of these types of reactions. An analytical approximation which is often used is valid for RG = 10 and normalized distances 0.1 ≤ L ≤ 1.5. The current-distance curves for an irreversible heterogeneous reaction occurring at the substrate while the tip process is diffusion-controlled can be calculated from

$$I_T(L) = \frac{i_T}{i_{T\infty}} = I_S(L)\left(1 - \frac{I_T^{ins}(L)}{I_T^{con}(L)}\right) + I_T^{ins}(L) \quad (12)$$

$$I_S = \frac{0.78377}{L(1 + 1/\Lambda)} + \frac{0.68 + 0.3315 exp(-1.0672/L)}{1 + F(L, \Lambda)} \quad (13)$$

where I_T^{ins} and I_T^{con} are normalized current related to the insulating substrate, and diffusion-controlled regeneration of redox mediator, respectively. The analytical approximations for these two values are [3]:

$$I_T^{con}(L) = \frac{0.78377}{L} + 0.3315 exp\left(\frac{-1.0672}{L}\right) + 0.68 \quad (14)$$

$$I_T^{ins}(L) = \frac{1}{\left\{0.15 + \frac{1.5385}{L} + 0.58 exp(-1.14/L) + 0.0908 exp\left[(L - 6.3)/1.017L\right]\right\}} \quad (15)$$

In this place it is important to highlight that the value of I_T^{ins} depends on the RG value too [100]. For the various RG value were calculated the parameters of this equation and these are able to read in [2].

Back to the Eq. 12 where it is necessary to define the remaining variables. I_s is the kinetically controlled substrate current. Λ is defined as $\Lambda = k_f d / D$, where k_f is the heterogeneous rate constant, and D is the diffusion coefficient. The function F depends on the L and Λ, is defined as follows: $F(L, \Lambda) = \left(\frac{11}{\Lambda} + 7.3\right) / (110 - 40L)$.

Notwithstanding above presented equations (Eq. 12 and 13) were derived for one step heterogeneous electron transfer reactions, it has been proofed that are applicable to more complicated substrate kinetics (e.g., liquid–liquid interfacial charge transfer, ET through self-assembled monolayers and mediated ET in living cells [2]).

On the other hand, study of the heterogeneous electron transfer reactions at SECM is possible to view from the opposite. For the finite heterogeneous kinetics at the tip and diffusion-controlled mediator regeneration at the substrate, an approximate was recently obtained for I_T as a function of tip potential E, and L [101]:

$$I_T(E, L) = \frac{0.78377}{L(\theta + 1/\kappa)} + \frac{0.68 + 0.3315exp(-1.0672/L)}{\theta\left[1 + \frac{\pi}{\kappa\theta}\frac{2\kappa\theta + 3\pi}{4\kappa\theta + 3\pi^2}\right]} \tag{16}$$

where

$$\kappa = \pi\lambda exp[-\alpha F(E - E^{o'})/RT] / (4I_T^{con}) \tag{17}$$

$$\theta = 1 + exp[F(E - E^{o'})/RT]D_o/D_R$$

$$\lambda = k^o a/D$$

$E^{o'}$ is the formal potential, α is the transfer coefficient, $I_T^{con}(L)$ is the normalized tip current for the same L the diffusion-controlled positive feedback at a conductive substrate, as defined by Eq. 14.

3.3.3. GC mode

The above-presented approximation models are related to the feedback mode of SECM determinations. The application of GC mode is connected with some difficulties in the quantitative interpretation of the measured results [17]:

- if the active region of the sample is to large, no steady-state situation is established. The local concentration depends not only on the position of the UME but also on the time that has passed since the onset of the reaction.
- the moving of probe disturbs the macroscopic diffusion layer of the sample through convection. This effect hinders the diffusion of reagents to the sample region underneath the UME.
- an enhancement of the current according to the principle of positive feedback can reach significant values if $d < 3a$.

Minimization of the quantitative issues is possible due to working with as small as possible UMEs (small RG). In these cases, the disturbance of the sample diffusion layer is minimized and feedback effects can also be neglected at small values of d. The lateral resolution is always poorer in GC mode than in the corresponding feedback experiments [17]. However, the GC mode offers a much higher sensitivity than feedback mode because the flux of reagents coming from the sample is measured essentially without a background signal [102]. This makes the GC mode appropriate for the investigation of immobilized enzymes and cells.

3.3.4. Numerical models

The analytical approximations presented above are best fits to numerical simulations of the diffusion problems for relatively simple and well defined electrochemical systems, e.g., an inlaid disk electrode approaching a flat, infinite, and uniformly reactive substrate surface. In most quantitative SECM experiments, the use of such approximations could be justified.

However, no analytical approximations are available for more complicated processes and system geometries, and so one has to resort to computer simulations.

The above mentioned models suggested the disk shaped tips. In real experiments, it is difficult to assume the ideal disk shaped tip. Production of tips, which may be similar to ideal model, is very difficult, especially when they have to be nanometer-sized. Characterization of the non-disk shaped tips (sphere cups, spheres, rings, ring disks and etched electrodes), experimental approach curves were obtained and then compared to simulated ones [3].

The solution of the diffusion phenomenon has been developed mostly as two-dimensional axi-symmetric diffusions. Even for an idealized situation (i.e. a flat, planar substrate, strictly perpendicular to the axis of the well-shaped disk tip) only numerical solutions can be obtained. The most important step is the validation of calculated data relate to the real measurements. Similar questions arise when the topography of substrate is complicated and/or its surface reactivity is highly non-uniform.

The most popular method for mathematical calculations of diffusion effect by SECM has been presented for the first time by *Fulian et al.* [13, 14]. They introduced the boundary element method (BEM), which is more suitable for problems with regions of complex or rapidly changing geometries. They used BEM to simulate the current responses for different SECM situations, such as a non-disk tip approaching a flat substrate; a flat disk tip over a hemispherical or a spherical cap-shaped substrate or a tilted substrate; a lateral scan of a flat disk tip over an insulating/conductive boundary. The next improvement and extension of this method, mostly related to the specific experiment simulation, has been published by many authors [103-105].

4. Conclusions

The SECM is the widely used technique in material and process investigation. Especially effects connected with electron transfer are studied. Micro and nano sized tips used in these experiments allow to measure nano and sub-nano currents levels and this way allow to describe the very small particles / surfaces. Significant progress has been made over the last five years in both quantitative (e.g. kinetic experiments) and qualitative (e.g. topographic imaging) applications of SECM to studies of structures and processes on the nanoscale. Combination of nanoparticles and SECM shifts the possible usage of this technique to the new horizon. This chapter summarizes the basic knowledge in the field of SECM in relation to study of nanoscale systems, specifically nanoparticles.

Acknowledgements

Financial support from the following projects NanoBioMetalNet CZ.1.07/2.4.00/31.0023 and NanoBioTECell GA CR P102/11/1068 is highly acknowledged.

Author details

David Hynek[1,2], Michal Zurek[1,2], Petr Babula[1], Vojtech Adam[1,2] and Rene Kizek[1,2*]

*Address all correspondence to:

1 Central European Institute of Technology, Brno University of Technology, Brno, Czech Republic

2 Department of Chemistry and Biochemistry, Faculty of Agronomy, Mendel University in Brno, Brno, Czech Republic

References

[1] Bard AJ, Denuault G, Lee C, Mandler D, Wipf DO. Scanning electrochemical microscopy - A new technique for the characterization and modification of surfaces. Accounts Chem Res. 1990 Nov;23(11):357-63.

[2] Sun P, Laforge FO, Mirkin MV. Scanning electrochemical microscopy in the 21st century. Phys Chem Chem Phys. 2007;9(7):802-23.

[3] Mirkin MV, Fan FRF, Bard AJ. Scanning electrochemical microscopy. 13. Evaluation of the tip shapes of nanometer size microelectrodes. J Electroanal Chem. 1992 Jul; 328(1-2):47-62.

[4] Edwards MA, Martin S, Whitworth AL, Macpherson JV, Unwin PR. Scanning electrochemical microscopy: principles and applications to biophysical systems. Physiol Meas. 2006 Dec;27(12):R63-R108.

[5] Nagy G, Nagy L. Scanning electrochemical microscopy: a new way of making electrochemical experiments. Fresenius J Anal Chem. 2000 Mar-Apr;366(6-7):735-44.

[6] Roberts WS, Lonsdale DJ, Griffiths J, Higson SPJ. Advances in the application of scanning electrochemical microscopy to bioanalytical systems. Biosens Bioelectron. 2007 Oct;23(3):301-18.

[7] Souto RM, Gonzalez-Garcia Y, Gonzalez S. Characterization of coating systems by scanning electrochemical microscopy: Surface topology and blistering. Prog Org Coat. 2009 Aug;65(4):435-9.

[8] de la Escosura-Muniz A, Ambrosi A, Merkoci A. Electrochemical analysis with nanoparticle-based biosystems. Trac-Trends Anal Chem. 2008 Jul-Aug;27(7):568-84.

[9] Amemiya S, Guo JD, Xiong H, Gross DA. Biological applications of scanning electrochemical microscopy: chemical imaging of single living cells and beyond. Anal Bioanal Chem. 2006 Oct;386(3):458-71.

[10] Amemiya S, Bard AJ, Fan FRF, Mirkin MV, Unwin PR. Scanning Electrochemical Microscopy. *Annual Review of Analytical Chemistry*. Palo Alto: Annual Reviews 2008:95-131.

[11] Wittstock G, Burchardt M, Pust SE, Shen Y, Zhao C. Scanning electrochemical microscopy for direct imaging of reaction rates. Angew Chem-Int Edit. 2007;46(10): 1584-617.

[12] Mirkin MV, Nogala W, Velmurugan J, Wang YX. Scanning electrochemical microscopy in the 21st century. Update 1: five years after. Phys Chem Chem Phys. 2011;13(48): 21196-212.

[13] Fulian Q, Fisher AC, Denuault G. Applications of the boundary element method in electrochemistry: Scanning electrochemical microscopy. J Phys Chem B. 1999 May; 103(21):4387-92.

[14] Fulian Q, Fisher AC, Denuault G. Applications of the boundary element method in electrochemistry: Scanning electrochemical microscopy, part 2. J Phys Chem B. 1999 May;103(21):4393-8.

[15] Szunerits S, Pust SE, Wittstock G. Multidimensional electrochemical imaging in materials science. Anal Bioanal Chem. 2007 Oct;389(4):1103-20.

[16] Hamou RF, Biedermann PU, Erbe A, Rohwerder M. Numerical simulation of probing the electric double layer by scanning electrochemical potential microscopy. Electrochim Acta. 2010 Jul;55(18):5210-22.

[17] Lu X, Wang Q, Liu X. Review: Recent applications of scanning electrochemical microscopy to the study of charge transfer kinetics. Anal Chim Acta. 2007 Oct;601(1): 10-25.

[18] Inaba M, Siroma Z, Funabiki A, Ogumi Z, Abe T, Mizutani Y, et al. Electrochemical scanning tunneling microscopy observation of highly oriented pyrolytic graphite surface reactions in an ethylene carbonate-based electrolyte solution. Langmuir. 1996 Mar;12(6):1535-40.

[19] Ibe JP, Bey PP, Brandow SL, Brizzolara RA, Burnham NA, Dilella DP, et al. On the electrochemical etching of tips for scanning tunneling microscopy. J Vac Sci Technol A-Vac Surf Films. 1990 Jul-Aug;8(4):3570-5.

[20] Macpherson JV, Unwin PR. Combined scanning electrochemical-atomic force microscopy. Anal Chem. 2000 Jan;72(2):276-85.

[21] Bain CD, Troughton EB, Tao YT, Evall J, Whitesides GM, Nuzzo RG. Formation of monolayer films by the spontaneous assembly of organic thiols from solution onto gold. J Am Chem Soc. 1989 Jan;111(1):321-35.

[22] Chen D, Li JH. Interfacial design and functionization on metal electrodes through self-assembled monolayers. Surf Sci Rep. 2006 Dec;61(11):445-63.

[23] Eckermann AL, Feld DJ, Shaw JA, Meade TJ. Electrochemistry of redox-active self-assembled monolayers. Coord Chem Rev. 2010 Aug;254(15-16):1769-802.

[24] Kiani A, Alpuche-Aviles MA, Eggers PK, Jones M, Gooding JJ, Paddon-Row MN, et al. Scanning electrochemical microscopy. 59. Effect of defects and structure on electron transfer through self-assembled monolayers. Langmuir. 2008 Mar;24(6):2841-9.

[25] Tao F, Bernasek SL. Understanding odd-even effects in organic self-assembled monolayers. Chem Rev. 2007 May;107(5):1408-53.

[26] Yamada H, Ogata M, Koike T. Scanning electrochemical microscope observation of defects in a hexadecanethiol monolayer on gold with shear force-based tip-substrate positioning. Langmuir. 2006 Aug 29;22(18):7923-7.

[27] Liu B, Bard AJ, Mirkin MV, Creager SE. Electron transfer at self-assembled monolayers measured by scanning electrochemical microscopy. J Am Chem Soc. 2004 Feb; 126(5):1485-92.

[28] Alizadeh V, Mousavi MF, Mehrgardi MA, Kazemi SH, Sharghi H. Electron transfer kinetics of cytochrome c immobilized on a phenolic terminated thiol self assembled monolayer determined by scanning electrochemical microscopy. Electrochim Acta. 2011 Jul;56(17):6224-9.

[29] Bar G, Rubin S, Taylor TN, Swanson BI, Zawodzinski TA, Chow JT, et al. Patterned self-assembled monolayers of ferrocene and methyl terminated alkanethiols on gold: A combined electrochemical, scanning probe microscopy, and surface science study. J Vac Sci Technol A-Vac Surf Films. 1996 May-Jun;14(3):1794-800.

[30] Salamifar SE, Mehrgardi MA, Kazemi SH, Mousavi MF. Cyclic voltammetry and scanning electrochemical microscopy studies of methylene blue immobilized on the self-assembled monolayer of n-dodecanethiol. Electrochim Acta. 2010 Dec;56(2): 896-904.

[31] Wittstock G, Hesse R, Schuhmann W. Patterned self-assembled alkanethiolate monolayers on gold. Patterning and imaging by means of scanning electrochemical microscopy. Electroanalysis. 1997 Jun;9(10):746-50.

[32] Zhang LM, Zuo RX, Liu XH, Lu XQ. The study on the self-assembled monolayer (SAMs) of porphyrin with different chain length by scanning electrochemical microscopy (SECM). Chin Chem Lett. 2006 May;17(5):661-4.

[33] Boldt FM, Baltes N, Borgwarth K, Heinze J. Investigation of carboxylic-functionalized and n-alkanethiol self-assembled monolayers on gold and their application as pH-sensitive probes using scanning electrochemical microscopy. Surf Sci. 2005 Dec; 597(1-3):51-64.

[34] Sanders W, Vargas R, Anderson MR. Characterization of carboxylic acid-terminated self-assembled monolayers by electrochemical impedance spectroscopy and scanning electrochemical microscopy. Langmuir. 2008 Jun;24(12):6133-9.

[35] Uosaki K, Wano H. Dynamic monitoring of electrochemical oxidative adsorption and reductive desorption of self-assembled monolayer of hexanethiol on Au(111) surface in KOH ethanol solution by in situ scanning tunneling microscopy. Abstr Pap Am Chem Soc. 2002 Aug;224:U449-U.

[36] Uosaki K, Wano H. Electrochemical formation and desorption of self-assembled monolayer of alkanethiol in ethanol solution monitored by in situ scanning tunneling microscopy. Abstr Pap Am Chem Soc. 2002 Apr;223:U413-U.

[37] Burshtain D, Mandler D. Studying the binding of Cd2+ by omega-mercaptoalkanoic acid self assembled monolayers by cyclic voltammetry and scanning electrochemical microscopy (SECM). J Electroanal Chem. 2005 Aug;581(2):310-9.

[38] Palecek E, Fojta M. Magnetic beads as versatile tools for electrochemical DNA and protein biosensing. Talanta. 2007 Dec;74(3):276-90.

[39] Quinn BM, Prieto I, Haram SK, Bard AJ. Electrochemical observation of a metal/insulator transition by scanning electrochemical microscopy. J Phys Chem B. 2001 Aug; 105(31):7474-6.

[40] Tel-Vered R, Bard AJ. Generation and detection of single metal nanoparticles using scanning electrochemical microscopy techniques. J Phys Chem B. 2006 Dec;110(50): 25279-87.

[41] Wijayawardhana CA, Wittstock G, Halsall HB, Heineman WR. Electrochemical immunoassay with microscopic immunomagnetic bead domains and scanning electrochemical microscopy. Electroanalysis. 2000 May;12(9):640-4.

[42] Wijayawardhana CA, Wittstock G, Halsall HB, Heineman WR. Spatially addressed deposition and imaging of biochemically active bead microstructures by scanning electrochemical microscopy. Anal Chem. 2000 Jan;72(2):333-8.

[43] Zhao C, Wittstock G. Scanning electrochemical microscopy for detection of biosensor and biochip surfaces with immobilized pyrroloquinoline quinone (PQQ)-dependent glucose dehydrogenase as enzyme label. Biosens Bioelectron. 2005 Jan;20(7):1277-84.

[44] Zhao CA, Wittstock G. Scanning electrochemical microscopy of quinoprotein glucose dehydrogenase. Anal Chem. 2004 Jun;76(11):3145-54.

[45] Wang YC, Hu R, Lin GM, Roy I, Yong KT. Functionalized Quantum Dots for Biosensing and Bioimaging and Concerns on Toxicity. ACS Appl Mater Interfaces. 2013 Apr;5(8):2786-99.

[46] Katz E, Willner I, Wang J. Electroanalytical and bioelectroanalytical systems based on metal and semiconductor nanoparticles. Electroanalysis. 2004 Jan;16(1-2):19-44.

[47] Zimin S, Vasin V, Gorlachev E, Buchin E, Naumov V. Investigations of PbSe layers after anodic electrochemical etching by scanning electron microscopy. In: Cantarero

A, Sailor M, Nassiopoulou A, Schmuki P, Canham L, eds. *Physica Status Solidi C: Current Topics in Solid State Physics*. Weinheim: Wiley-V C H Verlag Gmbh 2011.

[48] Liu GJ, Liu CY, Bard AJ. Rapid Synthesis and Screening of Zn(x)zCd(1-x)S(y)Se(1-y) Photocatalysts by Scanning Electrochemical Microscopy. J Phys Chem C. 2010 Dec; 114(49):20997-1002.

[49] Hutchings GJ, Brust M, Schmidbaur H. Gold - an introductory perspective. Chem Soc Rev. 2008 Sep;37(9):1759-65.

[50] Wain AJ. Imaging size effects on the electrocatalytic activity of gold nanoparticles using scanning electrochemical microscopy. Electrochim Acta. 2013 Mar;92:383-91.

[51] Sanchez-Sanchez CM, Vidal-Iglesias FJ, Solla-Gullon J, Montiel V, Aldaz A, Feliu JM, et al. Scanning electrochemical microscopy for studying electrocatalysis on shape-controlled gold nanoparticles and nanorods. Electrochim Acta. 2010 Nov;55(27): 8252-7.

[52] Sanchez-Sanchez CM, Solla-Gullon J, Vidal-Iglesias FJ, Aldaz A, Montiel V, Herrero E. Imaging Structure Sensitive Catalysis on Different Shape-Controlled Platinum Nanoparticles. J Am Chem Soc. 2010 Apr;132(16):5622-4.

[53] Markovic NM, Adzic RR, Vesovic VB. Structural effects in electrocatalysis - Oxygen reduction on the gold single-crystal electrodes with (110) and (111) orientations. J Electroanal Chem. 1984;165(1-2):121-33.

[54] Adzic RR, Markovic NM, Vesovic VB. Structural effects in electrocatalysis - Oxygen reduction on the Au(100) single-crystal electrode. J Electroanal Chem. 1984;165(1-2): 105-20.

[55] El-Deab MS, Sotomura T, Ohsaka T. Size and crystallographic orientation controls of gold nanoparticles electrodeposited on GC electrodes. J Electrochem Soc. 2005;152(1):C1-C6.

[56] Huang K, Anne A, Bahri MA, Demaille C. Probing Individual Redox PEGylated Gold Nanoparticles by Electrochemical-Atomic Force Microscopy. ACS Nano. 2013 May;7(5):4151-63.

[57] Wang J, Zhou FM. Scanning electrochemical microscopic imaging of surface-confined DNA probes and their hybridization via guanine oxidation. J Electroanal Chem. 2002 Nov;537(1-2):95-102.

[58] Wang J, Song FY, Zhou FM. Silver-enhanced imaging of DNA hybridization at DNA microarrays with scanning electrochemical microscopy. Langmuir. 2002 Aug;18(17): 6653-8.

[59] Turcu F, Schulte A, Hartwich G, Schuhmann W. Imaging immobilised ssDNA and detecting DNA hybridisation by means of the repelling mode of scanning electrochemical microscopy (SECM). Biosens Bioelectron. 2004 Nov;20(5):925-32.

[60] Turcu F, Schulte A, Hartwich G, Schuhmann W. Label-free electrochemical recognition of DNA hybridization by means of modulation of the feedback current in SECM. Angew Chem-Int Edit. 2004;43(26):3482-5.

[61] Carano M, Lion N, Abid JP, Girault HH. Detection of proteins on poly(vinylidene difluoride) membranes by scanning electrochemical microscopy. Electrochem Commun. 2004 Dec;6(12):1217-21.

[62] Carano M, Lion N, Girault HH. Copper staining/labeling and Scanning Electrochemical Microscopy readout of proteins on poly(vinylidene difluoride) membranes. Chimia. 2005;59(3):105-8.

[63] Carano M, Lion N, Girault HH. Detection of proteins on membranes and in microchannels using copper staining combined with scanning electrochemical microscopy. J Electroanal Chem. 2007 Jan;599(2):349-55.

[64] Cortes-Salazar F, Busnel JM, Li F, Girault HH. Adsorbed protein detection by scanning electrochemical microscopy. J Electroanal Chem. 2009 Oct;635(2):69-74.

[65] Glidle A, Yasukawa T, Hadyoon CS, Anicet N, Matsue T, Nomura M, et al. Analysis of protein adsorption and binding at biosensor polymer interfaces using X-ray photon spectroscopy and scanning electrochemical microscopy. Anal Chem. 2003 Jun; 75(11):2559-70.

[66] Kirchner CN, Szunerits S, Wittstock G. Scanning electrochemical microscopy (SECM) based detection of oligonucleotide hybridization and simultaneous determination of the surface concentration of immobilized oligonucleotides on gold. Electroanalysis. 2007 Jun;19(12):1258-67.

[67] Wilhelm T, Wittstock G. Analysis of interaction in patterned multienzyme layers by using scanning electrochemical microscopy. Angew Chem-Int Edit. 2003;42(20): 2247-50.

[68] Wilhelm T, Wittstock G. Generation of periodic enzyme patterns by soft lithography and activity imaging by scanning electrochemical microscopy. Langmuir. 2002 Nov; 18(24):9485-93.

[69] Chen PC, Chen RLC, Cheng TJ, Wittstock G. Localized Deposition of Chitosan as Matrix for Enzyme Immobilization. Electroanalysis. 2009 Apr;21(7):804-10.

[70] Wilhelm T, Wittstock G, Szargan R. Scanning electrochemical microscopy of enzymes immobilized on structured glass-gold substrates. Fresenius J Anal Chem. 1999 Sep-Oct;365(1-3):163-7.

[71] Wu ZQ, Jia WZ, Wang K, Xu JJ, Chen HY, Xia XH. Exploration of Two-Enzyme Coupled Catalysis System Using Scanning Electrochemical Microscopy. Anal Chem. 2012 Dec;84(24):10586-92.

[72] Wittstock G, Yu KJ, Halsall HB, Ridgway TH, Heineman WR. Imaging of immobilized antibody layers with scanning electrochemical microscopy. Anal Chem. 1995 Oct;67(19):3578-82.

[73] Le HQA, Sauriat-Dorizon H, Korri-Youssoufi H. Investigation of SPR and electrochemical detection of antigen with polypyrrole functionalized by biotinylated single-chain antibody: A review. Anal Chim Acta. 2010 Jul;674(1):1-8.

[74] Holmes JL, Davis F, Collyer SD, Higson SPJ. A new application of scanning electrochemical microscopy for the label-free interrogation of antibody-antigen interactions. Anal Chim Acta. 2011 Mar;689(2):206-11.

[75] Holmes JL, Davis F, Collyer SD, Higson SPJ. A new application of scanning electrochemical microscopy for the label-free interrogation of antibody-antigen interactions: Part 2. Anal Chim Acta. 2012 Sep;741:1-8.

[76] Wittstock G. Modification and characterization of artificially patterned enzymatically active surfaces by scanning electrochemical microscopy. Fresenius J Anal Chem. 2001 Jun;370(4):303-15.

[77] Lu XQ, Wang TX, Zhou XB, Li Y, Wu BW, Liu XH. Investigation of Ion Transport Traversing the "Ion Channels" by Scanning Electrochemical Microscopy (SECM). J Phys Chem C. 2011 Mar;115(11):4800-5.

[78] Kim J, Izadyar A, Nioradze N, Amemiya S. Nanoscale Mechanism of Molecular Transport through the Nuclear Pore Complex As Studied by Scanning Electrochemical Microscopy. J Am Chem Soc. 2013 Feb;135(6):2321-9.

[79] Wilburn JP, Wright DW, Cliffel DE. Imaging of voltage-gated alamethicin pores in a reconstituted bilayer lipid membrane via scanning electrochemical microscopy. Analyst. 2006;131(2):311-6.

[80] Zhan W, Bard AJ. Scanning electrochemical microscopy. 56. Probing outside and inside single giant liposomes containing Ru(bpy)(3)(2+). Anal Chem. 2006 Feb;78(3): 726-33.

[81] Amemiya S, Ding ZF, Zhou JF, Bard AJ. Studies of charge transfer at liquid vertical bar liquid interfaces and bilayer lipid membranes by scanning electrochemical microscopy. J Electroanal Chem. 2000 Mar;483(1-2):7-17.

[82] Cannes C, Kanoufi F, Bard AJ. Cyclic voltammetry and scanning electrochemical microscopy of ferrocenemethanol at monolayer and bilayer-modified gold electrodes. J Electroanal Chem. 2003 Apr;547(1):83-91.

[83] Tsionsky M, Zhou JF, Amemiya S, Fan FRF, Bard AJ, Dryfe RAW. Scanning electrochemical microscopy. 38. Application of SECM to the study of charge transfer through bilayer lipid membranes. Anal Chem. 1999 Oct;71(19):4300-5.

[84] Burt DP, Cervera J, Mandler D, Macpherson JV, Manzanares JA, Unwin PR. Scanning electrochemical microscopy as a probe of Ag+ binding kinetics at Langmuir phospholipid monolayers. Phys Chem Chem Phys. 2005;7(15):2955-64.

[85] Slevin CJ, Liljeroth P, Kontturi K. Measurement of the adsorption of drug ions at model membranes by scanning electrochemical microscopy. Langmuir. 2003 Apr; 19(7):2851-8.

[86] Bergner S, Vatsyayan P, Matysik FM. Recent advances in high resolution scanning electrochemical microscopy of living cells - A review. Anal Chim Acta. 2013 May; 775:1-13.

[87] Li XM, Geng QH, Wang YY, Si ZK, Jiang W, Zhang XL, et al. Fabrication of active horseradish peroxidase micropatterns with a high resolution by scanning electrochemical Microscopy. Electroanalysis. 2007 Aug;19(16):1734-40.

[88] Roberts WS, Davis F, Collyer SD, Higson SPJ. Construction and interrogation of enzyme microarrays using scanning electrochemical microscopy - optimisation of adsorption and determination of enzymatic activity. Analyst. 2011;136(24):5287-93.

[89] Cougnon C, Gohier F, Belanger D, Mauzeroll J. In Situ Formation of Diazonium Salts from Nitro Precursors for Scanning Electrochemical Microscopy Patterning of Surfaces. Angew Chem-Int Edit. 2009;48(22):4006-8.

[90] Cougnon C, Mauzeroll J, Belanger D. Patterning of Surfaces by Oxidation of Amine-Containing Compounds Using Scanning Electrochemical Microscopy. Angew Chem-Int Edit. 2009;48(40):7395-7.

[91] Janin M, Ghilane J, Randriamahazaka H, Lacroix JC. Microelectrodes modification through the reduction of aryl diazonium and their use in scanning electrochemical microscopy (SECM). Electrochem Commun. 2009 Mar;11(3):647-50.

[92] Schwamborn S, Stoica L, Neugebauer S, Reda T, Schmidt HL, Schuhmann W. Local Modulation of the Redox State of p-Nitrothiophenol Self-Assembled Monolayers Using the Direct Mode of Scanning Electrochemical Microscopy. ChemPhysChem. 2009 May;10(7):1066-70.

[93] Coates M, Cabet E, Griveau S, Nyokong T, Bedioui F. Microelectrochemical patterning of gold surfaces using 4-azidobenzenediazonium and scanning electrochemical microscopy. Electrochem Commun. 2011 Feb;13(2):150-3.

[94] Yong KT, Swihart MT, Ding H, Prasad PN. Preparation of Gold Nanoparticles and their Applications in Anisotropic Nanoparticle Synthesis and Bioimaging. Plasmonics. 2009 Jun;4(2):79-93.

[95] Fedorov RG, Mandler D. Local deposition of anisotropic nanoparticles using scanning electrochemical microscopy (SECM). Phys Chem Chem Phys. 2013;15(8): 2725-32.

[96] Li F, Edwards M, Guo JD, Unwin PR. Silver Particle Nucleation and Growth at Liquid/Liquid Interfaces: A Scanning Electrochemical Microscopy Approach. J Phys Chem C. 2009 Mar;113(9):3553-65.

[97] Malel E, Colleran J, Mandler D. Studying the localized deposition of Ag nanoparticles on self-assembled monolayers by scanning electrochemical microscopy (SECM). Electrochim Acta. 2011 Aug;56(20):6954-61.

[98] Treichel DA, Mirkin MV, Bard AJ. Scanning electrochemical microscopy. 27. Application of a simplified treatment of an irreversible homogeneous reaction following electron-transfer to the oxidative dimerization of 4-nitrophenolate in acetonitrile. J Phys Chem. 1994 Jun;98(22):5751-7.

[99] Lefrou C, Cornut R. Analytical Expressions for Quantitative Scanning Electrochemical Microscopy (SECM). ChemPhysChem. 2010 Feb;11(3):547-56.

[100] Amphlett JL, Denuault G. Scanning electrochemical microscopy (SECM): An investigation of the effects of tip geometry on amperometric tip response. J Phys Chem B. 1998 Dec;102(49):9946-51.

[101] Sun P, Mirkin MV. Kinetics of electron-transfer reactions at nanoelectrodes. Anal Chem. 2006 Sep;78(18):6526-34.

[102] Wittstock G, Wilhelm T. Characterization and manipulation of microscopic biochemically active regions by scanning electrochemical microscopy (SECM). Anal Sci. 2002 Nov;18(11):1199-204.

[103] Sklyar O, Wittstock G. Numerical simulations of complex nonsymmetrical 3D systems for scanning electrochemical microscopy using the boundary element method. J Phys Chem B. 2002 Aug;106(30):7499-508.

[104] Sklyar O, Kueng A, Kranz C, Mizaikoff B, Lugstein A, Bertagnolli E, et al. Numerical simulation of scanning electrochemical Microscopy experiments with frame-shaped integrated atomic force microscopy-SECM probes using the boundary element method. Anal Chem. 2005 Feb;77(3):764-71.

[105] Sklyar O, Trauble M, Zhao CA, Wittstock G. Modeling steady-state experiments with a scanning electrochemical microscope involving several independent diffusing species using the boundary element method. J Phys Chem B. 2006 Aug;110(32):15869-77.

Chapter 7

Electrocatalysis and the Production of Nanoparticles

Metin Açıkyıldız, Ahmet Gürses,
Mehtap Ejder Korucu and Kübra Güneş

Additional information is available at the end of the chapter

http://dx.doi.org/10.5772/58340

1. Introduction

Electrochemistry has much in common with surface science, but the application of the principles of catalytic activity to the reactions taking place in an electrochemical environment is not straightforward. Electrochemical reactions of practical interest imply at least one step where an electron is transferred between species coming from the solution side or the electrode surface. As a result, electrochemical reactions occurring at the interfaces are governed by the interaction of the reactant both with the solvent and with the electrode [1].

It is undeniable that electrocatalysis is a very important subject. This is not only for the obvious situations, for example, increasing the amount of energy obtained from a fuel cell, but also for every kind of electrochemical reaction which is going to be developed industrially, because the cost of producing a certain amount of a substance is simply proportional to the total amount of electricity which is needed, and this depends upon the overpotentials at the electrode–solution interfaces [2].

Electrocatalysis is one of the most actual fields of contemporary electrochemistry. Although Bowden and Rideal are the true originators of studies of rates on a series of different materials, Horiuti and Polanyi have wrote the first paper which began the theory of electrocatalysis [2]. In the 1920s, Bowden and Rideal made a series of measurements of hydrogen evolution overpotential as a function of current density for a number of metals and then the electrocatalysis term was coined in 1936 by Kobozev, but in the next 30 years little attention was paid to this phenomenon [3].

Electrocatalysis is the field of catalysis that deals with the catalysis of redox reactions. Redox reactions not only play a key role in a multitude of devices, in particular electrochemical power sources such as fuel cells, batteries, and electrolysis cell, but also in many naturally occurring processes, especially those related to energy conversion in living systems. There is growing

INTECH
open science | open minds

© 2014 The Author(s). Licensee InTech. This chapter is distributed under the terms of the Creative Commons Attribution License (http://creativecommons.org/licenses/by/3.0), which permits unrestricted use, distribution, and reproduction in any medium, provided the original work is properly cited.

awareness that the way that mankind will secure future availability of sustainable energy will depend heavily on electrochemistry, as it allows the storage and consumption of electrons in chemical bonds at ambient conditions. It is the task of electrocatalysis to make this process as efficient, reversible, robust, and cost-effective as possible [4].

Electrocatalysis and heterogeneous catalysis are closely related in that they involve well-controlled sequences of elementary bond-breaking and making processes and share many common mechanistic principles in the transformation of molecules over supported metal and metal oxide catalysts [5].

Electrocatalysis is not limited to heterogeneous catalysts: molecular catalysts, either in solution or immobilized on a conductive but otherwise inert support, may also catalyze redox reactions. Therefore, the field of electrocatalysis bridges many areas of catalysis [4].

In electrocatalysis, the high performance of electrodes is the result of some combinations of surface reactivity, electronic and ionic conductivity and electron hole pair separation and facile mass transport of molecules, which is furnished by the architectural design of both electrodes in the cell, i.e., the arrangement of material in space. In electrochemical reactions, the electron is a reactant that scales with the electrified surface area. This fact poses the issue of the optimum pore size of electrode materials compatible with the highest electrocatalytic activity. In this respect, it should be noted that micropores (less than 5 nm) are not useful because electrochemical reactions require ionic conducting media for charge balance, and the mass transport rate of ionic species into micropores becomes several orders of magnitude lower than in the bulk of the solution. Therefore, in electrocatalysis surface accessibility is maximized by using materials with meso or macropores structures. Accordingly, since some decades ago, this issue has become of considerable interest from the standpoint of electrochemical research [6].

Traditionally, for laboratory and small-scale plants, platinum was the first choice as electrode material for both cathodic and anodic processes. Platinum metal in a shape of bars, foils (continuous and latter perforated) or gauze, was used over a long period. Nevertheless, platinum was not accustomed in commercial use, due to its disadvantages (too expensive, extensive corrosion when used for $Cl_{2(g)}$ production, etc.) [3].

Literature survey made from 2005 to 2013 shows that there are a lot of studies focused on development of alternative materials and their usages at the different electrocatalytic applications. These materials can be classified as noble metals, alloys or metallic composites; polymer based metallic composites, carbon-supported metallic composites, carbon-based materials etc.

In a few specific studies, the noble metals were used alone and in these researches, the efficiency of nanostructured Pd film [7], single crystalline platinum [8], porous gold [9], and Pt particles [10] were investigated in terms of electrocatalytic reactions and electroanalysis.

Figure 1 shows the scanning electron microscopy (SEM) and transmission electron microscopy (TEM) images of the Au porous textile-like sheet arrays (AuPTSAs) developed by one-step electrodeposition with the assistance of *N*-methylimidazole, without any template, seed, or surfactant. It was determined that the AuPTSAs which displayed remarkable improvement of the electrocatalytic activity over polycrystalline Au nanoparticles for methanol oxidation.

http://dx.doi.org/10.5772/58340

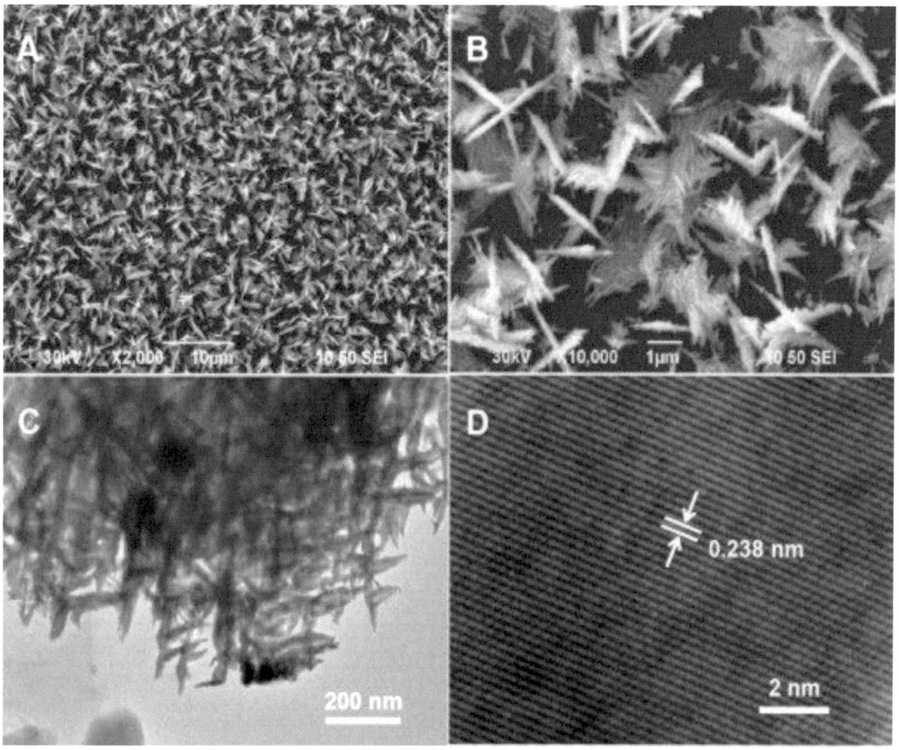

Figure 1. Low (A, C) and high (B, D) magnification of SEM (A, B) and TEM (C, D) images of the typical AuPTSAs using the applied potential of −0.4 V for 600 s in the electrolysis solution containing 10 mM $HAuCl_4$, 1.25 M *N*-methylimidazole, and 0.5 M H_2SO_4 [9]

In addition, alloys or metallic composites such as Co_3O_4 and Ag/Co_3O_4 [11], Au and Au-Pd core-shell nanoparticles [12], electrodeposited Pd coatings on gold [13], titanium supported Pt [14], nanosized Pt/IrO_2 [15], Ni, Pt, and Ni-Pt nanoparticles [16], Pt supported on nano-tungsten carbide [17], nanostructured Pt/SnO_2-SbO_x-RuO_2 [18], Au-Pt [19, 20], Pt and Pt–Pd core-shell nanoparticles [21], Pt-Sn [22], SnO_2-Au nanoparticles [23], Fe-containing Se/Ru composite [24], titanium supported nanocrystallite Pt and Pt_x-Sn_y [25], $Pt_{25}Cu_{75}$ and $Pt_{20}Cu_{20}Co_{60}$ nanoparticles and Pt-Carbon [26], platinum modified by ruthenium and tin [27], manganese oxide nanorods modified Au, Pt and glassy carbon [28], nano TiO_2 supported Pt [29], bimetallic PdNi nanoparticles [30], transition metal modified ruthenium-selenium [31], Pt-Sn nanoparticles [32], titanium oxides supported Pt [33], and $Ru_xCr_ySe_z$ [34] were used as catalysts.

Figure 2 shows the SEM images of the titanium supported nanocrystallite platinum and platinum-tin catalysts prepared by electrodeposition techniques. It was determined that the modification of Ti with Pt and/or PtSn (i.e. Pt/Ti and Pt-Sn/Ti electrodes) had many advantages compared to Pt/Pt and Pt-Sn/Pt electrodes and the combination effect of Pt, Sn and Ti improves the catalytic activity and the stability of the prepared electrode through complete oxidation of the intermediate product of methanol oxidation. It was dedicated that the introduction of a small amount of Sn deposited with Pt improved the catalytic activity and the stability of prepared electrode and the optimum molar ratio of Pt:Sn was found as 8:1 for binary catalysts.

Figure 2. SEM patterns of the (a) Ti, (b) Pt/Ti (galvanostatic deposition), (c) Pt/Ti (potentiostatic deposition) and (d) Pt-Sn/Ti [25].

Poly(3,4-ethylene-dioxythiophene) coated platinum [35], myoglobin immobilized $LiMn_2O_4$-chitosan [36], MnO_2-polypyrrole composites [37] were used as polymer based metallic composite catalysts.

It was recently found in a work [36], the new electrochemical biosensor constructed by immobilization of bovine myoglobin and $LiMn_2O_4$ in a thin film of chitosan on the surface of carbon ionic liquid electrode has a good electrocatalytic activity toward trichloroacetic acid with a lower detection limit. In addition, it was determined that the obtained $LiMn_2O_4$ microspheres as the cathode material for lithium ion batteries had high discharge capacity and good cycling stability. The SEM images of α-MnO_2 precursor and $LiMn_2O_4$ microspheres calcined at 750 ºC were shown in Figure 3.

The carbon supported mainly nano-sized metallic composites given in Table 1 emerge a critical importance both qualitatively and quantitatively. In order to achieve fine dispersion, high utilization and stable nanoscale metallic particles, catalyst-supporting strategies have been explored. Compared to bulk metal catalysts, supported catalysts show higher activity and stability. Carbon particles are frequently used as catalyst supports because of their relative stability in both acidic and basic media, good electric conductivity and high specific surface

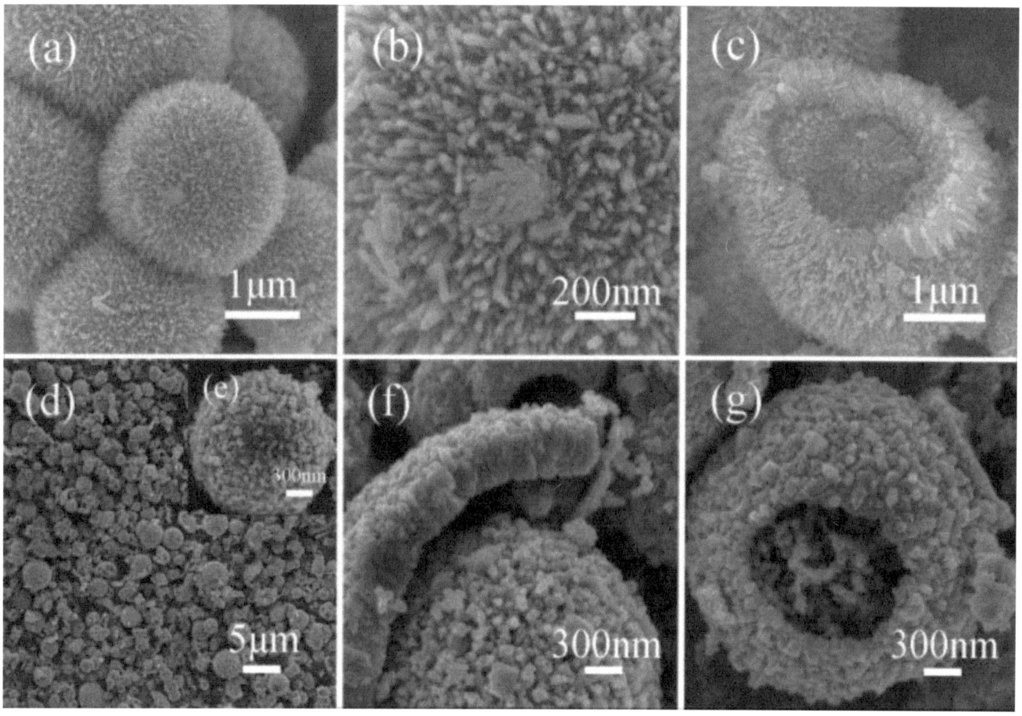

Figure 3. The SEM images of α-MnO_2 precursor (a-c), and $LiMn_2O_4$ microspheres calcined at 750 °C (d-g) [36].

area [38]. The carbon black, nanostructured carbon materials and mesoporous carbon are the most used carbon based catalysts supports.

Catalysts	References
Carbon supported monometallic materials	
Pt	[39-54]
Pd	[49,55]
Ru	[51, 56, 57]
Au	[58]
Ir	[59]
Carbon supported binary materials	
Pt-Ru	[47, 50-54, 60-63]
Pt-Sn	[53, 61, 64-67]
Pt-Pd	[53, 68-70]
Pt-Co	[71, 72]
Pt-CuO	[73]
Pt-MoO	[74]
Pt-Rh	[53]

Catalysts	References
Carbon supported binary materials	
Pt-Sb	[75]
Pt-Cr	[61]
Pd-Co	[55, 76, 77]
Pd-Fe	[55]
Pd-V	[78]
Au-Zn	[79]
Au-Pt	[58]
Au-Pd	[58]
Ag-WC	[80]
Ru-Se	[56, 81]
Ir-Sn	[59]
Ni-TiO_2	[82]
Carbon supported ternary materials	
Pt-Co-Ru	[83]
Ru-Fe-Se	[56]
Pt-Cr-Ru	[84]
Carbon supported quaternary materials	
Pt-Mn-Cu-X (X: Fe, Co, Ni, and Sn)	[85]
Pt-Mn-Mo-X (X: Fe, Co, Ni, Cu and Sn)	[85]
Single-walled carbon nanotube supported	
Pt	[86, 87]
Pd	[88]
Multi-walled carbon nanotube supported	
Pt	[86, 87, 89]
Pd	[90]
Pt-Pd	[91]
Pt-Ru	[92]
Pt-Au	[93]
Pd-Ni	[94]
Ni-Fe	[95]
Pt-M (M: Fe, Co, and Ni)	[96]
Pt nanoparticles on carbonized polyaniline nanotubes	[97]
Carbon nanotubes and active carbon supported Pt	[98]
Pt supported by multi-walled carbon nanotubes coated with sulfated TiO_2	[99]
PtRu alloy catalyst supported on graphene-carbon black	[100]
Reduced graphene/Pd nanocomposite modified with glassy carbon	[101]
Pt supported on reduced graphene oxide-ordered mesoporous carbon	[102]

http://dx.doi.org/10.5772/58340

Catalysts	References
Cd-doped cobalt nanoparticles encapsulated in graphite shell	[103]
Pt and nickel oxide nanoparticles at glassy carbon	[104]
Carbon xerogels supported Pt	[105, 106]
Carbon aerogel supported Pt	[107]
Pt-Ru and Pt-Ru supported on H_2O_2 treated carbon black	[108]
Pt deposited on unsupported and carbon supported Ru nanoparticles	[109]
Oxidized glassy carbon supported Pt/Ru nanoparticles	[110]
Pt monolayers deposited on carbon-supported Ru and Rh nanoparticles	[111]
Pt nanoparticles supported on functionalized graphene	[112]
Polyoxometalate-deposited Pt/C	[113]
Pt–SnO_2 nanoparticles supported on graphitized mesoporous carbon	[114]
Fe associated with pyridinic nitrogen	[115]
Active carbon supported Ir-V nanoparticle	[116]
Silicotungstic acid stabilized Pt-Ru nanoparticles supported on carbon nanofibers	[117]
Polypyrrole/carbon supported Pt	[118]
Au and MnOx nanoparticles deposited on glassy carbon	[119]
Pt/carbon hollow nanospheres	[120]

Table 1. Catalysts from carbon supported metallic composites

One of the most important applications related to the electrocatalysis is fuel cells. The performance of methanol fuel cells is limited due to both the CO poisoning of the anode surface and the low catalytic activity of Pt-based anodic catalyst. Adsorbed CO has been identified as a key surface poison and the later pathways have been attracted great interest because their optimization can improve energy conversion efficiency at low potentials. Thus, the Pt(111) surface is expected to facilitate the methanol oxidation through the direct pathway and exhibits a good CO tolerance [39].

The Pt octahedron nanoparticles on supported carbon black were produced by Liu et al [39]. The TEM and HRTEM images of the Pt octahedral and tetrahedral nanoparticles were given in Figure 4. The authors have determined that the defect and step sites introduce on the surface of the Pt octahedron nanoparticles by applying a number of potential cycles. Further cycling lead to the formation of Pt(100) terraces. The enhanced catalytic activity observed after a number of potential cycles was due to the formation of the defect and step sites, and the rapid decay in the long-term performance was ascribed to the formation of Pt(100) terraces. They stated that the surface structure significantly affects the methanol decomposition process.

Porous graphene [121], graphitic carbon nitride [122], nitrogen-containing multi-walled carbon nanotube modified glassy carbon electrodes [123], and nitrogen-doped graphene [124] can be grouped as carbon-based catalysts. The SEM images from two of these catalysts are shown in Figure 5. Figure 5A shows the interconnected 3D porous network of graphene sheets on the electrochemically reduced grapheme oxide modified electrode surface. After chitosan

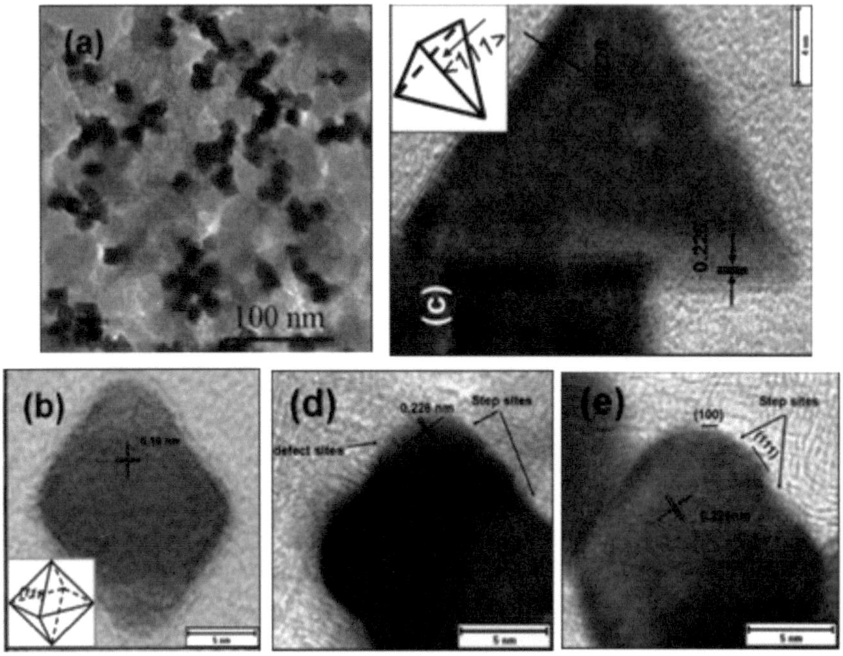

Figure 4. (a) TEM image of the Pt octahedron nanoparticles on supported carbon black. (b) HRTEM image of a single Pt octahedron nanoparticle; inset: the geometrical model of the Pt octahedron nanoparticle. (c) HRTEM image of a single Pt tetrahedral nanoparticle; inset: the geometrical model of the Pt tetrahedral nanoparticle. (d) The typical HRTEM image of a single Pt octahedron nanoparticle after 50 cycles. (e) The typical HRTEM image of a single Pt octahedron nanoparticle after 100 cycles [39].

addition to this electrode, it was argued that the electrode is a good material platform promising for construction of the third-generation enzyme biosensor, biofuel cells and bio electrochemical devices.

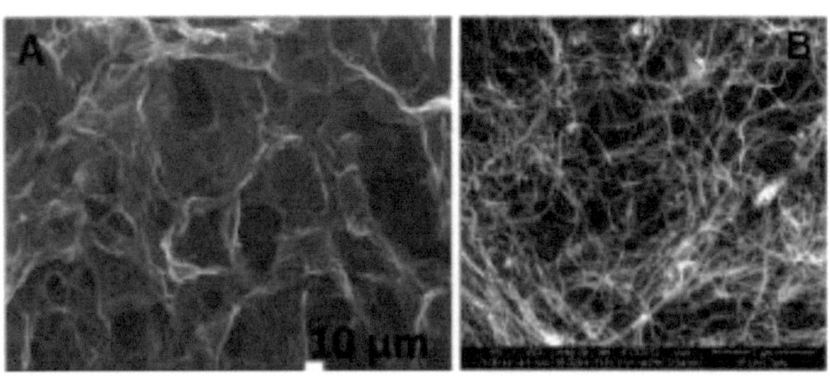

Figure 5. (A) SEM images the surface of the electrochemically reduced graphene oxide modified electrode [121] and (B) nitrogen-doped multi-walled carbon nanotubes deposited on glassy carbon [123]

Figure 5B shows the surface morphology of the nitrogen-doped multi-walled carbon nanotubes deposited on glassy carbon. It was determined that the nitrogen doped carbon nanotube materials demonstrate a high electrocatalytic activity for O_2 reduction in alkaline solution.

The specific catalysts such as glucose oxidase/graphene/gold nanocomposites [125], The heat-treated cobalt–porphyrin/carbon [126], SiO_2 supported catalyst powders [127], palladium doped glassy carbon, nitrogen implanted glassy carbon and argon implanted glassy carbon [128], hybrid nanocomposites modified glassy carbon electrode [129], hemoglobin immobilized on poly (styrene-alternative-maleic acid)/multiwalled carbon nanotubes blends [130], active carbon-supported cupper phthalocyanine [131], Pt nanoparticles supported by zeolite-type mesoporous $Cs_{2.5}H_{0.5}PW_{12}O_{40}$ solid super-acid [132], carbon-ceramic supported bimetallic Pt-Ni nanoparticles [133], Ni nano-pottery structure [134], Pt nanoparticles modified with ethylidyne [135], poly(3,4-ethylenedioxythiophene)-modified screen-printed electrodes [136], Pt nanoparticles on nitrogen-doped magnetic carbon nanoparticles [137], carbon supported gold, carbon powder and carbon powder grafted with anthraquinone derivatives [138] were not included to the above groups.

Figure 6 indicates the SEM images of Ni nano-pottery structures produced by electrodeposition in a nanoporous alumina template. The researchers used these nanostructures as cathode catalyst for hydrogen evolution reaction (HER) in alkaline water electrolysis. Their catalytic activity was compared to that of the Ni nano-rod and film and the result shows that the HER activity was greatly enhanced when using the Ni nano-pottery structure which mainly resulted from both larger and higher numbers of surface reactive sites.

Figure 7 summarizes the electrocatalytic applications in which are used electrode materials described and classified above. From this figure, it can be seen that the catalytic activity tests of electrode materials are generally performed with oxygen reduction and methanol oxidation reactions.

The mechanism of the electrochemical oxygen reduction reaction is quite complicated and involves many intermediates, primarily depending on the natures of the electrode material, catalyst, and electrolyte [139]. It can be presented as follows in acidic or alkali medium.

In acidic media;

$$O_{2(g)} + 4H^{+}_{(aq)} + 4\bar{e} \rightarrow H_2O_{(l)}$$

$$O_{2(g)} + 2H^{+}_{(aq)} + 2\bar{e} \rightarrow H_2O_{2(aq)}$$

$$H_2O_{2(aq)} + 2H^{+}_{(aq)} + 2\bar{e} \rightarrow 2H_2O_{(l)}$$

In alkali media;

$$O_{2(g)} + H_2O_{(l)} + 4\bar{e} \rightarrow 4OH^{-}_{(aq)}$$

$$O_{2(g)} + H_2O_{(l)} + 2\bar{e} \rightarrow HO_2^{-}{}_{(aq)} + OH^{-}_{(aq)}$$

$$HO_2^{-}{}_{(aq)} + H_2O_{(l)} + 2\bar{e} \rightarrow 3OH^{-}_{(aq)}$$

The reduction pathways such as the 1, 2, and 4-electron reduction pathways have unique significance, depending on the applications. In fuel cell processes, the 4-electron direct pathway is highly preferred [139].

With growing crisis in energy conservation and environmental pollution, a particular emphasis has been placed on sustainable conversion of environment-friendly feedstock to valuable matters and energy. Fuel cell technology provides a sustainable strategic manner for energy conversion with high-efficiency and cleanliness, but it has to rely on simple small-molecular high energy-containing fuels such as hydrogen, methanol, methanoate, and ethanol [80].

Electrochemical reactions involved in the methanol fuel cell include an anode reaction for oxidizing of fuel and a cathode reaction for reducing of hydrogen ion and oxygen. These reactions are as follows:

Anode reaction: $CH_3OH_{(aq)+}H_2O_{(l)} \rightarrow 6H^{++}6\bar{e}^{+}CO_{2(g)}$

Cathode reaction: $3/2\ O_{2(g)+}6H^{+}{}_{(aq)+}6\bar{e} \rightarrow 3H_2O_{(l)}$

Overall reaction: $CH_3OH_{(aq)+}3/2\ O_{2(g)} \rightarrow 2H_2O_{(l)}+CO_{2(g)}$

As shown in the above reactions, methanol and water react with each other to produce carbon dioxide, six hydrogen ions, and six electrons at the anode. At the cathode, the hydrogen ions, electrons, and oxygen react to produce water. The overall reaction consists of the production of water and carbon dioxide via the reaction between methanol and oxygen. Through these reactions, a large portion of energy corresponding to the heat of combustion of methanol is converted to an electrical energy. In order to facilitate these reactions, both the anode and the cathode of the methanol fuel cell comprise catalysts [140].

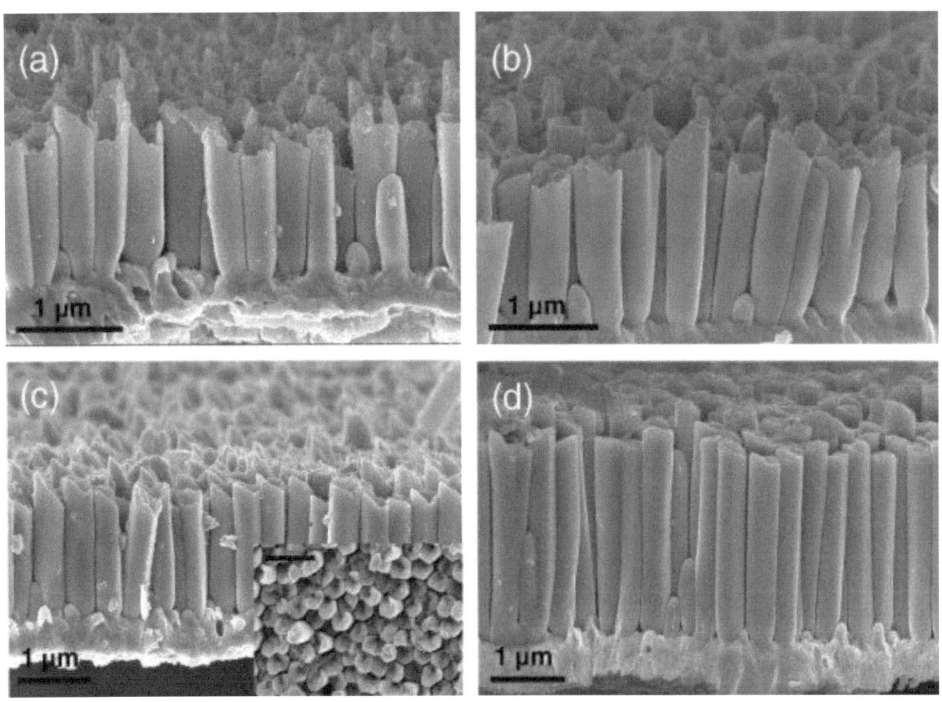

Figure 6. SEM images of Ni nanostructure obtained at different deposition times (a) 150, (b) 300, (c) 350, and (d) 600 s [134].

http://dx.doi.org/10.5772/58340

Most of electrocatalysis appears to be centered on fuel cells in which are carrying out the oxidation of hydrogen, oxygenates, or hydrocarbon molecules to CO_2 and the reduction of oxygen to water and more specifically proton exchange membrane (PEM) [5].

The oxygen reduction reaction (ORR) is an important cathode reaction in fuel cells. However, this reaction is kinetically slow and the cathodic polarization in fuel cells is usually over 0.3 V, accounting for the major part of the voltage loss in a single cell. So far, the best catalyst for the ORR has been Pt, which is expensive and limited in natural resource. Promoting the activity of Pt and searching for alternatives are two basic approaches to solve this problem [14].

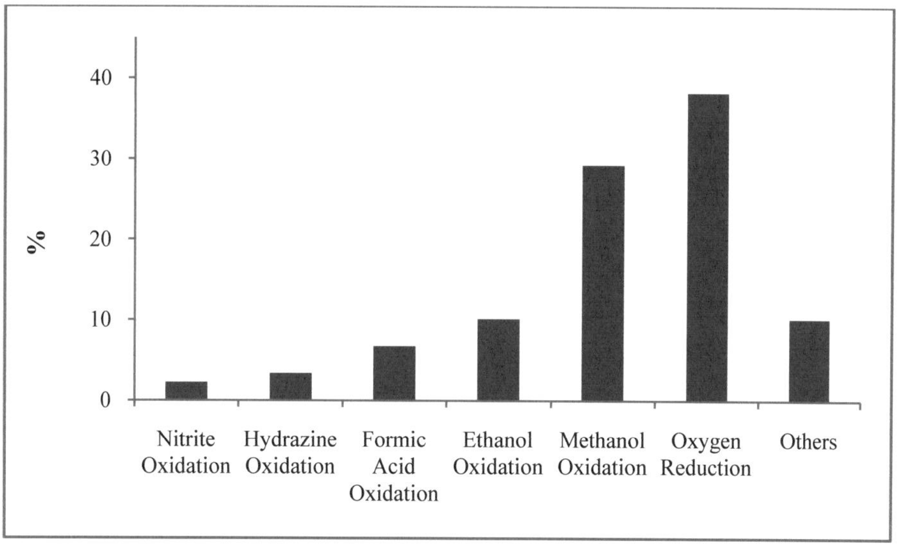

Figure 7. The testing reactions for the catalysts and their percentages of application (2005-2013)

The electrochemical reduction of a proton source to produce hydrogen (H_2) gase is a reaction of both fundamental and economic significance. Electrocatalysis of this reaction can be achieved homogeneously via protonation and electrochemical (or photo-chemical) reduction of a suitable Brønsted base in solution. Advances in this area have afforded molecular electrocatalysts that employ cobalt, molybdenum, nickel, or iron, instead of platinum that is currently the preferred electrocatalyst for proton reduction in water. Given the interest of developing H_2-production electrocatalysts based on cheap and abundant materials, synthetic models of the iron–iron hydrogenase enzymes have been the subject of numerous studies [141].

Methanol, ethanol, and 2-propanol are the main alcohols used directly as fuels in the direct alcohols fuel cells (DAFCs). Methanol is nonrenewable, volatile and a flammable substance; moreover, it has a high toxicity. The use of methanol may result in major issues if applied to portable devices. On the other hand, ethanol is less toxic and has a higher energy density compared to methanol [103].

2. Improvements in electrocatalytic materials and nanoparticles

It should be noted that ready-to-use electrocatalytic materials are complex systems where the resultant performance, i.e., activity, selectivity, and stability, depends on many factors. Even a single-supported nanoparticle in contact with an electrolyte represents a multi parametric system: different facets, surface defects, interactions with the support material, surface and bulk composition, size effects, specific properties of the electrochemical interface, electrolyte composition, and other parameters control its catalytic properties [4].

The rates of electrocatalytic reactions depend upon the surface structure and composition of the electrode as well as on the applied potential. The former includes the chemical composition of the electrode surface, its long-range geometry (with possible surface reconstructions or relaxations) and the in situ surface atomic arrangement on a short range scale whereas the later follows the changes that result in the atomic structure as well as the electronic structure of the metal. Recent developments in electrocatalysis have been driven by progress in ab initio theory of catalysis, development of new experiment methods to study materials and surfaces and the synthesis of new catalytic materials. The discovery of new nanoscale materials with unique surface structures and geometric architectures often provide unexpected surface properties as well as enhanced electrocatalytic performance [142].

The common criteria for a high performance catalyst are: (i) a narrow nanoscale size distribution; (ii) a uniform composition throughout the nanoparticles; (iii) a fully alloyed degree and (iv) high dispersion on carbon support [38]. In principle the catalytic activity can be increased either by a fine dispersion of the nanoparticles (NPs) or by modulating the electronic properties of the catalyst NPs by interaction with support chemical or morphological defects; as an example, a way to enhance the durability of the catalyst support assembly in the case of oxygen reduction is to strengthen the catalyst NPs support interaction by introducing into the substrate defects that can act as trapping sites for anchoring the catalyst NPs [128].

Most current studies on electrocatalysis based on the use of ensembles of nanostructured metal materials to understand the structure-function relationship. An important aspect of nanostructured metal materials is their structure-dependent catalytic activity. It has been shown that the electrocatalytic activity of nanostructured metallic materials is extremely sensitive to their sizes. Due to differences in electrochemical activity at different crystal surfaces, the shape of the nanoparticles also greatly alters their functionality. There had been numerous research efforts reported to control the shape of metal nanoparticles to obtain optimized activity in electrocatalysis [143].

Nanoparticles are widely investigated for their catalytic activity due to their high surface area to bulk atomic ratious and the achievable high surface area at low bulk concentration and associated lower cost. Studies of the reactivity of nanoparticles as catalysts for electrochemical reactions involve the adsorption or growth of nanomaterials onto solid substrates, e.g. carbon materials, oxide materials, metal substrates or molecular templates, which allow formation of an electronic contact required to control electrochemical reactions. The potential is then applied through the substrate material. Although the substrate is chosen with a low activity for the specific catalyzed reaction, the potential distribution at the interface and the electrochemical results might be influenced through the substrate double layer. To decouple substrate–nanoparticle and substrate–electrolyte interaction from the electrochemical results is therefore not straightforward [12].

Nanoparticles have been used extensively in the fields of physical, chemical and material sciences in the past few years. These materials have a high specific surface area, large pore size and unique features, such as surface plasmon absorption, improved magnetic property, high reactivity, and enhanced catalytic activity. They are also expected to show advantages for use as electrocatalysts due to their large active surface areas, small catalyst loading and abilities to prevent aggregation between particles. Among the nanoparticles, metal nanoparticles are objects of great interest in modern chemistry and material researches, where they find application in such diverse fields as photochemistry, nanoelectronics, optics, catalysis, and electrocatalysis. In fact, often enough these particles do possess physical as well as chemical properties, which are distinct both from the bulk phase and from isolated atoms and molecules. Use of noble metal nanoparticles seems to be inevitable in fuel cells. So far only platinum micro/nanoparticles are chosen as the main anode and cathode material in fuel cells. However, transition metals (Ni, Cr, and...) alloyed with platinum are also found to be efficient catalysts to decrease the CO poisoning effect in the electro oxidation of fuels. Further these catalysts are effective to reduce the cost compared to pure platinum based catalysts [144].

Generally, Pt has long been regarded as the best electrocatalyst for both anode and cathode in fuel cells. However, large-scale commercial application of fuel cells has been hindered largely by the high price of the Pt catalysts due to the limited reserve of Pt in nature, poor kinetics at the anode caused by reaction intermediate adsorption on Pt catalyst and the degradation of cathode performance because of poisoning of Pt catalyst resulting from diffusion of fuel molecules from the anode across membrane and CO poisoning. As a result, various catalysts such as Pt-based alloys, transition-metal macro cycles and Pt-based catalysts have been researched to solve the problems above and the latter presents much more activity and better long-term stability under fuel-cell operation conditions in comparison with others [102].

There is no doubt that many researchers would agree that the oxygen reduction reaction is more important than any other in respect to electrocatalysis. This is because it gives rise to the principal over potential in fuel cells and fuel cells are in a stage of rapid development so that anything related to increasing the efficiency of them is economically important [2]. Overpotential value related to the charge transfer processes at the anode or cathode is the ones that could be lowered by the proper selection of electrode material. This is the issue of the electrocatalysis.

Just to show the tremendous savings which could be obtained by improved electrocatalysis of the oxygen reduction reaction (and any corresponding one with the same types of i_0) it can be started with a familiar equation (Eq 1), derived from Tafel's Law, and said that the change in overpotential at a given current density in the irreversible region with two different i_0's are given (roughly of course) by:

$$\Delta\eta = 2\frac{RT}{F} ln \frac{(i_0)_1}{(i_0)_2} \quad (1)$$

where; $\Delta\eta$ is the overpotential (V), 2RT/F is the so-called "Tafel slope" (V), i_0 is the current density (A/m^2). There is only one assumption in this equation and that is that this value for transfer coefficient (α) in the relevant Tafel equation for the reaction is about one-half, which is a widely observed value (though for certain rate determining steps, it may indeed be substantially increased, thus reducing the overpotential) [2].

Of course, everybody knows that "platinum is a good catalyst" and this phrase has often given rise to the impression that one has to look for a substance which will be, in a universal way, "a good catalyst". There are several reasons why platinum has been brought out and focused upon and two of them are that it is an easily available substance which can catalyze substances at highly anodic potentials without being decomposed. In this sense, it has wide applicability. Indeed, one of the difficulties in finding other catalysts, better than platinum, is that, although one might make theoretical calculations which would indicate a given substance to be a better catalyst than platinum it may not withstand the anodic potentials which are often met in, *e.g.*, the reduction of oxygen and undergo decomposition itself, thus making it impractical to use [2].

Platinum was found to be the electrode material on top of which hydrogen evolves with minimum overpotential. From purely academic, the problem how to lower the overpotential (thus lowering the product price) gets on importance with the growth of technical processes of electrolysis. Performed on a large scale, their economy depends greatly on the electricity consumption per unit product, i.e. on the value of cell voltage [3].

3. Applications of electrocatalytic materials

The goal of the applied electrocatalysis is to improve the electrode activity for the wanted reaction, but at the same time to prevent (or slow down) side reactions. Technical reasons request the selected electrode material to be stable (without corrosion and wear), economical reasons favor cheaper materials, while environmental concern is in favor of nonpolluting electrode materials [3].

The transition from one-metal-only electrocatalyst to a two or more metal (or their compounds) catalysts was not at all easy. It took almost 30 years from the first rhodium electrodeposited on titanium electrocatalyst till the first titanium-based anodes in the cathodic protection of rebar in concrete. In this process, rhodium and platinum electrodeposited on titanium (1958), paint/thermal decomposition coatings (1960), first mixed metal oxide coatings (1965), Pt/Ir coated first titanium based electrodes (1968), and first non-precious metal alloys (1974) were produced and titanium-based anodes in electrogalvanizing (1984) and in the cathodic protection applications (1986) were used. Development of electrocatalysts without precious metals started in the middle 1970's and then extended over a long period [3].

http://dx.doi.org/10.5772/58340

The strategy lying behind this important switch from one type of electrocatalysts, made out of rather expensive and limited in quantities precious metals, to the new type of compound catalysts engineered trivial components, but promising unrestricted possibilities was fully correct [3].

Pt and other group VIII metals are known to be very active in the electrocatalytic oxidation of alcohols and the reduction of oxygen in fuel cells, automotive exhaust catalysis, and hydrogenolysis, and hydrogenation catalysis in the conversion of petroleum and renewable resources. This is predominantly the result of the well-established Sabatier's Principle which suggests that the metals in middle of the periodic table demonstrate optimal metal-adsorbate bond strengths necessary to balance surface reaction steps and product desorption steps [5].

Gold first became recognizable as a catalyst when clusters deposited on several metal oxides exhibited oxidation towards carbon oxides at low temperature. Nowadays this effect is thoroughly investigated due to techniques and tools allowing studying the size and shape of the particles and their correlation with catalytic activities. Gold nanoparticles (AuNPs) catalyze a number of reactions either suspended in solution or deposited on the electrode surface [145].

The knowledge obtained from model experiments can be successfully used to design "real-world electrocatalysts". For example, one of the main modern paradigms of the proton exchange membrane fuel cells (PEMFCs) electrocatalyst preparation is to produce high surface area nanoparticulate materials where the catalyst nanoparticles are supported on an electronically conducting carbon support. As stability issues strongly dominate in selection of electrocatalyst material, Pt-based materials are used at the cathode side of present state-of-the-art PEMFCs [4].

In general, a fuel cell is an electrochemical device which requires electrocatalysts at the anode oxidize the fuel to produce electrons and protons and at the cathode reduce oxygen and consume the protons and electrons produced (as shown in Fig. 8).

The vast majority of polymer electrolyte membrane fuel cell work has been performed using Nafion® as the electrolyte [146]. The proton conducting membrane allows the proton to rapidly transfer from the anode to the cathode whereas the wire which connects the two electrodes allows for electron transfer. The fuel oxidation at the anode and the reduction of oxygen which occurs at the cathode are shown [142].

The differences in the rates of reaction at the anode and cathode dictate the potential at which the cell can operate and the maximum current density. On the anode side, the challenge is to efficiently catalyze the oxidation of hydrogen or oxygenate fuels such as methanol, formic acid or ethanol. Unfortunately, the fuel sources are often contaminated by CO or other species that can poison the electrode surface. On the cathode side, the requirement is to effectively reduce oxygen, usually from air and to prevent the dissolution and loss of catalyst. The most active electrocatalytic materials are typically comprised of expensive transition metals nanoparticles such as Pt that are too costly to provide viable commercial solutions. Core–shell nanoparticles have of a very thin shell of the active Pt catalyst supported onto a second metal. To synthesize these systems, bimetallic nanoparticles are often leached in acid to remove the excess of the electro-dissolving non-noble element leaving the protective layer composed of the Pt or the Pt

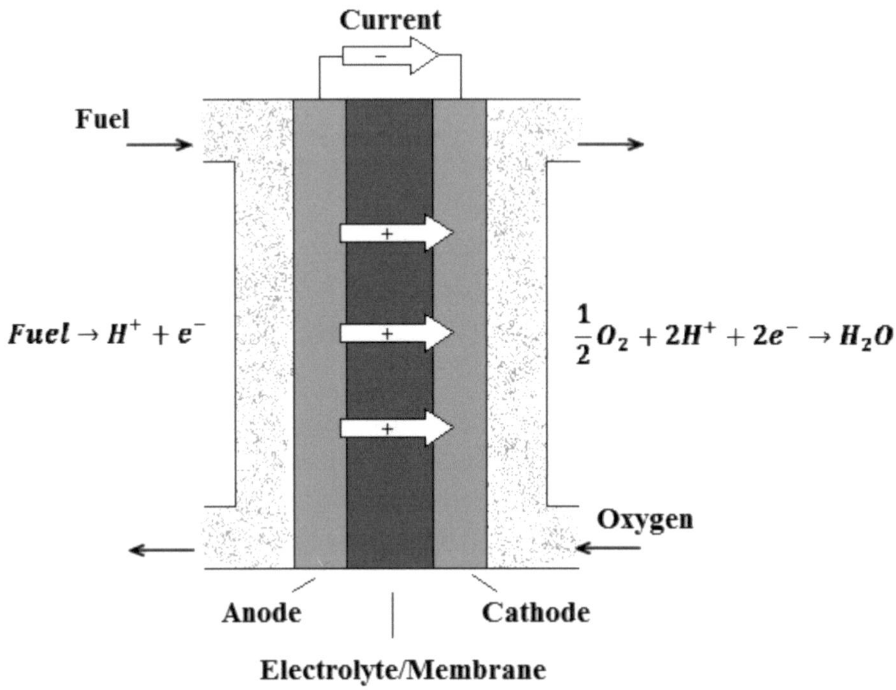

Figure 8. Schematic illustration of a fuel cell

enriched surface phase. In other words, the electro-dissolution creates de-alloyed Pt bimetallic catalysts that are composed of an inexpensive core element with platinum present only in the nanoparticle shell. This leaves the composition of the material in active usage and close to the nominal (started) alloyed/bimetallic specification [142].

4. Conclusion

Electrocatalysts are of ubiquitous importance and can be found in a large array of research fields and applications, including corrosion science, development of electroanalytical sensors, waste water treatment, electro-organic synthesis and energy conversion devices (e.g. batteries, fuel cells and solar cells). In the context of the increasing global energy demand and continuous depletion of fossil fuel-reserves, the exploitation of renewable carbon-neutral energy sources, such as sunlight, is one of the greatest scientific challenges of this century. The production of hydrogen through water splitting, either electrochemically or photochemically, seems to be a promising and sustainable solution to the storage of energy in a chemical and stable form, before converting it back into electricity, for example in a hydrogen-driven fuel cell. Electrocatalysis is the key enabling technology for these processes.

In recent years, research in electrocatalysis has experienced a significant progress mainly through the application of very modern research techniques and the use of improved electrode materials. The widespread interest in nanostructured materials mainly originates from the fact that their properties (optical, electrical, mechanical and chemical performance) are usually different from those of the bulk materials. In addition, the development of high resolution instrumentation and new developments in analytical chemistry and electrochemical techniques permits the detailed characterization of atomic distribution, before, during, and after a reaction takes place, giving unprecedented information about the status of the catalyst during the reaction, and most importantly the time evolution of the exposed catalytic surfaces at the atomistic level.

In order to reduce the usage of precious metals and, thus, cost, the best possible performance has to be extracted from a given amount of catalyst. This involves mainly catalyst preparation, pretreatment, electrode preparation and, in the case of supported catalysts, also the choice of a suitable support. Tremendous progress has been made in developing preparation methods, carbon supports and down-selection approaches for catalysts in recent years. Several simple and tunable methods have shown a superior ability to synthesize state-of-the-art catalysts. Some new carbon materials have been demonstrated to be feasible as catalyst supports, although their applications still face some challenges in terms of synthesis, metal loading and electrode preparation. Combination of these advanced metal loading methods and excellent carbon supports could bring about a breakthrough in the exploration for a new generation of catalysts in the near future.

In addition, we predict that the efforts for the development of new catalysts with different characteristics will increasingly continue in the future too. Further improvements considering the exploration of these materials will help to commercialization of the fuel cells. We feel that there is a need for further works in the investigation of the effectiveness of these materials in the alternative areas as chemical catalyst, chemical sensor, and biological biosensor.

Author details

Metin Açıkyıldız[1], Ahmet Gürses[2], Mehtap Ejder Korucu[3] and Kübra Güneş[2]

1 Kilis Aralık University, Faculty of Science and Art, Dep. of Chemistry, Kilis, Turkey

2 Atatürk University, K.K. Education Faculty, Dep. of Chemistry Education, Erzurum, Turkey

3 Kafkas University, Faculty of Science and Literature, Dep. of Chemistry, Kars, Turkey

References

[1] Santos E. and Schmickler W. (2010) Recent Advances in Theoretical Aspects of Electrocatalysis, Theory and Experiment in Electrocatalysis Modern Aspects of Electrochemistry, 50, 25-88

[2] Bockris, J.O'M. (2005) A primer on electrocatalysis, Journal of the Serbian Chemical Society, 70 (3) 475-487.

[3] Jordanov S.H., Paunović P., Popovski O., Dimitrov A., Slavkov D., (2004) Electrocatalysts in the Last 30 Years – From Precious Metals to Cheaper but Sophisticated Complex Systems, Bulletin of the Chemists and Technologists of Macedonia, 23(2), 101-112

[4] Bandarenka A.S., Koper M.T.M., (2013) Structural and electronic effects in heterogeneous electrocatalysis: Toward a rational design of electrocatalysts, Journal of Catalysis, 308, 11-24.

[5] Wieckowski A. and Neurock M., (2011) Contrast and Synergy between Electrocatalysis and Heterogeneous Catalysis, Advances in Physical Chemistry.

[6] Arvia J., Salvarezza R.C. and Triaca W.E., (2004) Noble Metal Surfaces and Electrocatalysis.Review and Perspectives, Journal of New Materials for Electrochemical Systems, 7, 133-143.

[7] Xu S., Zhang H., Huang F., Wang P., Xia Y., Li Z., Huang W., (2013) Electrochemical fabrication of a cauliflower-like nanostructured Pd film from pure Pd and its applications in electrocatalysis and electroanalysis, Electrochimica Acta, 107, 537-541.

[8] Sitta E., Gómez-Marín A.M., Aldaz A., Feliu J.M., (2013) Electrocatalysis of H_2O_2 reduction/oxidation at model platinum surfaces, Electrochemistry Communications, 33, 39-42.

[9] Feng J-J., Lv Z-Y., Qin S-F., Li A-Q., Fei Y., Wang A-J., (2013) N-methylimidazole-assisted electrodeposition of Au porous textile-like sheet arrays and its application to electrocatalysis, Electrochimica Acta, 102, 312-318.

[10] Trojànek A., Langmaier J., Samec Z., (2006) Electrocatalysis of the oxygen reduction at a polarised interface between two immiscible electrolyte solutions by electrochemically generated Pt particles, Electrochemistry Communications, 8, 475–481.

[11] Pan L., Tang J., Wang F., (2013) Solvothermal synthesis of sheet-like Co_3O_4 and Ag/Co_3O_4 nanocomposites and their electrocatalysis performances, Materials Chemistry and Physics, 140, 391-397.

[12] Gründer Y., Fabian M.D., Booth S.G., Plana D., Fermín D.J., Hill P.I., Dryfe R.A.W., (2013) Solids at the liquid–liquid interface: Electrocatalysis with pre-formed nanoparticles, Electrochimica Acta, 110, 809-815.

[13] Erikson H., Liik M., Sarapuu A., Marandi M., Sammelselg V., Tammeveski K., (2013) Electrocatalysis of oxygen reduction on electrodeposited Pd coatings on gold, Journal of Electroanalytical Chemistry, 691, 35–41.

[14] Sun Y., Lu J., Zhuang L., (2013) Pt/Ti catalyst prepared by simultaneous etching-displacement method and the electrocatalysis toward oxygen reduction reaction, Journal of Electroanalytical Chemistry, 688, 189–195.

[15] Cruz J.C., Baglio V., Siracusano S., Ornelas R., Arriaga L.G., Antonucci V., Aricó A.S., (2012) Nanosized Pt/IrO_2 electrocatalyst prepared by modified polyol method for application as dual function oxygen electrode in unitized regenerative fuel cells, International Journal of Hydrogen Energy, 37, 5508-5517.

[16] Domínguez-Crespo M.A., Ramírez-Meneses E., Torres-Huerta A.M, Garibay-Febles V., Philippot K., (2012) Kinetics of hydrogen evolution reaction on stabilized Ni, Pt and Ni-Pt nanoparticles obtained by an organometallic approach, International Journal of Hydrogen Energy, 37, 4798-4811.

[17] Elezovíc N.R., Babic B.M., Gajíc-Krstajíc Lj. Ercius P., Radmilovíc V.R., Krstajíc N.V., Vračar Lj.M., (2012) Pt supported on nano-tungsten carbide as a beneficial catalyst for the oxygen reduction reaction in alkaline solution, Electrochimica Acta, 69, 239–246.

[18] Frolova L., Lyskov N., Dobrovolsky Y., (2012) Nanostructured $Pt/SnO_2–SbO_x–RuO_2$ electrocatalysts for direct alcohol fuel cells, Solid State Ionics, 225, 92–98.

[19] Bi X., Wang R., Ding Y., (2011) Boosting the performance of Pt electro-catalysts toward formic acid electro-oxidation by depositing sub-monolayer Au clusters, Electrochimica Acta, 56, 10039–10043.

[20] Lee J.K., Lee J., Han J., Lim T.H., Sung Y.E., Tak Y., (2008) Influence of Au contents of AuPt anode catalyst on the performance of direct formic acid fuel cell, Electrochimica Acta, 53, 3474-3478.

[21] Long N.V., Ohtakid M., Hienc T.D., Randya J., Nogamia M., (2011) A comparative study of Pt and Pt–Pd core–shell nanocatalysts, Electrochimica Acta, 56, 9133– 9143.

[22] Antolini E., E.R. Gonzalez E. (2010) The electro-oxidation of carbon monoxide, hydrogen/carbon monoxide and methanol in acid medium on Pt-Sn catalysts for low-temperature fuel cells: A comparative review of the effect of Pt-Sn structural characteristics, Electrochimica Acta, 56, 1–14.

[23] Chen W., Ny D., Chen S.,(2010) SnO_2–Au hybrid nanoparticles as effective catalysts for oxygen electroreduction in alkaline media, Journal of Power Sources, 195, 412–418.

[24] Delacôte C., Lewerac A., Pisarekd M., Kuleszac P.J., Zelenaye P., Alonso-Vantea N., (2010) The effect of diluting ruthenium by iron in Ru_xSe_y catalyst for oxygen reduction, Electrochimica Acta, 55, 7575–7580.

[25] Hassan H.B., (2009) Electrodeposited Pt and Pt-Sn nanoparticles on Ti as anodes for direct methanol fuel cells, Journal of Fuel Chemistry and Technology, 37, 346-354.

[26] Neyerlin K.C., Srivastava R., Yua C., Strasser P., (2009) Electrochemical activity and stability of dealloyed Pt–Cu And Pt–Cu–Co electrocatalysts for the oxygen reduction reaction (ORR), Journal of Power Sources, 186, 261-267.

[27] Macdonald J.P., Gualtieri B., Runga N., Teliz E., Zinola C.F., (2008) Modification of platinum surfaces by spontaneous deposition: Methanol oxidation electrocatalysis, International Journal of Hydrogen Energy, 33, 7048-7061.

[28] Mohammad A.M., Awad M.I., El-Deab M.S., Okajima T., Ohsaka T., (2008) Electrocatalysis by nanoparticles: Optimization of the loading level and operating Ph for the oxygen evolution at crystallographically oriented manganese oxide nanorods modified electrodes, Electrochimica Acta, 53, 4351–4358.

[29] Rajalakshmi N., Lakshmi N., Dhathathreyan K.S., (2008) Nano titanium oxide catalyst support for proton exchange membrane fuel cells, International Journal of Hydrogen Energy, 33, 7521–7526.

[30] Ramos-Sànchez G., Yee-Madeira H., Solorza-Feria O., (2008) PdNi electrocatalyst for oxygen reduction in acid media, International Journal of Hydrogen Energy, 33, 3596–3600.

[31] Cheng H., Yuan W., Scott K., Browning D.J., Lakeman J.B., (2007) The catalytic activity and methanol tolerance of transition metal modified-ruthenium–selenium catalysts, Applied Catalysis B: Environmental, 75, 221–228.

[32] Lycke D.R., Gyenge E.L., (2007) Electrochemically assisted organosol method for Pt-Sn nanoparticle synthesis and in situ deposition on graphite felt support: Extended reaction zone anodes for direct ethanol fuel cells, Electrochimica Acta, 52, 4287–4298.

[33] Vračar L.M., Krstajić N.V., Radmilović V.R., JakšićM.M., (2006) Electrocatalysis by nanoparticles–oxygen reduction on Ebonex/Pt electrode, Journal of Electroanalytical Chemistry, 587, 99–107.

[34] Suàrez-Alcàntara K., Rodríguez-Castellanos A., Dante R., Solorza-Feria O., (2006) $Ru_xCr_ySe_z$ electrocatalyst for oxygen reduction in a polymer electrolyte membrane fuel cell, Journal of Power Sources, 157, 114–120.

[35] Monge-Romero I.C., Suárez-Herrera M.F., (2013) Electrocatalysis of the hydroquinone-benzoquinone redox couple at platinum electrodes covered by a thin film of poly(3,4-ethylenedioxythiophene), Synthetic Metals, 175, 36-41.

[36] Gu Y., Tang Z., Deng Y., Wang L., (2013) Hierarchical $LiMn_2O_4$ microspheres for high rate lithium ion batteries, and direct electrochemistry and electrocatalysis, Electrochimica Acta 94, 165-171.

[37] Ding K.Q., Cheng F.M., (2009) Cyclic voltammetrically prepared MnO_2–PPy composite material and its electrocatalysis towards oxygen reduction reaction (ORR), Synthetic Metals, 159, 2122-2127.

[38] Liu H., Song C., Zhang L., Zhang J., Wang H., Wilkinson D.P., (2006) A review of anode catalysis in the direct methanol fuel cell, Journal of Power Sources, 155, 95–110.

[39] Liu Y., Wang Y., Zhang J., Shi S., Feng P., Wang T., (2009) Observation of surface structural changes of Pt octahedron nanoparticles and its effect in electrocatalysis oxidation of methanol, Catalysis Communications, 10, 1244-1247.

[40] Sun W., Li L., Lei B., Li T., Ju X., Wang X., Li G., Sun Z., (2013) Fabrication of graphene–platinum nanocomposite for the direct electrochemistry and electrocatalysis of myoglobin, Materials Science and Engineering C, 33, 1907-1913.

[41] Calvillo L., Gangeri M., Perathoner S., Centi G., Moliner R., Làzaro M.J., (2011) Synthesis and performance of platinum supported on ordered mesoporous carbons as catalyst for PEM fuel cells: Effect of the surface chemistry of the support, International Journal of Hydrogen Energy, 36, 9805-9814.

[42] Ma J., Sun H., Su F., Chen Y., Tang Y., Lu T., Zheng J., (2011) Ethanol electrooxidation on carbon-supported Pt nanoparticles catalyst prepared using complexing self-reduction method, International Journal of Hydrogen Energy, 36, 7265-7274.

[43] Bai Z., Yanga L., Zhangb J., Li L., Hua C., Lva J., Guoa Y., (2010) High-efficiency carbon-supported platinum catalysts stabilized with sodium citrate for methanol oxidation, Journal of Power Sources, 195, 2653–2658.

[44] Fu Z., Li W., Zhang W., Sun F., Zhou Z., Xiang X., (2010) Preparation and activity of carbon-supported porous platinum as electrocatalyst for methanol oxidation, International Journal of Hydrogen Energy, 35, 8101–8105.

[45] Brimaud S., Coutanceau C., Garnier E., Lèger J.M., Gèrard F., Pronier S., Leoni M., (2007) Influence of surfactant removal by chemical or thermal methods on structure and electroactivity of Pt/C catalysts prepare by water-in-oil microemulsion, Journal of Electroanalytical Chemistry, 602, 226–236.

[46] Liu Z., Gan L.M., Hong L., Chen W., Lee J.Y., (2005) Carbon-supported Pt nanoparticles as catalysts for proton exchange membrane fuel cells, Journal of Power Sources, 139, 73–78.

[47] Babu P.K., Chung J.H., Oldfield E., Wieckowski A., (2008) CO surface diffusion on platinum fuel cell catalysts by electrochemical NMR, Electrochimica Acta, 53, 6672-6679.

[48] Joo J.B., Kim P., Kim W., Kim J., Yi J., (2006) Preparation of mesoporous carbon templated by silica particles for use as a catalyst support in polymer electrolyte membrane fuel cells, Catalysis Today, 111, 171–175.

[49] Liu Z., Hong L., Tham M.P., Lim T.H., Jiang H., (2006) Nanostructured Pt/C and Pd/C catalysts for direct formic acid fuel cells, Journal of Power Sources, 161, 831–835.

[50] Rojas S., García-García F.J., Järas S., María V. Martínez-Huerta M.V., García Fierro J.L., Boutonnet M., (2005) Preparation of carbon supported Pt and PtRu nanoparticles from microemulsion Electrocatalysts for fuel cell applications, Applied Catalysis A: General, 285, 24–35.

[51] Kawaguchi T., Sugimoto W., Murakami Y., Takasu Y., (2005) Particle growth behavior of carbon-supported Pt, Ru, PtRu catalysts prepared by an impregnation reductive-pyrolysis method for direct methanol fuel cell anodes, Journal of Catalysis, 229, 176–184.

[52] Carmo M., Paganin V.A., Rosolen J.M., Gonzalez E.R., (2005) Alternative supports for the preparation of catalysts for low-temperature fuel cells: the use of carbon nanotubes, Journal of Power Sources, 142, 169–176.

[53] Ávila-García I., Ramírez C., Hallen López J.M., Arce Estrada E.M., (2010) Electrocatalytic activity of nanosized Pt alloys in the methanol oxidation reaction, Journal of Alloys and Compounds, 495, 462–465.

[54] Diaz V., Ohanian M., Zinola C.F., (2010) Kinetics of methanol electrooxidation on Pt/C and PtRu/C catalysts, International Journal of Hydrogen Energy, 35, 10539-10 546.

[55] Alexeyeva N., Sarapuu A., Tammeveski K., Vidal-Iglesias F.J., Solla-Gullon J., Feliu J.M., (2011) Electroreduction of oxygen on Vulcan carbon supported Pd nanoparticles and Pd–M nanoalloys in acid and alkaline solutions, Electrochimica Acta, 56, 6702–6708.

[56] Gonzàlez-Huerta R.G., Chàvez-Carvayar J.A., (2006) Solorza-Feria O., Electrocatalysis of oxygen reduction on carbon supported Ru-based catalysts in a polymer electrolyte fuel cell, Journal of Power Sources, 153, 11–17.

[57] Bron M., Bogdanoff P., Fiechter S., Tributsch H., (2005) Enhancement of oxygen electroreduction activity via surface modification of carbon supported ruthenium nanoparticles: A new class of electrocatalysts, Journal of Electroanalytical Chemistry, 578, 339–344.

[58] Atwan M.H., Charles L.B. Macdonald C.L.B., Derek O. Northwood D.O., Gyengec E.L., (2006) Colloidal Au and Au-alloy catalysts for direct borohydride fuel cells: Electrocatalysis and fuel cell performance, Journal of Power Sources, 158, 36–44.

[59] Cao L., Sun G., Li H., Xin Q., (2007) Carbon-supported IrSn catalysts for direct ethanol fuel cell, Fuel Cells Bulletin, 11, 12-16.

[60] Guo J., Sun G., Wang Q., Wang G., Zhou Z., Tang S., Jiang L., Zhou B., Xin Q., (2006) Carbon nanofibers supported Pt–Ru electrocatalysts for direct methanol fuel cells, Carbon, 44, 152–157.

[61] Lèger J.M. (2005) Preparation and activity of mono-or bi-metallic nanoparticles for electrocatalytic reactions, Electrochimica Acta, 50, 3123–3129.

[62] Sarma L.S., Lin T.D., Tsai Y.W., Chen J.M., Hwang B.J., (2005) Carbon-supported Pt–Ru catalysts prepared by the Nafion stabilized alcohol-reduction method for application in direct methanol fuel cells, Journal of Power Sources, 139, 44–54.

[63] Deivaraj T.C., Jim Yang Lee J.Y., (2005) Preparation of carbon-supported PtRu nanoparticles for direct methanol fuel cell applications–a comparative study, Journal of Power Sources, 142, 43–49.

[64] Yi L., Liu L., Wang X., Liu X., Yi W., Wang X., (2013) Carbon supported Pt-Sn nanoparticles as anode catalyst for direct borohydride-hydrogen peroxide fuel cell: Electrocatalysis and fuel cell performance, Journal of Power Sources, 224, 6-12.

[65] Almeida T.S., Palma L.M., Leonello P.H., Morais C., Kokoh K.B., De Andrade A.R., (2012) An optimization study of PtSn/C catalysts applied to direct ethanol fuel cell: Effect of the preparation method on the electrocatalytic activity of the catalysts, Journal of Power Sources, 215, 53-62.

[66] Jeyabharathi C., Venkateshkumar P., Mathiyarasua J., Phani K.L.N., (2008) Platinum–Tin bimetallic nanoparticles for methanol tolerant oxygen-reduction activity, Electrochimica Acta, 54, 448-454.

[67] De Souza R.F.B., Parreira L.S., Rascio D.C., Silva J.C.M., Teixeira-Neto E., Calegaro M.L., Spinace E.V., Neto A.O., Santos M.C., (2010) Study of ethanol electro-oxidation in acid environment on Pt_3Sn/C anode catalysts prepared by a modified polymeric precursor method under controlled synthesis conditions, Journal of Power Sources, 195, 1589–1593.

[68] Lopes T., Antolini E., Gonzalez E.R., (2008) Carbon supported Pt–Pd alloy as an ethanol tolerant oxygen reduction electrocatalyst for direct ethanol fuel cells, International Journal of Hydrogen Energy, 33, 5563-5570.

[69] Bayrakçeken A., Cangül B., Zhang L.C., Aindow M., Erkey C., (2010) PtPd/BP2000 electrocatalysts prepared by sequential supercritical carbon dioxide deposition, International Journal of Hydrogen Energy, 3 5, 11669-11680.

[70] Kadirgan F., Beyhan S., Atilan T., (2009) Preparation and characterization of nano-sized Pt–Pd/C catalysts and comparison of their electro-activity toward methanol and ethanol oxidation, International Journal of Hydrogen Energy, 34, 4312-4320.

[71] Huang Q., Yang H., Tang Y., Lu T., Akins D.L., (2006) Carbon-supported Pt–Co alloy nanoparticles for oxygen reduction reaction, Electrochemistry Communications, 8, 1220–1224.

[72] Dubau L., Maillard F., Chatenet M., André J., Rossinot E., (2010) Nanoscale compositional changes and modification of the surface reactivity of Pt_3Co/C nanoparticles during proton-exchange membrane fuel cell operation, Electrochimica Acta, 56, 776–783.

[73] Amin R.S., Abdel Hameed R.M., El-Khatib K.M., El-Abd H., Souaya E.R., (2012) Effect of preparation conditions on the performance of nano Pt–CuO/C electrocatalysts

for methanol electro-oxidation International Journal of Hydrogen Energy, 37, 18870-18881.

[74] Elezović N.R., Gajić-Krstajić Lj.M., Vračar Lj.M, Krstajić N.V., (2010) Effect of chemisorbed CO on MoO_x-Pt/C electrode on the kinetics of hydrogen oxidation reaction, International Journal of Hydrogen Energy, 35, 12878-12887.

[75] Peng B., Wang J.Y., Zhang H.Y., Lin Y.H., Cai W.B., (2009) A versatile electroless approach to controlled modification of Sb on Pt surfaces towards efficient electrocatalysis of formic acid, Electrochemistry Communications, 11, 831-833.

[76] Lima F.H.B., Castro J.F.R., Santos L.G.R.A., Ticianelli E.A., (2009) Electrocatalysis of oxygen reduction on carbon-supported Pt–Co nanoparticles with low Pt content, Journal of Power Sources, 190, 293-300.

[77] Wang W., Zheng D., Du C., Zou Z., Zhang X., Xia B., Yang H., Akins D.L., (2007) Carbon-supported Pd-Co bimetallic nanoparticles as electrocatalysts for the oxygen reduction reaction, Journal of Power Sources, 167, 243–249.

[78] Ang S.Y., Walsh D.A., (2010) Palladium–vanadium alloy electrocatalysts for oxygen reduction: Effect of heat treatment on electrocatalytic activity and stability, Applied Catalysis B: Environmental, 98, 49–56.

[79] He P., Wang X., Fu P., Wang H., Yi L., (2011) The studies of performance of the Au electrode modified by Zn as the anode electrocatalyst of direct borohydride fuel cell, International Journal of Hydrogen Energy, 36, 8857-8863.

[80] Gong X.B., You S.J., Wang X.H., Gan Y., Zhang R.N., Ren N.Q., (2013) Silver-tungsten carbide nanohybrid for efficient electrocatalysis of oxygen reduction reaction in microbial fuel cell, Journal of Power Sources 225, 330-337

[81] Cheng H., Yuan W., Scott K., (2006) The influence of a new fabrication procedure on the catalytic activity of ruthenium–selenium catalysts, Electrochimica Acta, 52, 466–473.

[82] Abdel Aal A., Hassan H.B., (2009) Electrodeposited nanocomposite coatings for fuel cell application, Journal of Alloys and Compounds, 477, 652-656.

[83] Escudero-Cid R., Hernàndez-Fernàndez P., Pèrez-Flores J.C., Rojas S., Garcia-Rodríguez S., Fatàs E., Ocón P., (2012) Analysis of performance losses of direct methanol fuel cell with methanol tolerant PtCoRu/C cathode electrode, International Journal of Hydrogen Energy, 37, 7119-7130.

[84] Pèrez G., Pastor E., Zinola C.F., (2009) A novel Pt/Cr/Ru/C cathode catalyst for direct methanol fuel cells (DMFC) with simultaneous methanol tolerance and oxygen promotion, International Journal of Hydrogen Energy, 34, 9523-9530.

[85] Ammam M., Easton E.B., (2012) Quaternary PtMnCuX/C (X=Fe, Co, Ni, and Sn) and PtMnMoX/C (X=Fe, Co, Ni, Cu and Sn) alloys catalysts: Synthesis, characterization

and activity towards ethanol electrooxidation, Journal of Power Sources, 215, 188-198.

[86] Alexeyeva N., Tammeveskia K., Lopez-Cuderob A., Solla-Gullónb J., Feliu J.M., (2010) Electroreduction of oxygen on Pt nanoparticle/carbon nanotube nanocomposites in acid and alkaline solutions, Electrochimica Acta, 55, 794–803.

[87] Chen Y., Zhang G., Ma J., Zhou Y., Tang Y., Lu T., (2010) Electro-oxidation of methanol at the different carbon materials supported Pt nano-particles, International Journal of Hydrogen Energy, 35, 10109-10117.

[88] Guo D.J., Li H.L., (2005) High dispersion and electrocatalytic properties of palladium nanoparticles on single-walled carbon nanotubes, Journal of Colloid and Interface Science, 286, 274–279.

[89] Gao G., Yang G., Xu M., Wang C., Xu C., Li H., (2007) Simple synthesis of Pt nanoparticles on noncovalent functional MWNT surfaces: Application in ethanol electrocatalysis, Journal of Power Sources, 173, 178–182.

[90] Chen L., Hu G., Zou G., Shao S., Wang X., (2009) Efficient anchorage of Pd nanoparticles on carbon nanotubes as a catalyst for hydrazine oxidation, Electrochemistry Communications, 11, 504-507.

[91] Golikand A.N., Asgari M., Lohrasbi E., (2011) Study of oxygen reduction reaction kinetics on multi-walled carbon nano-tubes supported Pt-Pd catalysts under various conditions, International Journal of Hydrogen Energy, 36, 13317-13324.

[92] Hernàndez-Fernàndez P., Nuño R., Fatàs E., Fierro J.L.G., Ocón P., (2011) MWCNT-supported PtRu catalysts for the electrooxidation of methanol: Effect of the functionalized support, International Journal of Hydrogen Energy, 36, 8267-8278.

[93] Xiao F., Mo Z., Zhao F., Zeng B., (2008) Ultrasonic-electrodeposition of gold–platinum alloy nanoparticles on multi-walled carbon nanotubes–ionic liquid composite film and their electrocatalysis towards the oxidation of nitrite, Electrochemistry Communications, 10, 1740–1743.

[94] Li R., Wei Z., Huang T., Yu A. (2011) Ultrasonic-assisted synthesis of Pd–Ni alloy catalysts supported on multi-walled carbon nanotubes for formic acid electrooxidation, Electrochimica Acta, 56, 6860–6865.

[95] Wang J., Dong Z., Huang J., Li J., Jin X., Niu J., Sun J., Jin J., Ma J., (2013) Filling carbon nanotubes with Ni–Fe alloys via methylbenzene-oriented constant current electrodeposition for hydrazine electrocatalysis, Applied Surface Science 270, 128-132.

[96] Hsieh C.T., Lin J.Y., (2009) Fabrication of bimetallic Pt–M (M=Fe, Co, and Ni) nanoparticle/carbon nanotube electrocatalysts for direct methanol fuel cells, Journal of Power Sources, 188, 347-352.

[97] Gavrilov N., Dašić-Tomić M., Pašti I., Ćirić-Marjanović G., Mentus S., (2011) Carbonized polyaniline nanotubes/nanosheets-supported Pt nanoparticles: Synthesis, characterization and electrocatalysis, Materials Letters, 65, 962–965.

[98] Huang H., Zhang W., Li M., Gan Y., Chen J., Kuang Y., (2005) Carbon nanotubes as a secondary support of a catalyst layer in a gas diffusion electrode for metal air batteries, Journal of Colloid and Interface Science, 284, 593–599.

[99] Guo D.J., Qiu X.P., Chen L.Q., Zhu W.T., (2009) Multi-walled carbon nanotubes modified by sulfated TiO_2–A promising support for Pt catalyst in a direct ethanol fuel cell, Carbon, 47, 1680-1685.

[100] Woo S., Lee J., Park S.K., Kim H., Chung T.D., Piao Y., (2013) Enhanced electrocatalysis of PtRu onto graphene separated by Vulcan carbon spacer, Journal of Power Sources, 222, 261-266.

[101] Zhang Y., Zhao Y., Yuan S., Wang H., He C., (2013) Electrocatalysis and detection of nitrite on a reduced graphene/Pd nanocomposite modified glassy carbon electrode, Sensors and Actuators B, 185, 602-607.

[102] Tang J., Wang T., Sun X., Hu Y., Xie Q., Guo Y., Xue H., He J., (2013) Novel synthesis of reduced graphene oxide-ordered mesoporous carbon composites and their application in electrocatalysis, Electrochimica Acta, 90, 53-62.

[103] Barakat, N.A.M,. Abdelkareem, M.A., and Kim, H.Y., (2013) Ethanol electro-oxidation using cadmium-doped cobalt/carbon nanoparticles as novel non precious electrocatalyst, Applied Catalysis A: General, 455, 193-198.

[104] El-Nagar G.A., Mohammad A.M., El-Deab M.S., El-Anadouli B.E., (2013) Electrocatalysis by design: Enhanced electrooxidation of formic acid at platinum nanoparticles–nickel oxide nanoparticles binary catalysts, Electrochimica Acta, 94, 62-71.

[105] Alegre C., Gàlvez M.E., Baquedano E., Pastor E., Moliner R., La´zaro M.J., (2012) Influence of support's oxygen functionalization on the activity of Pt/carbon xerogels catalysts for methanol electro-oxidation, International Journal of Hydrogen Energy, 37, 7180-7191.

[106] Job N., Marie J., Lambert S., Berthon-Fabry S., Achard P., (2008) Carbon xerogels as catalyst supports for PEM fuel cell cathode, Energy Conversion and Management, 49, 2461-2470.

[107] Kim H.J., Kim W.I, Park T.J., Park H.S., Suh D.J., (2008) Highly dispersed platinum–carbon aerogel catalyst for polymer electrolyte membrane fuel cells, Carbon, 46, 1393-1400.

[108] Carmo M., Linardi M., Poco J.G.R., (2008) H_2O_2 treated carbon black as electrocatalyst support for polymer electrolyte membrane fuel cell applications, International Journal of Hydrogen Energy, 33, 6289-6297.

[109] Kuk S.T., Wieckowski A., (2005) Methanol electrooxidation on platinum spontaneously deposited on unsupported and carbon-supported ruthenium nanoparticles, Journal of Power Sources, 141, 1–7.

[110] Roth C., Papworth A.J., Hussain I., Nichols R.J., Schiffrin D.J., (2005) A Pt/Ru nanoparticulate system to study the bifunctional mechanism of electrocatalysis, Journal of Electroanalytical Chemistry, 581, 79–85.

[111] Lima F.H.B., Gonzalez E.R., (2008) Electrocatalysis of ethanol oxidation on Pt monolayers deposited on carbon-supported Ru and Rh nanoparticles, Applied Catalysis B: Environmental, 79, 341-346.

[112] Kou R., Shao Y., Wang D., H. Engelhard M.H., Kwak J.H., Wang J., Viswanathan V.V., Wang C., Lin Y., Wang Y., Aksay I.A., Liu J., (2009) Enhanced activity and stability of Pt catalysts on functionalized graphene sheets for electrocatalytic oxygen reduction, Electrochemistry Communications, 11, 954-957.

[113] Choi S.M., Seo M.H., Kim H.J., Lim E.J., Kim W.B., (2010) Effect of polyoxometalate amount deposited on Pt/C electrocatalysts for CO tolerant electrooxidation of H_2 in polymer electrolyte fuel cells, International Journal of Hydrogen Energy, 35, 6853-6862.

[114] Cui X., Cui F., He Q., Guo L., Ruan M., Shi J., (2010) Graphitized mesoporous carbon supported Pt–SnO_2 nanoparticles as a catalyst for methanol oxidation, Fuel, 89, 372–377.

[115] Kothandaraman R., Nallathambi V., Artyushkova K., Scott Calabrese Barton S.C., (2009) Non-precious oxygen reduction catalysts prepared by high-pressure pyrolysis for low-temperature fuel cells, Applied Catalysis B: Environmental, 92, 209-216.

[116] Li B., Qiao J., Zheng J., Yang D., Ma J., (2009) Carbon-supported Ir–V nanoparticle as novel platinum-free anodic catalysts in proton exchange membrane fuel cell, International Journal of Hydrogen Energy, 34, 5144-5151.

[117] Maiyalagan T., (2009) Silicotungstic acid stabilized Pt–Ru nanoparticles supported on carbon nanofibers electrodes for methanol oxidation, International Journal of Hydrogen Energy, 34, 2874-2879.

[118] Zhao H., Li L., Yang J., Zhang Y., (2008) Nanostructured polypyrrole/carbon composite as Pt catalyst support for fuel cell applications, Journal of Power Sources, 184, 375–380.

[119] El-Deab M.S., Ohsaka T., (2007) Electrocatalysis by design: Effect of the loading level of Au nanoparticles–MnO_x nanoparticles binary catalysts on the electrochemical reduction of molecular oxygen, Electrochimica Acta, 52, 2166–2174.

[120] Song Y.Y., Li Y., Xia X.H., (2007) One-step pyrolysis process to synthesize dispersed Pt/carbon hollow nanospheres catalysts for electrocatalysis, Electrochemistry Communications, 9, 201–205.

[121] Cui M., Xu B., Hu C., Shao H.B., Qu L., (2013) Direct electrochemistry and electrocatalysis of glucose oxidase on three-dimensional interpenetrating, porous graphene modified electrode, Electrochimica Acta, 98, 48-53.

[122] Zhai H.S., Cao L., Xia X.H., (2013) Synthesis of graphitic carbon nitride through pyrolysis of melamine and its electrocatalysis for oxygen reduction reaction, Chinese Chemical Letters, 24, 103–106.

[123] Vikkisk M., Kruusenberg I., Joost U., Shulga E., Tammeveski K., (2013) Electrocatalysis of oxygen reduction on nitrogen-containing multi-walled carbon nanotube modified glassy carbon electrodes, Electrochimica Acta, 87, 709–716.

[124] Lin Z., Waller G.H., Liu Y., Liu M., Wong C.P., (2013) 3D Nitrogen-doped graphene prepared by pyrolysis of graphene oxide with polypyrrole for electrocatalysis of oxygen reduction reaction, Nano Energy, 2, 241-248.

[125] Cao X., Ye Y., Li Y., Xu X., Yu J., Liu S., (2013) Self-assembled glucose oxidase/graphene/gold ternary nanocomposites for direct electrochemistry and electrocatalysis, Journal of Electroanalytical Chemistry, 697, 10-14.

[126] Yamanaka I., Ichihashi R., Iwasaki T., Nishimura N., Murayama T., Ueda W., Takenaka S., (2013) Electrocatalysis of Heat-treated Cobalt–Porphyrin/Carbon for Hydrogen Peroxide Formation, Electrochimica Acta, 108, 321-329.

[127] Caravaca A., Lucas-Consuegra A. de, Ferreira V.J., Figueiredo J.L., Faria J.L., Valverde J.L., Dorado F., (2013) Coupling catalysis and gas phase electrocatalysis for the simultaneous production and separation of pure H_2 and C_2 hydrocarbons from methane and natural gas, Applied Catalysis B: Environmental, 142–143, 298-306.

[128] Perini L., Durante C., Favaro M., Agnoli S., Granozzi G., Gennaro A., (2014) Electrocatalysis at Palladium Nanoparticles: Effect of the Support Nitrogen Doping on the Catalytic Activation of Carbon-Halogen Bond, Applied Catalysis B, Environmental, 144, 300-307.

[129] Song Y., Liu H., Wang Y., Wang L., (2013) A novel bi-protein bio-interphase of cytochrome c and glucose oxidase: Electron transfer and electrocatalysis, Electrochimica Acta, 93, 17-24.

[130] Baghayeri M., Zare E.N., Namadchian M., (2013) Direct electrochemistry and electrocatalysis of hemoglobin immobilized on biocompatible poly (styrene-alternative-maleic acid)/functionalized multiwall carbon nanotubes blends, Sensors and Actuators B: Chemical, 188, 227-234.

[131] Ding L., Qiao J., Dai X., Zhang J., Zhang J., Tian B., (2012) Highly active electrocatalysts for oxygen reduction from carbon-supported copper-phthalocyanine synthesized by high temperature treatment, International Journal of Hydrogen Energy, 37, 14103-14113.

[132] Dsoke S., Kolary-Zurowska A., Zurowski, A., Mignini P., Kulesza P.J., Marassi R., (2011) Rotating disk electrode study of $Cs_2.5H_{0.5}PW_{12}O_{40}$ as mesoporous support for

Pt nanoparticles for PEM fuel cells electrodes, Journal of Power Sources, 196, 10591–10600.

[133] Habibi B., Delnavaz N., (2011) Carbon-ceramic supported bimetallic Pt-Ni nanoparticles as an electrocatalyst for oxidation of formic acid, International Journal of Hydrogen Energy, 36, 9581-9590.

[134] Yi Y., Lee J.K., Lee H.J., Uhm S., Nam S.C, Lee J., (2009) A single-step approach to create nano-pottery structures for efficient water electrocatalysis, Electrochemistry Communications, 11, 2121-2124.

[135] Solla-Gullón J., Gómez, Aldaz A., Pèrez J.M., (2008) A combination of SERS and electrochemistry in Pt nanoparticle electrocatalysis: Promotion of formic acid oxidation by ethylidyne, Electrochemistry Communications, 10, 319–322.

[136] Su W.Y., Cheng S.H., (2008) Electrocatalysis and sensitive determination of cysteine at poly (3,4-Ethylenedioxythiophene)-modified screen-printed electrodes, Electrochemistry Communications, 10, 899–902.

[137] Choi B., Yoon H., Park I.S., Jang J., Sung Y.E., (2007) Highly dispersed Pt nanoparticles on nitrogen-doped magnetic carbon nanoparticles and their enhanced activity for methanol oxidation, Carbon, 45, 2496–2501.

[138] Lobyntseva E., Kallio T., Alexeyeva N., Tammeveski K., Kontturi K., (2007) Electrochemical synthesis of hydrogen peroxide: Rotating disk electrode and fuel cell studies, Electrochimica Acta, 52, 7262–7269.

[139] Song C. and Zhang J., (2008) Electrocatalytic Oxygen Reduction Reaction, in: Zhang J. (Ed.), PEM Fuel Cell Electrocatalysts and Catalyst Layers, Fundamentals and Applications. Springer-Verlag, London, 89-134.

[140] Lee S-H., Pak C-H., Chang H., Shin J-K., (2009) Catalyst for cathode in fuel cell, US Patent 7566514 B2.

[141] Quentel F., Gloaguen F., (2013) Kinetic and thermodynamic aspects of the electrocatalysis of acid reduction in organic solvent using molecular diiron-dithiolate compounds, Electrochimica Acta, 110, 641-645.

[142] Braunchweig B., Hibbitts D., Neurock M., Wieckowski A., (2013) Electrocatalysis: A direct alcohol fuel cell and surface science perspective, Catalysis Today, 202, 197-209.

[143] Guo D.J. and Ding Y., (2012) Porous Nanostructured Metals for Electrocatalysis, Electroanalysis, 24 (11), 2035–2043.

[144] Habibi B. and Dadashpour E., (2013) Carbon-ceramic supported bimetallic Pt-Ni nanoparticles as an electrocatalyst for electrooxidation of methanol and ethanol in acidic media, International Journal of Hydrogen Energy, 38, 5425-5434.

[145] Kaminska I., Opallo M., Łaszcz A., Czerwinski A., Niedziolka-Jonsson J., (2013) (Bio)electrocatalysis at tin-doped indium oxide nanoparticulate film decorated with gold, Electrochimica Acta, 106, 165–171.

[146] Wasmus S. and Küver A., (1999) Methanol oxidation and direct methanol fuel cells: a selective review, Journal of Electroanalytical Chemistry, 461, 14–31.

Chapter 8

Preventing of Cathode Passivation/Deposition in Electrochemical Treatment Methods – A Case Study on Winery Wastewater with Electrocoagulation

Murat Eyvaz, Ercan Gürbulak, Serdar Kara and
Ebubekir Yüksel

Additional information is available at the end of the chapter

http://dx.doi.org/10.5772/58580

1. Introduction

Electrochemical methods and processes have been applied for many years in environmental applications such as water/wastewater treatment, recovery of metals, electroplating and qualitative/quantitative analysis in various aqueous media. Among these processes, electrocoagulation (EC) has gained many interest due to providing simple, reliable and cost effective operation for the treatment of wastewaters without and need for additional chemicals, and thus the secondary pollution. EC is declared an environment-friendly technique since the 'electron' is the main reagent and does not require addition of the reagents/chemicals. This will minimize the sludge generation to a great extent and eventually eliminate some of the harmful chemicals used as coagulants in the conventional effluent treatment methods. EC process can effectively destabilize small colloidal particles and generates lower quantity of sludge compared to other processes [1]. This technique uses a direct current (DC) source between metal electrodes immersed in polluted water [2]. In this method, soluble metal electrodes (such as iron and aluminium mostly) form metal hydroxides when subjected to a suitable current. The metal hydroxides act as coagulants and lead to the removal of various contaminants [3]. The pros and cons of the EC are tabulated in Table 1

EC is an efficient technique since adsorption of hydroxide on mineral surfaces are a 100 times greater on 'in situ' rather than on preprecipitated hydroxides when metal hydroxides are used as coagulant [3]. Besides, the 'electron' is the main reagent and does not require addition of the reagents/chemicals, which will minimize the sludge generation to a great extent and eventually eliminate some of the harmful chemicals used as coagulants in the conventional

INTECH
open science | open minds

© 2014 The Author(s). Licensee InTech. This chapter is distributed under the terms of the Creative Commons Attribution License (http://creativecommons.org/licenses/by/3.0), which permits unrestricted use, distribution, and reproduction in any medium, provided the original work is properly cited.

effluent treatment methods. EC process can effectively destabilize small colloidal particles and generates lower quantity of sludge compared to other processes [1]. EC has been successfully applied to the treatment of a lot of wastewaters including either organic or inorganic pollutants as well as drinking waters due to its benefits: environmental compatibility, versatility, energy efficiency, safety, selectivity, amenability to automation, and cost effectiveness. Additionally, electrochemical based systems allow controlled and rapid reactions, smaller systems become viable and, instead of using chemicals and micro-organisms, the systems employ only electrons to facilitate water treatment [3, 5]. However, the main drawback of conventional EC (DC-EC) is inevitable formation of an impermeable oxide film on the cathode, which results in lower removal performance of pollutants and higher operating costs (due to higher energy consumption) [3, 6].

In the EC process, an electric field is applied to the medium for a short time, and the treated dispersion transferred to an integrated clarifier system where the water–contaminant mixture separates into a floating layer, a mineral-rich sediment, and clear water. The aggregated mass settles down due to gravitational force. The clear water can be extracted by conventional methods [4]. Generally, DC power supply is traditionally employed to generate an electric field and ion transportation between the immersed sacrificial electrodes in the EC reactor. However, the major threat of the EC process using DC is that an impermeable oxide film may be formed on the cathode during the electrolysis. This leads a *"cathode passivation"*, which decreases the ionic transfer between the anode and cathode directly, hindering the metal dissolution and indirectly preventing metal hydroxide formation. Electrolytic dissolution of anode and electrolytic deposition of cathode in the EC cell employing DC power supply are schematically shown in Fig. 1. The DC-EC technology is inherent with the formation of an impermeable oxide layer on the cathode as well as deterioration of the anode due to oxidation. These limitations of the DC-EC process have been minimized to some extent by the addition of parallel plate sacrificial electrodes in the cell configuration. However, many have preferred the use of alternating current (AC) power supply in EC process [4, 7]. According to the reference [4], the AC cyclic energization is believed to retard the normal mechanisms of electrode attack that are experienced in DC-EC system, and thus, ensure reasonable electrode life. In addition to that, since the AC electric fields in an AC-EC separator do not cause electrophoretic transport of the charged particles due to the frequent change of polarity, it can induce dipole–dipole interactions in a system containing nonspherical charged species. As a result, the AC electric fields may also disrupt the stability of balanced dipolar structures existing in such a system. This is, however, not possible in a DC-EC separator using DC electric fields.

To prevent the main disadvantage of the EC, cathode passivation, AC can be preferred as power supply or anode and cathode can be replaced periodically with each other in DC mode. However, the latter option is not feasible for continuous operations in practical. The alternating pulse current (APC) method was proposed with experimental results in detail for the first time by Mao et al. [8] by using AC power. However, the rectangular wave produced by "a time relay from already existing DC power supply" was declared by Eyvaz et al. [9] for the first time in EC applications.

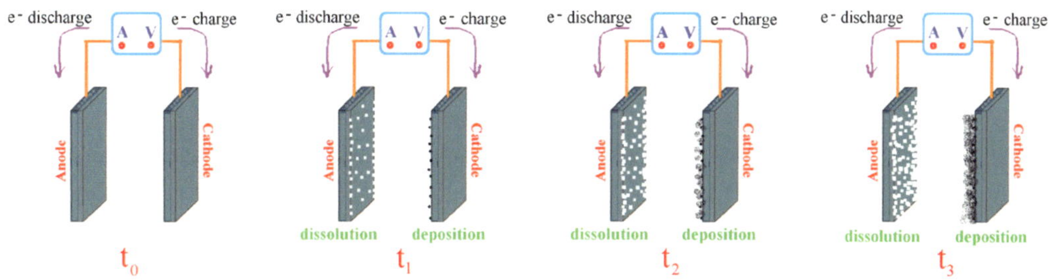

Figure 1. Schematic representation of electrode surface change during the EC treatment using DC supply (t_0–t_3 are treatment times of continuous process in order).

Mao et al. [8] proposed a novel current feed style in electrocoagulation aiming at preventing the cathode passivation by using AC in synthetic oily wastewater. They investigated the effects of APC on the aluminium electrodes' surfaces by scanning electron microscope (SEM) and X-ray photoelectron spectroscopy (XPS), not observing passivation of Al, also achieving uniform dissolution of anode and cathode. Similar to APC, periodic reversal current feed was obtained by a simple time relay device integrated with a DC power supply by Eyvaz et al. [9]. They used this system to prevent electrode passivation of aluminium electrodes in EC of textile dye solutions. Their results indicated that APC was found superior to DC and higher removal efficiencies in shorter operation times and longer fill-and-draw periods could be gained by APC. Removal efficiencies increased in APC system after optimum operation time belongs to DC system as well. In another research containing the comparison of AC and DC, Vasudevan et al. [10] reported that under the identical experimental conditions, similar cadmium removal efficiencies (97.5% and 96.2%) were achieved. However, the benefits of reducing energy consumption were two fold with AC (0.665 and 1.236 kWh/m^3 for AC and DC, respectively).

According to the research of Keshmirizadeh et al. [11], equal removal efficiencies were obtained in direct current and alternating pulse current. In the APC mode, the water recovery was very significant, measuring as high as 0.92 m^3/m^3 wastewater. For DC mode, the water recovery was less than 0.5 m^3/m^3 of wastewater.The APC mode was found to be more efficient than the DC mode with a lower anode over-voltage, slower anode polarization and passivity, and lower tank voltage. The operating time was 3–25% less when APC mode was used, based on initial Cr(VI) concentration of 50–1000 mg/L, respectively. Because of the reduction in operating time, less energy was consumed, which made the APC mode more cost effective. Application of APC eliminated uneven wear (dissolution) of electrodes; typically, the anode material dissolved and electroreduction products stuck to the cathodes when DC mode was used. When the APC mode was employed, electrocoagulation produced a highly dense or compact sludge at the reactor bottom, resembling dense clay soil layers. It also produced more clear supernatant. The APC mode minimized waste and increased sludge stability.

More recently, periodic electrode reversal methode (PREC) in EC was optimized by response surface methodology (RSM) for color removal of synthetic Methyl Orange wastewater by Pi et al. [12]. Color removals of 97 % with PREC and 82 % with conventional EC were gained in the optimal RSM conditions. It is concluded that EC with PREC can effectively retard cathodic passivation by resulted in lower energy and electrode material consumptions.

Although many researches on treatment of synthetic or real industrial wastewaters with EC are available, very few researches have been carried out on the economical applicability of AC electrocoagulation. Therefore, in this chapter, the effects of power supply type (DC or AC) on EC performance were investigated both technically and economically. An adjustable time relay plugged into the DC power supply was employed to obtain APC to avoid additional AC power supply cost. Winery wastewater was selected as the model electrolyte solution due to its high strength pollutant capacity and APC, an (almost) new method, was applied to the winery wastewaters for the first time with this study.

2. Brief description of electrocoagulation

EC is a complicated process involving many chemicals and physical phenomena that use consumable electrodes to supply ions into the wastewater stream. In an EC process the coagulating ions are produced 'in situ' and it involves three successive stages: (i) formation of coagulants by electrolytic oxidation of the 'sacrificial electrode',(ii) destabilization of the contaminants, particulate suspension, and breaking of emulsions and (iii) aggregation of the destabilized phases to form flocs. The destabilization mechanism of the contaminants, particulate suspension, and breaking of emulsions have been described in broad steps and may be summarized as: (1) Compression of the diffuse double layer around the charged species by the interactions of ions generated by oxidation of the sacrificial anode, (2) Charge neutralization of the ionic species present in wastewater by counter ions produced by the electroche-mi-cal dissolution of the sacrificial anode. These counter ions reduce the electrostatic interparticle repulsion to the extent that the van der Waals attraction predominates, thus causing coagulation. A zero net charge results in the process. (3) Floc formation; the floc formed as a result of coagulation creates a sludge blanket that entraps and bridges colloidal particles still remaining in the aqueous medium. The solid oxides, hydroxides and oxyhydroxides provide active surfaces for the adsorption of the polluting species. EC has been successfully employed in removing metals, suspended particles, clay minerals, organic dyes, and oil & grease from a variety of industrial effluents [3]. A brief literature review of EC efficiency on the treatment of different waters/wastewaters is presented in Table 2.

In this process, a potential is applied to the metal anodes, typically fabricated from either iron or aluminium, which causes two separate reactions: (1) Fe/Al is dissolved from the anode generating corresponding metal ions, which almost immediately hydrolyze to polymeric iron or aluminium hydroxide. These polymeric hydroxides are excellent coagulating agents. The consumable (sacrificial) metal anodes are used to continuously produce polymeric hydroxides in the vicinity of the anode. Coagulation occurs when these metal cations combine with the negative particles carried toward the anode by electrophoretic motion. Contaminants present in the wastewater stream are treated either by chemical reactions and precipitation or physical and chemical attachment to colloidal materials being generated by the electrode erosion. They are then removed by electro-flotation, or sedimentation and filtration. Thus, rather than adding coagulating chemicals as in conventional coagulation process, these coagulating agents are generated in situ. (2) Water is also electrolyzed in a parallel reaction, producing small bubbles of oxygen at the anode and hydrogen at the cathode. These bubbles attract the flocculated particles and, through natural buoyancy, float the flocculated pollutants to the surface [3]. The most important reactions are summarised in Fig. 2.

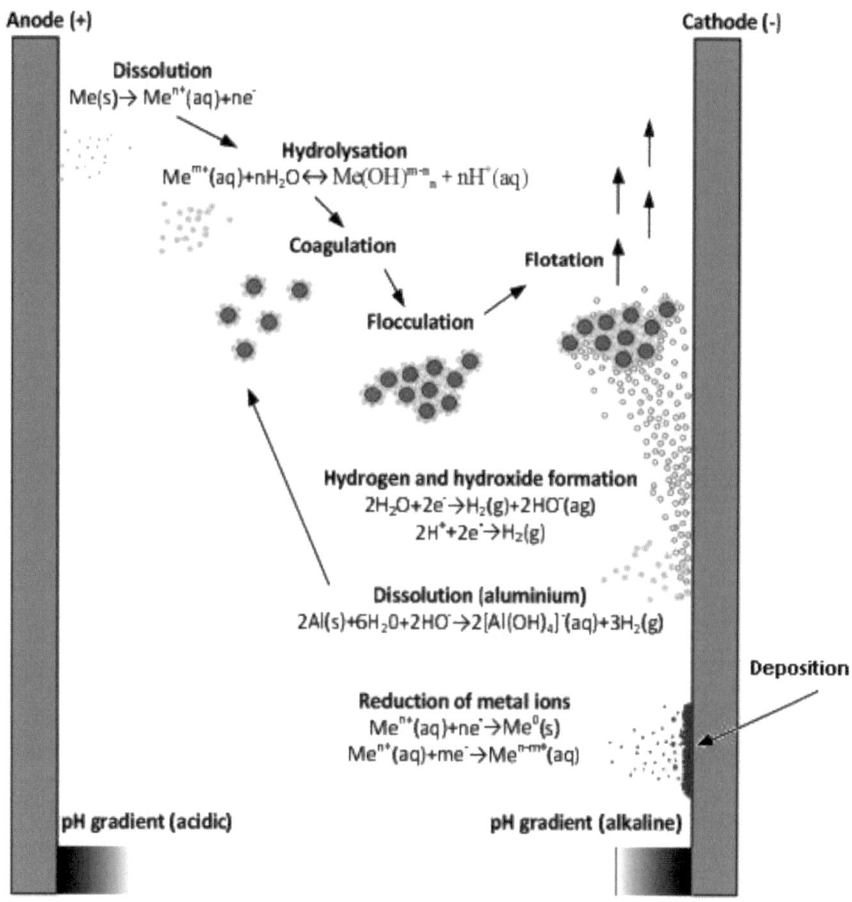

Figure 2. Schematic representation of typical reactions during the EC treatment using DC supply [13].

The (EC) process involves generation of coagulants in situ by dissolving sacrificial anodes such as aluminium or iron upon application of a DC. When iron electrode is used as anodes upon oxidation in an electrolytic system, it produces iron hydroxide, $Fe(OH)_n$ where n=2 or 3 [14, 15]. The major disadvantage of EC compared to chemical coagulation (usually ferric or aluminium chloride/sulfate) is that high conductivity water is required. This fact is especially relevant for drinking water treatment, as conductivity can not be enhanced by salts due to total dissolved solids (TDSs) limitations in drinking water [16]. Two mechanisms for the production of metal hydroxide have been proposed when iron electrodes are used [17, 18]:

Mechanism I:

$$\text{Anode: } 2Fe \rightarrow 2Fe^{2+} + 4e^{-} \tag{1}$$

$$2Fe^{2+} + 5H_2O + 1/2O_2 \rightarrow 2Fe(OH)_{3(s)} + 4H^{+} \tag{2}$$

$$\text{Cathode: } 4H_2O + 2e^{-} \rightarrow 4OH^{-} + 2H_{2(g)} \tag{3}$$

$$\text{Overall reaction: } 2Fe + 5H_2O + 1/2O_2 \rightarrow 2Fe(OH)_{3(s)} + 4H_{2(g)} \tag{4}$$

Mechanism II:

$$\text{Anode: } Fe \rightarrow Fe^{2+} + 2e^{-} \tag{5}$$

$$Fe^{2+} + 2OH^{-} \rightarrow Fe(OH)_{2(s)} \tag{6}$$

$$\text{Cathode: } 2H_2O + 2e^{-} \rightarrow H_2 + 2OH^{-} \tag{7}$$

$$\text{Overall reaction: } Fe + 2H_2O \rightarrow Fe(OH)_{2(s)} + H_{2(g)} \tag{8}$$

When aluminium electrodes in the EC process are used as an anode and a cathode, the main reactions are at the anode as follows [19-22]:

$$\text{Anode:} Al \rightarrow Al^{3+} + 3e^{-} \tag{9}$$

$$2H_2O \rightarrow O_{2(g)} + 4H+ + 4e^{-} \tag{10}$$

$$\text{Cathode:} 3H_2O + 3e^{-} \rightarrow 3OH^{-} + 3/2\ H_{2(g)} \tag{11}$$

$$2Al + 6H_2O + 2OH- \rightarrow 2Al(OH)^{-4} + 3\ H_{2(g)} \tag{12}$$

$$\text{Overall reaction:} Al^{3+} + 3H_2O \rightarrow Al(OH)_{3(s)} + 3H^+ \quad (13)$$

The performance of the EC process depends on many operational parameters such as pH of the solution, applied current to the reactor, conductivity of the (water/wastewater) solution, electrolysis time as well as electrode spesifications such as arrangement of electrode, electrode shape, distance between the electrodes, etc. Main operational parameters influencing the EC efficiency is schematically given in Fig. 3. The effects of each parameters in details can be found elsewhere in the literature. However, in this study, effects of six of the parameters such as pH of the winery wastewater (by adjusting the pH with acid or base), arrangement of the sacrificial electrodes (parallel or serial connection to power supply), electrolysis time (duration of applying voltage to the wastewater), current density (applied current to the unit area of active electrode surface) and especially power supply type (by changing the polarity of the anodes and cathodes) were investigated.

Figure 3. Schematic display of the various operating parameters influencing the EC process performance.

2.1. Brief description of alternating pulse current electrocoagulation

Usually, DC is used in EC systems. In this case, an impermeable oxide layer may form on the cathode material as well as corrosion formation on the anode material due to oxidation. This prevents the effective current transfer between the anode and cathode, so the performance of EC reactor declines. These disadvantages of DC have been diminished by the addition of parallel plate sacrificial electrodes in the EC unit configuration. However, many have preferred the use of AC in EC unit [7]. It is believed that the cyclic energization between the anode–cathode in AC system delays the cathode passivation and anode deterioration that are experienced in DC system, and thus, ensure reasonable electrode life [4]. A hypothesis for the lower electrode consumption with AC is that since DC only flows in one direction, there may be irregular wear on the plates due to the onslaught of the current and subsequent oxidation occurring in the same preferential points of the electrode. In the case of AC, the cyclical energization retards the normal mechanisms of attack on an electrode and makes this attack more uniform, thus ensures longer electrode life [23].

There are a few studies in which (AC) has been tested although typically DC has been used in EC systems. Vasudevan *et al.* [24] studied the removal of fluoride from water with DC and AC EC systems. They observed similar removal efficiencies with both technologies. However, energy consumption was slightly lower with AC technology. Eyvaz *et al.* [9] used APC in their study. APC enhanced removal efficiency compared to DC current. Pollutant removal decreases over the course of time with DC systems, possibly due to passivation of electrodes, whereas in an APC system this was not observed. Polarity reversal has also been suggested by other authors to reduce passivation of electrodes [25].

Schematic display of current waves of AC and APC systems are given in Fig. 4. An adjustable time relay plugged into the DC power supply was employed to obtain APC (It also represents AC in our study). According to EC unit with time relay system, turn on and turn off modes switch to positive pole to negative pole or reversion to it. For example, when the time relay is turned on in an EC reactor including two electrodes namely 1 (anode) and 2 (cathode), electrode 1 is then converted to cathode while electrode 2 is becoming anode. When the time relay is turned off, only DC system is in circuit, electrode 1 becomes anode this time. Current wave in real AC is shown as sine wave in Fig. 4.

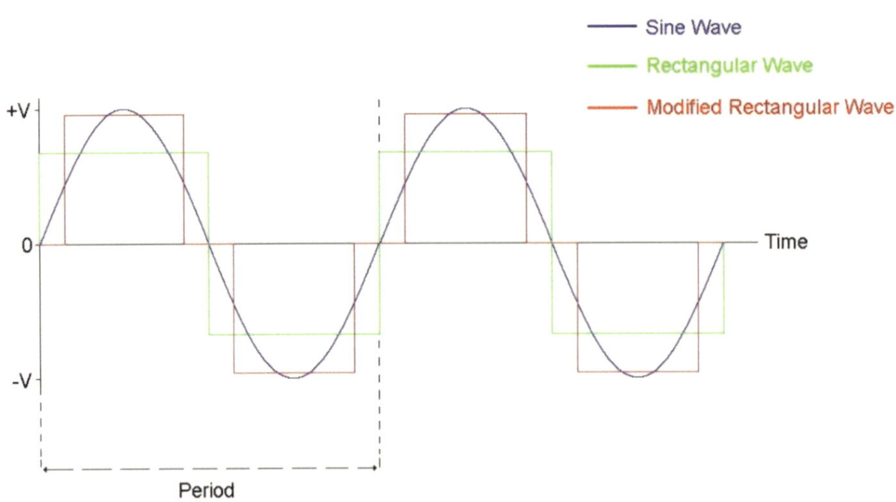

Figure 4. Schematic display of current waves of AC and APC systems.

3. Winery industry, winery wastewaters and treatment methods

Wine production processes generate organic and inorganic pollutions mostly associated with solid wastes and liquid effluents. The liquid effluents usually referred as "winery wastewater" are mainly originated inwashing operations during grape harvesting, pressing and first fermentation phases of wine processing [26-29]. Winery wastewater is produced in significant volumes around the world [30-32]. Each winery is also unique in wastewater generation, highly variable, 0.8 to 14 L per litre of wine [33-35] and is characterized by high contents of organic material and nutrients, high acidity and large variations in the seasonal flow production [36, 37] and is generated mainly as the result of cleaning practices in winery, such as washing operations during crushing and pressing grapes, rinsing of fermentations tanks, barrels washing, bottling and purges from the cooling process. As a consequence of the working period and the winemaking technologies, volumes and pollution loads greatly vary over the year [35]. The organic matter might reach during vintage periods up to >30,000 mg/L chemical oxygen demand (COD), and the high sodium adsorption ratio (SAR>15) make such water inadequate to be disposed to common sewage systems [37,38].

Although different varieties of grapes and strains of yeasts result in different types of wine and consequently winery wastewaters with different characteristics, in general, the typical raw winery wastewater presents a pH between 3 and 4, COD ranges from 320 to 296,119 mg O_2/L and BOD5 values around 125–130,000 mg O_2/L [32, 39]. The main organic compounds present in this kind of wastewaters are soluble sugars (fructose and glucose), organic acids (tartaric, lactic and acetic), alcohols (glycerol and ethanol) and high-molecular-weight compounds, such as polyphenols, tannins and lignin [39, 40].

Wine distillery wastewater, the product of the distillation of ethanol, wine and waste biological material, produces large volumes of liquid that involves unacceptable environmental risks [41-47]. The disposal of the untreated waste from the wine sector causing salination and eutrophication of water resources; waterlogging and anaerobiosis and loss of soil structure with increased vulnerability to erosion [33, 35].

Several winery wastewater treatments are available, and among them biological treatment methods have been recognized as a reasonable alternative way for a significant degradation of wastewater with high organic content, however, the presence of recalcitrant compounds for the microorganisms frequently makes impossible the complete treatment of a winery wastewater [27, 29]. The winery wastewater treatment technologies can be sorted as natural evaporation in ponds, evaporation–condensation with or without combustion, direct dispersion on soil as a fertilizer and intensification of the natural evaporation capacity of the ponds by means of sprinklers and panels as physicochemical methods; aerobic or anaerobic treatment, trickling filters, lagoons as biological methods [30, 47-49]. These methods classified as schematically in Fig. 5.

Advantages	Drawbacks
☑ Needs simple equipments, designable for virtually any size, ☑ Cost-effective, and easily operable, ☑ The start-up and operating costs are relatively low, ☑ Low maintenance cost with no moving parts, ☑ No chemical addition is required, ☑ Minimum chance of secondary pollution, ☑ Needs low current, and it can be run even by green processes, such as, solar, windmills and fuel cells, ☑ Effectively destabilize small colloidal particles, ☑ Produces minimal sludge, ☑ Lowering the sludge disposal cost, ☑ Gas bubbles (produced during electrolysis) can enhance flotation, ☑ Removes many species that chemical coagulation cannot remove, ☑ More readily filterable and can be utilized as a soil additive, ☑ Contains metal oxides that pass the leachability test, ☑ Gives palatable, clear, colorless and odorless water.	☒ The 'sacrificial anodes' need to be replaced periodically. ☒ Requires a minimum conductivity depending on reactor design, limiting its use with water containing low dissolved solids. ☒ In case of the removal of organic compounds, some toxic chlorinated organic compound may be formed in situ if chlorides are also present. ☒ Wastewater with high humic and fluvic acid content may be amenable to the formation of trihalomethanes. ☒ If phenols and algal metabolic and decomposition products are present, chlorine may lead to bad taste and odor. ☒ An impermeable oxide film may be formed on the cathode that may interfere with the performance of the EC cell. ☒ The cost of operating EC may be high in those are as where the cost of electricity is high.

Table 1. Advantages and drawbacks of the EC process.

COD and BOD of the winery wastewater can be removed significantly by biological treatment. However, the color of the wastewater remains dark brown as that before the treatment because of the non-biodegradable colored compounds such as melanoidins that can be degraded only 6–7% by biological treatment [47, 49-51]. EC process produces coagulants such as iron or aluminium (Al) hydroxides having a considerable sorption capacity by anodic dissolution and also pollutants are removed simultaneously by deposition on cathode electrode or by flotation due to the hydrogen gas produced at the cathode [47, 52-55]. Because the wastewater is not enriched with anions, the sludge produced in EC process is more compact than the sludge generated by chemical coagulation [47, 55]. Besides this, EC process has many advantages like simple equipment, easy operation, a shortened reactive retention time and less sludge amount when compared chemical coagulation [2, 47].

In recent years EC technique has been applied to the wastewaters generated from food industry such as distillery and fermentation [1, 49, 51, 56-58], dairy [59-61], potato chips manufacturing [21], pasta and cookie process [62], poultry slaughterhouse [63-65], and yeast [20, 66]. Although in literature, there are a lot of studies including various treatment methods shown in Fig. 5, there have been very few research [46, 47, 67] conducted about technical and economic analysis of EC process on winery industry wastewaters. Therefore, the purpose of this work is to examine the treatment performance of EC process employing Fe and Al electrodes on winery wastewater when investigating the APC on the overall EC efficiency.

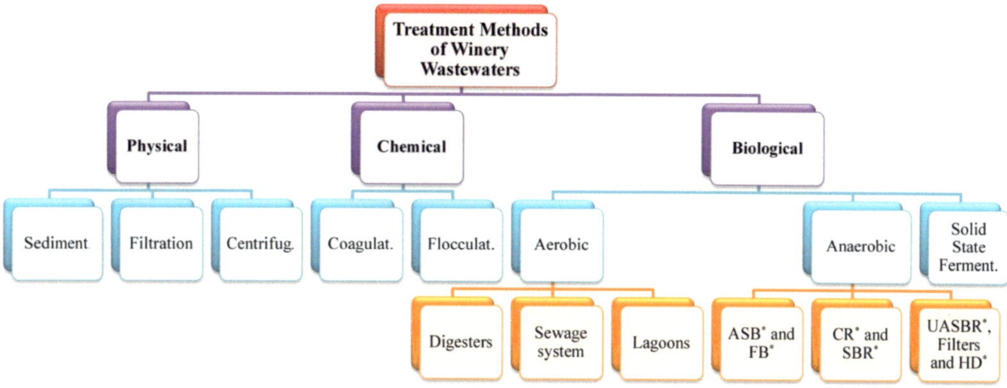

Figure 5. Schematic representation of various methods used in the treatment of winery wastewaters, *: ASB: Anaerobic sludge beds, FB: Fluidized beds, CR: Contact reactor, SBR: Sequencing batch reactor, UASBR: Upflow anaerobic sludge blanket reactor, HD: Hybrid digesters.

Pollutants	Electrode material	Operational parameters investigated	Summary of the work	Ref.
Winery	Aluminium, iron, stainless steel	Initial pH, current density and electrolysis time	• When Fe electrodes were used under optimal conditions, the removal efficiencies of COD, color, and turbidity were calculated as 46.6, 80.3, and 92.3%, respectively. They were found as 48.5% for COD, 97.2% for color and 98.6% for turbidity, when Al electrodes were used. • A new approach combining electrochemical methods with ultrasound in the strong electromagnetic field resulted in significantly better removal efficiencies for majority of the measured parameters compared to the biological methods, advanced oxidation processes or electrocoagulation.	[47], [67]
Paper-pulp mill	Aluminium, iron	Initial pH, temperature, current density, treatment time.	• Temperature has negative effect on the removal efficiency. • Al–Al has a high efficiency in the color removal and Fe–Fe is effective in the COD and Phenol removal. • Pimaric-type acids were removed with higher efficiency than abietic-type resin acids. • EC had no significant effect on bacterial toxicity despite a high removal efficiency of resin acids and copper. • The sludge aptitude to settling is better with Fe electrodes than with Al electrodes.	[68-70]
Textile effluents, dyes	Aluminium, iron, stainless steel	Initial pH, current density, anode-cathode polarization	• Anode–cathode polarization reduces the reaction for removing TOC and dye from aqueous solutions.	[9], [71], [72]

Pollutants	Electrode material	Operational parameters investigated	Summary of the work	Ref.
		period, power supply type, electrode material, electrode connection mode.	• EC employing SS electrodes was more economical; consumed less material and produced less sludge, and pH of the medium was more stabilized than EC with Fe electrodes.	
Boron removal	Aluminium, iron	Initial pH, current density, initial boron concentration, treatment time, temperature	• 98 % removal of boron from produced water was achieved. • Adsorption is chemisorption and endothermic. • The effect of water purification for higher boron concentrationsin the solution is better than for low ones. • For the iron and aluminium electrodes, pH = 6 is the most suitable value, and the aluminium anode is the best one for the boron removal. • The highest current density gave the quickest treatment for boron removal from synthetically prepared waters containing boron equivalent industrial wastewaters.	[73-75]
Landfill leachate	Aluminium, iron	Electrode material, current density, initial pH, operating time, Cl^- concentration	• Aluminum supplies more COD removal (56%) than iron electrode (35%) at the end of the 30 min operating time. • Coagulation and EC treatment mainly affected hydrophobic molecules and after treatment 30% of the initial BDOC quantity was removed. • Electrolysis (and as a consequence EC) increased the amount of hydrophilic organic compounds of lower apparent molecular weight. • Under conditions of iron electrode, 4.96 mA/cm^2 current density, 2319 mg/L Cl- concentration, 90 min electrolysis time and unchanged the raw pH (6.4-7.3), the removal efficiencies of COD, NH_3-N, TP, BOD_5 and turbidity are 49.8, 38.6, 82.2, 84.4 and 69.7%, respectively.	[76-78]
Drinking water	Aluminium, iron, stainless steel	Anode metal type, NOM source, initial NOM concentration, co-occurring solutes, initial fluoride concentration, electrode connection type.	• Between the three metals tested, iron was the least costly and most available material, it presented greater DOC removal, it showed no passivation layer and linear voltage ramp, and residual metal met guideline values. • Removal of fluoride was better for bipolar connection than for monopolar connection. • The operating costs for monopolar and bipolar connections were 0.38 and 0.62 US$/$m^3$, respectively, for the initial fluoride concentration of 10 mg/L.	[16], [79], [80]

Pollutants	Electrode material	Operational parameters investigated	Summary of the work	Ref.
			• The residual arsenic concentration was maintained below the limiting value recommended during a period of 16 h of continuous mode operation for EC-MF system.	
Olive mill	Aluminium, iron	Electrolysis time, current density, chloride concentration, initial pH, settling time, electrode material and polarization, amount of hydrogen peroxide, addition of coagulant-aid	• EC can remove more than 70% of COD, polyphe-nols and dark color. • EC treatment makes good solid matter and turbidity removal efficiency, 71% and 75%, respectively. • EC in the absence of coagulant aid and oxidant is not too efficient for the treatment of this type of wastewater.	[81-83]
Oily wastewaters	Aluminium, iron, steel	electrode connection mode, current density, initial oil concentration, pH, NaCl dosage	• The best performance was obtained using mild steel MP electrode system. • EC process operated under the optimal conditions involves a total cost of 0.46 US$ per cubic meter of treated oily bilge water. • The oil removal efficiency showed its best values at high current density values, high initial oil concentration with an emulsion of pH around 7. • Sacrifice anode like Fe found to be more effective than Al for the removal of sulfide species and organic matters.	[55], [84], [85]
Poultry slaughterhouse, manure	Aluminium, iron	Stirring speed, current density,	• It has been possible to decrease COD of poultry slaughterhouse wastewater about 2170 mg/L to a less than 300 mg/L in a matter of 30 min. under stirring speed of 150 rpm, initial pH 3 and a current density of 1.0 mA/cm^2 conditions. • Aluminum electrode performed better in reducing the COD, with a removal efficiency as 93% in 25 at low initial pH, such as 3, and current density of 150 A/m^2 .On the other hand, iron electrode was more successful in removing oil-grease with 98% efficiency, irrespective of the initial pH. From economic point of view, iron electrode is clearly preferable; the total operating cost is between 0.3 and 0.4 $\$/m^3$, which is nearly half that of aluminum electrode.	[63], [64], [76]

Pollutants	Electrode material	Operational parameters investigated	Summary of the work	Ref.
			• Under the optimal conditions, about 90% of COD and 92% of residual color could be effectively removed from the UASB effluent with the further contribution of the EC technology used as a post-treatment unit.	
Electroplating/ metal	Aluminium, iron	Electrode material, current density, wastewater pH, conductivity, initial metal concentration	• The Fe–Fe and Fe–Al electrode combinations were more effective for the removal of Cu, Cr and Ni from metal plating wastewater. • At the current density of 25 mA/cm2 with a total energy consumption of 49 kWh/m^3 , more than 96% removal value was achieved for all studied metals except Mn which was 72.6%.	[86-88]

Table 2. cont. A brief literature review of EC efficiency on the treatment of different waters/wastewaters containing various organic/inorganic pollutants.

4. Materials and methods

4.1. Materials

4.1.1. Winery wastewater used in this study

The wastewater used in this work was taken from an equalization tank of a wine factory located in the city of Tekirdağ (in TURKEY), producing approximately 350 m^3 of wastewater daily. The characteristics of the wastewater are presented in Table 3.

Parameter	Value
pH	5.2
COD, mg/L	20,400 ± 1,100
BOD_5, mg/L	11,120 ± 1,055
TOC, mg/L	4,230 ± 940
TSS, mg/L	1,045 ± 85
Turbidity, NTU	1,600 ± 510
Color, Pt-Co	5,300 ± 100
Conductivity, µS/cm	2,800 ± 93

Table 3. Characteristics of the winery wastewater used in this study

http://dx.doi.org/10.5772/58580

4.1.2. EC setup and electrode connection modes

EC reactor was made from plexiglas reactor with dimensions of 130 × 130 × 120 mm and operated in batch mode. Al and Fe electrodes with effective area of 143 cm^2 were used and the distance between the electrodes was 20 mm. Electrodes were connected to a digital DC power supply (Maksimel, Ankara, Turkey) in various electrode connection modes of which details are given below. A time relay (Siemens Sirius, Germany) was used with DC power supply to change polarity of the electrodes when performing APC experiments.

4.1.3. Monopolar electrodes in parallel connections (MP-P)

To improve the performances of an EC it may be necessary to interchange the polarity of the electrode intermittently. However, a two-electrode EC cell is not suitable for wastewater treatment, because for a workable rate of metal dissolution the use of electrodes with large surface area is required. Performance improvement has been achieved by using EC cells with monopolar electrodes either in parallel or series connections [3]. In the MP-P system, anodes and cathodes are in parallel connection, the current is divided between all the electrodes in relation to the resistance of the individual cells. Hence, a lower potential difference is required in parallel connection, when compared with serial connections. MP-P connection mode is given in Fig. 6.

4.1.4. Monopolar electrodes in serial connections (MP-S)

In the MP-S system, each pair of sacrificial electrodes is internally connected with each other, because the cell voltages sum up, a higher potential difference is required for a given current. MP-S connection mode is given in Fig. 7.

4.1.5. Bipolar electrodes in serial connections (BP-S)

In the BP-S system, there is no electrical connection between inner electrodes, only the outer electrodes are connected to the power supply. Outer electrodes are monopolar and inner ones are bipolar. This connection mode has simple setup with and has less maintenance cost during operation. BP-S connection mode is given in Fig. 8.

4.1.6. Time relay device

An adjustable time relay (3RP1525-1BW30 Siemens Sirius Time Relay 20>240VAc/Dc) plugged into the DC power supply was employed to obtain APC. It represents AC in our study. According to EC unit with time relay system, turn on and turn off modes switch to positive pole to negative pole or reversion to it. For example, when the time relay is turned on in an EC reactor including two electrodes namely 1 (anode) and 2 (cathode), electrode 1 is then converted to cathode while electrode 2 is becoming anode. When the time relay is turned off, only DC system is in circuit, electrode 1 becomes anode this time.

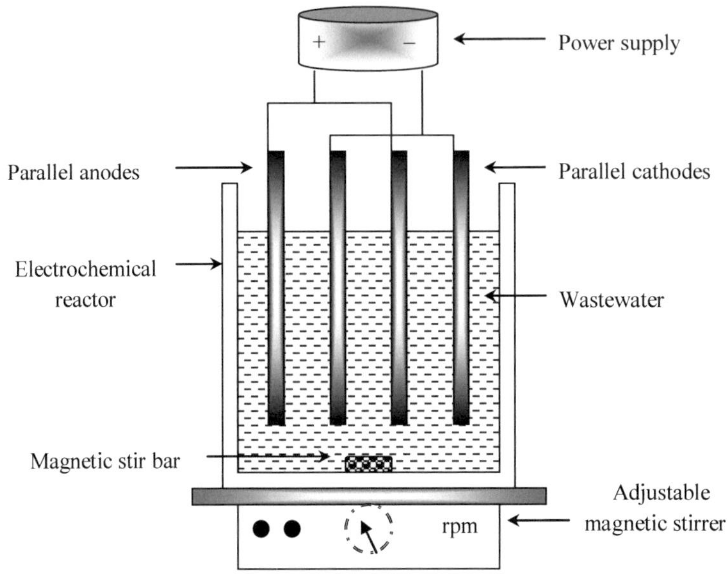

Figure 6. EC reactor with MP-P electrodes [89].

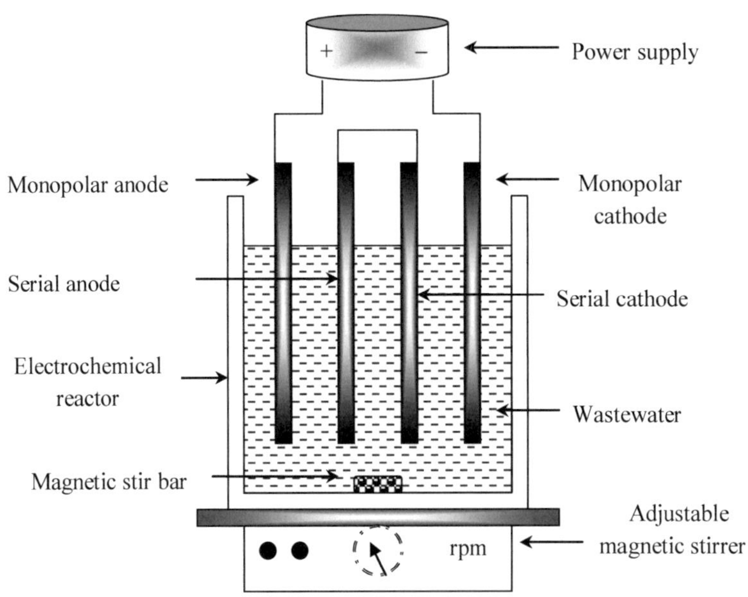

Figure 7. EC reactor with MP-S electrodes [89].

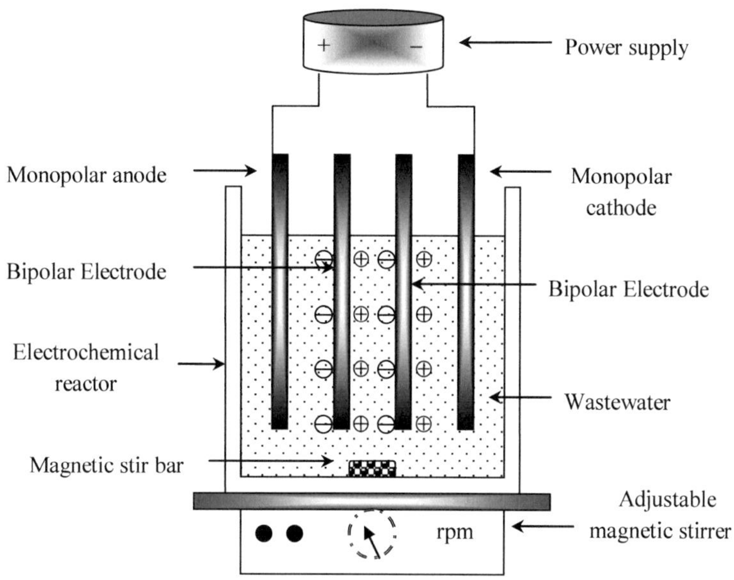

Figure 8. EC reactor with BP-S electrodes [89].

4.2. Methods

All experiments were performed at constant temperature of 25°C. In each run, 1,500 mL of winery wastewater was placed into the reactor. Magnetic stirring (250 rpm, Velp Are) was applied to provide a homogenous solution in the reactor. Conductivity was 2.800 μS/cm that was the conductivity of the wastewater itself where no supported electrolyte was added. The current and pH were adjusted to the desired value before the process. After each run, electrode surfaces were removed by dipping for 1 min in a solution prepared by mixing 100 cm^3 of HCl solution (36.5%) and 200 cm^3 of hexamethylenetetramine aqueous solution (2.80%) [71] and washed thoroughly with demineralized water to remove any solid residues on the surfaces, dried and re-weighted. The solution was filtered through a filter paper (Whatman 40 ashless-NJ, USA) after each run and then analyzed. The solid residue was dried until constant weight was obtained for the calculation of sludge amounts. The experiments were performed in three replications used to compute the mean value and standard deviations. Therefore large amounts of data were collected; analyzed and figured for pH, current density and time experiments; only on effects of COD removal efficiencies; so, all of the graphics could not be presented here due to the limited space in the chapter. Electrode and energy consumptions, sludge formations and operating costs are given in Tables. Economic data used for the evaluation of the total operating costs dye given for the first quarter of 2013, Turkey market, in Table 4.

Item with warranty period	Cost
Power supply and installing, \$, 5 years	10,000
EC tank and installing, \$, 10 years	500
Maintenance and depreciation, \$/m^3	0.005
Electricity, \$ /kWh	0.17
Labor costs, \$/m^3	0.1
Aluminium electrode, \$/kg	0.5
Iron electrode, \$/kg	0.5
Chemicals (acid, salt, etc.), \$/m^3	0.04
Sliudge disposal cost, \$/kg	0.012

Table 4. Economic factors used in the total operating cost calculations.

Measurements of COD and total suspended solids (TSS) were performed according to the procedure of Standard Methods (2005). The pH and conductivity of solutions were measured using a multi meter (Hach Lange HQ40d-Düsseldorf, Germany). An UV spectrophotometer (HACH Co., model DR5000-Düsseldorf, Germany) was employed to measure color and turbidity of the wastewater. The initial pH was adjusted to a desired value using NaOH (Merck-Darmstadt, Germany) or H_2SO_4 (Merck-Darmstadt, Germany).

Pollutant removal efficiencies are calculated as follows:

$$\% \, Removal \ efficiency = \frac{C_0 - C}{C_0} \times 100 \tag{14}$$

where C is COD, color or turbidity value of treated aqueous solution (mg/L, Pt-Co or NTU) and C_0 is the initial relating concentrations (mg/L, Pt-Co or NTU).

5.Results and discussions

5.1. Determining the optimum experimental parameters to comparison DC & APC systems

5.1.1. Effects of initial pH

The EC process is highly dependent on the initial pH of the solution [3]. In aluminium case, precipitation mechanism of monomeric and polymeric $Al(OH)_3$ species at pH 4.0–6.5 and adsorption mechanism of $Al(OH)_3$ and polymeric $Al(OH)_3$ species at pH > 6.5 are effective on the removal of pollutants. However, in the iron case, good removal efficiency can be achieved on floc formation at pH 6–8 [89-91]. Five pH values (4, 5, 6, 7, 8, and 9) were selected to investigate the optimal pH at which maximum removal efficiencies, minimum electrode and energy consumptions were observed for three electrode connection systems as well as both of the electrode material, Fe and Al. The effects of pH have been investigated at constant current density of 40 mA/cm^2 and 60 min of operating time. Because there were a lot of parameters (such as pH, current density, operating time, electrode material, electrode arrangement, and current type) of which effects were investigated on the great number of process outputs (such as COD, turbidity and color removals; energy and electrode consumptions; sludge amount, and total operating cost), large amounts of data were collected, therefore, only some of them are given as figures; as for the others, they are presented in tables. However, effects of APC are analyzed and discussed in much more details in the following sections. The effects of initial pH on COD removal are featured in Fig. 9. In both cases of different electrode materials, no great differences are observed between connection modes. As seen from the figure, the highest COD removal efficiencies were observed at pH 5 for Al electrode and at pH 7 for Fe electrode, where also maximum turbidity and color removals as well as minimum energy and electrode consumptions with minimum sludge formations (Table 5) were achieved. According to the Fig. 9, COD removal efficiency increased when pH increased from 4 to 5 and then it decreased at higher pH values, until pH 9. In similar trend, Fe electrode shows maximum performance for COD removal at pH 7, (maxium turbidity and color removals were also obtained at these pHs: 5 and 7, as seen in Table 5). It was concluded that colloid particles were destabilized by the metal ions produced by anodic dissolution and these ions reacted with organic pollutions by adsorption or co-precipitation while they were precipitating in the form of hydroxides at these pH values [54, 71]. According to Table 5 where the optimum pHs are presented for both electrodes, Fe and Al electrodes show similar performances with all connection modes on the removal of color and turbidity. Additionally, almost equal amounts of sludge revealed. However, MP-S and BP-S systems exhibit high consumptions as the consequence of the serial connection requiring higher potential. When electrode consumptions are compared, more electrode material is consumed in iron case than that of aluminium. The lowest total operating cost was gained with MP-P mode as expected.

5.1.2. Effects of current density

Current density is the most important parameter for controlling the reaction rate within the reactor in all electrochemical processes. It is well known that the magnitude of current density determines the amount of Al or Fe ions released from the electrodes and the formation rate of $Me(OH)_n$ (coagulant production rate) [92, 93] and adjusts the rate and size of the bubble production, and hence affects the growth of flocs [3, 94, 95]. In this research, all the experiments were applied under pHs 5 and 7 for Al and Fe electrodes, respectively with 60 min of electrolysis time to examine current density effects. Fig. 10 depicts the current density effects on COD removal. The removal percentages reach maximum at 40-50 mA/cm^2 and stay constant or decrease at higher current densities. Increment in current density raised the formation of hydroxide flocs and promote the removal efficiency by coagulation but meanwhile it causes an increase in energy consumption. However, after a certain value of current density, cathode passivation occurred and dissolution of anode material stopped or reduced and also floc formation and removal efficiencies decreased. The turbidity and color removal efficiencies were also came in sight the same trend with current density as COD removal efficiency. Because the lower current density with higher removal efficiencies is preferable, depending on the results, 40 mA/cm^2 was chosen as the approppriate current density value for the following time experiments. The other results belong to this value are shown in Table 6.

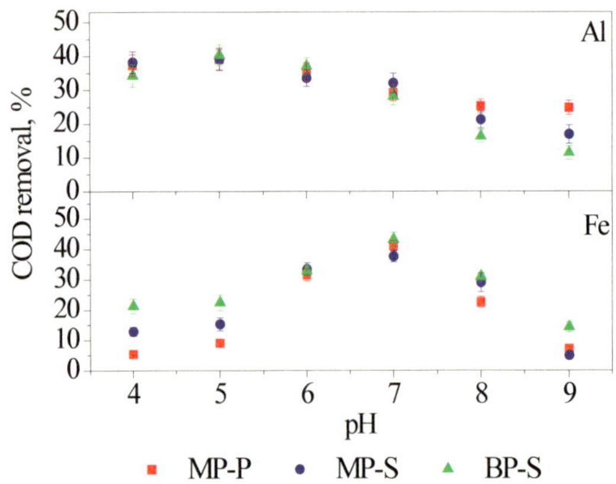

Figure 9. Effects of initial pH on removal efficiencies for Fe and Al electrodes with different electrode arrangements (current density of 40 mA/cm^2; operating time of 60 min).

	Al			Fe		
Parameter	MP-P	MP-S	BP-S	MP-P	MP-S	BP-S
Initial pH	5	5	5	7	7	7
Current density, mA/cm^2	40	40	40	40	40	40
Operating time, min	60	60	60	60	60	60
Initial voltage, V	28	84	85	28	84	85
Final voltage, V	29	85	87	29	85	87
Initial COD, mg/L	20,400	20,400	20,400	20,400	20,400	20,400
Initial color, Pt-Co	5,300	5,300	5,300	5,300	5,300	5,300
Initial turbidity, NTU	1,600	1,600	1,600	1,600	1,600	1,600
COD removal, %	39	39.1	40.3	41	37.6	43.1
Color removal, %	70	69.1	71	73.2	69.1	73.2
Turbidity removal, %	86.3	85.2	82.1	81	80.6	81
Energy consumption, kWh/kg COD	6.83	20.19	19.93	6.49	20.99	18.64
Energy consumption, KWh/m^3	36.20	107.34	109.25	36.20	107.34	109.25

http://dx.doi.org/10.5772/58580

		Al			Fe	
Energy cost, $/kg COD	1.16	2.35	3.39	1.16	3.57	3.17
Energy cost, $/m^3	6.15	18.25	18.57	6.15	18.25	18.57
Faraday (charge loading), coulomb/m^3	47.40	47.40	47.40	47.40	47.40	47.40
Electrode consumption, kg Al or Fe/m^3	0.40	0.45	0.45	0.88	0.89	0.87
Electrode cost, $/m^3	0.20	0.23	0.23	0.44	0.45	0.43
Sludge formation, kg/m^3	9.86	9.93	10.17	10.73	10.06	11.16
Sludge formation, kg/kg COD removed	1.86	1.87	1.87	1.92	1.97	1.90
Sludge disposal cost, $/kg	0.12	0.12	0.13	0.13	0.12	0.13
Operating cost, $/kg COD removed	0.83	2.35	2.32	0.82	2.47	2.19
Total operating cost, $/m^3	6.62	18.74	19.07	6.86	18.96	19.28

Table 5. Optimum results obtained from pH experiments.

5.1.3. *Effects of operating time*

Operating time is another important factor in EC process, which is necessary to provide sufficient current applied to the electrodes where the metal ions generated by the dissolution to form metal hydroxide species. Therefore reasonable electrolysis times should be applied in the EC reactor. To investigate the effects of operating time on the EC, optimum parameters obtained from the former pH and current density experiments were used: pHs 5 and 7 for Al and Fe, respectively; 40 mA/cm^2 of current density. Influence of the operating time on COD removal and the other results at the appropriate electrolysis time are presented in Fig. 11 and Table 7 respectively.

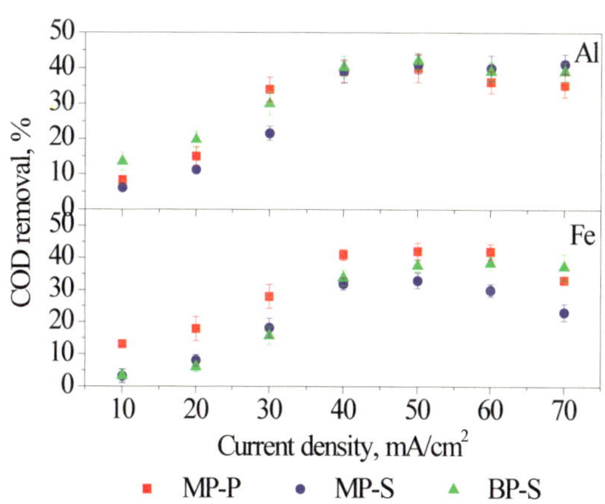

Figure 10. Effects of current density on removal efficiencies for Fe and Al electrodes with different electrode arrangements (initial pH of 5 for Al and 7 for Fe; operating time of 60 min).

	Al			Fe		
Parameter	MP-P	MP-S	BP-S	MP-P	MP-S	BP-S
Initial pH	5	5	5	7	7	7
Current density, mA/cm^2	40	40	40	40	40	40
Operating time, min	60	60	60	60	60	60
Initial voltage, V	28	84	85	28	84	85
Final voltage, V	29	85	87	29	85	87
Initial COD, mg/L	20,400	20,400	20,400	20,400	20,400	20,400
Initial color, Pt-Co	5,300	5,300	5,300	5,300	5,300	5,300
Initial turbidity, NTU	1,600	1,600	1,600	1,600	1,600	1,600
COD removal, %	39	39.1	40.3	41	37.6	43.1
Color removal, %	70	69.1	71	73.2	69.1	73.2
Turbidity removal, %	86.3	85.2	82.1	81	80.6	81
Energy consumption, kWh/kg COD	6.83	20.19	19.93	6.49	20.99	18.64
Energy consumption, KWh/m^3	36.20	107.34	109.25	36.20	107.34	109.25
Energy cost, $/kg COD	1.16	2.35	3.39	1.16	3.57	3.17
Energy cost, $/m^3	6.15	18.25	18.57	6.15	18.25	18.57
Faraday (charge loading), coulomb/m^3	47.40	47.40	47.40	47.40	47.40	47.40
Electrode consumption, kg Al or Fe/m^3	0.40	0.45	0.45	0.88	0.89	0.87
Electrode cost, $/m^3	0.20	0.23	0.23	0.44	0.45	0.43
Sludge formation, kg/m^3	9.86	9.93	10.17	10.73	10.06	11.16
Sludge formation, kg/kg COD removed	1.86	1.87	1.87	1.92	1.97	1.90
Sludge disposal cost, $/kg	0.12	0.12	0.13	0.13	0.12	0.13
Operating cost, $/kg COD removed	0.83	2.35	2.32	0.82	2.47	2.19
Total operating cost, $/m^3	6.62	18.74	19.07	6.86	18.96	19.28

Table 6. Optimum results obtained from current density experiments.

As seen from Fig.11, COD removal efficiencies of both electrodes in three connection modes increases until a certain operating time value, then, remain steady or decrease. Here, two explanation may be done: Firstly, for an electrolysis time beyond the optimum electrolysis time, the pollutant removal efficiency does not increase as sufficient numbers of flocs are available for the removal of the pollutant [1], secondly, removal efficiency does not increase, on the contrary, it decreases due to the anodic passivation and cathodic polarization which can impede the performance of EC [9, 96]. This situation is also valid in the case of the current density experiments. The optimum operating times (chosen as 90 and 105 min for Fe and Al, respectively), at which the maximum pollutant removals and minimum energy/electrode consumptions were gained, is presented with experimental results in Table 7. Because the removal efficiencies in the Table are close to each other, MP-P connection mode with minimum total operating cost is preferred for the follow-up experiments of the comparison of DC and APC.

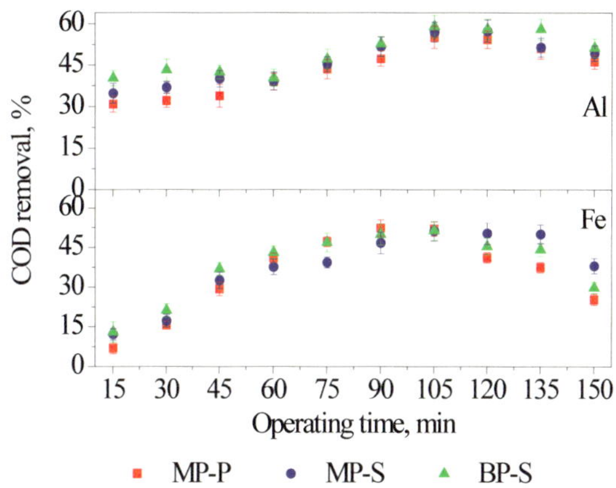

Figure 11. Effects of operating time on removal efficiencies for Fe and Al electrodes with different electrode arrangements (initial pH of 5 for Al and 7 for Fe; current density of 40 mA/cm²).

	Al			Fe		
Parameter	MP-P	MP-S	BP-S	MP-P	MP-S	BP-S
Initial pH	5	5	5	7	7	7
Current density, mA/cm^2	40	40	40	40	40	40
Operating time, min	105	105	105	90	90	90
Initial voltage, V	28	84	85	28	84	86
Final voltage, V	29	85	88	29	85	90

	Al			Fe		
Initial COD, mg/L	20,400	20,400	20,400	20,400	20,400	20,400
Initial color, Pt-Co	5,300	5,300	5,300	5,300	5,300	5,300
Initial turbidity, NTU	1,600	1,600	1,600	1,600	1,600	1,600
COD removal, %	55.1	57	59	52.4	46.7	49.9
Color removal, %	96.3	92.1	94.7	81.4	76.7	81.1
Turbidity removal, %	97.7	97	95.4	90.8	88.2	91.5
Energy consumption, kWh/kg COD	8.45	24.23	23.97	7.62	25.35	24.71
Energy consumption, KWh/m^3	63.36	187.85	192.29	54.30	161.01	167.68
Energy cost, $/kg COD	1.44	4.12	4.07	1.30	4.31	4.20
Energy cost, $/m^3	10.77	31.93	32.69	9.23	27.37	28.51
Faraday (charge loading), coulomb/m^3	82.94	82.94	82.94	71.10	71.10	71.10
Electrode consumption, kg Al or Fe/m^3	0.75	0.75	0.77	1.33	1.30	1.30
Electrode cost, $/m^3	0.37	0.37	0.39	0.66	0.65	0.65
Sludge formation, kg/m^3	13.49	13.87	14.31	13.52	12.33	12.98
Sludge formation, kg/kg COD removed	1.80	1.79	1.78	1.90	1.94	1.91
Sludge disposal cost, $/kg	0.16	0.17	0.17	0.16	0.15	0.16
Operating cost, $/kg COD removed	1.02	2.81	2.77	0.95	2.97	2.89
Total operating cost, $/m^3	11.45	32.62	33.39	10.20	28.31	29.46

Table 7. Optimum results obtained from operating time experiments.

5.2. Comparison of DC & APC systems

As encountered as a problem, cathode passivation, in the current density and operating time experiments, it may be described schematically as given in Fig. 12 similar to the results given in Table 8. To compare DC and APC system, MP-P connection mode selected for both electrode material. Current density and operating time experiments were repeated with time relay integrated with DC power supply to generate polarization between anodes and cathodes at certain intervals. Time relay was set to 300 Hz^{-1} [9] meaning anode–cathode polarization period of 10 min. The results are depicted in Figs. 13 and 14 for current density and operating time, respectively. As seen in the figures, removal efficiencies are not stopped or decreased at or after a certain current density or time value, moreover, at the same values APC sounds superior to DC. To make a local comparison between APC and DC, the results of APC are presented at optimum experimental conditions that formerly determined for DC system employed Fe or

http://dx.doi.org/10.5772/58580

Al electrodes in MP-P connection mode in Table 8. Moreover, few reports in literature about power supply effects on EC process are presented with overall results in Table 9.

	Al		Fe	
Parameter	DC	APC	DC	APC
Initial pH	5	5	7	7
Current density, mA/cm^2	40	40	40	40
Operating time, min	105	105	90	90
Initial voltage, V	28	28	28	28
Final voltage, V	29	29	29	29
Initial COD, mg/L	20,400	20,400	20,400	20,400
Initial color, Pt-Co	5,300	5,300	5,300	5,300
Initial turbidity, NTU	1,600	1,600	1,600	1,600
COD removal, %	55.1	77	52.4	75
Color removal, %	96.3	99	81.4	99
Turbidity removal, %	97.7	99	90.8	99
Energy consumption, kWh/kg COD	8.45	6.05	7.62	5.32
Energy consumption, KWh/m^3	63.36	63.36	54.30	54.30
Energy cost, $/kg COD	1.44	1.03	1.30	0.91
Energy cost, $\$/m^3$	10.77	10.77	9.23	9.23
Faraday (charge loading), $coulomb/m^3$	82.94	82.94	71.10	71.10
Electrode consumption, kg Al or Fe/m^3	0.75	0.85	1.33	1.36
Electrode cost, $\$/m^3$	0.37	0.43	0.66	0.68
Sludge formation, kg/m^3	13.49	18.10	13.52	18.16
Sludge formation, kg/kg COD removed	1.80	1.72	1.90	1.78
Sludge disposal cost, $/kg	0.16	0.22	0.16	0.22
Operating cost, $/kg COD removed	1.02	0.74	0.95	0.67
Total operating cost, $\$/m^3$	11.45	11.55	10.20	10.27

Table 8. Comparison of DC and APC at the same experimental conditions.

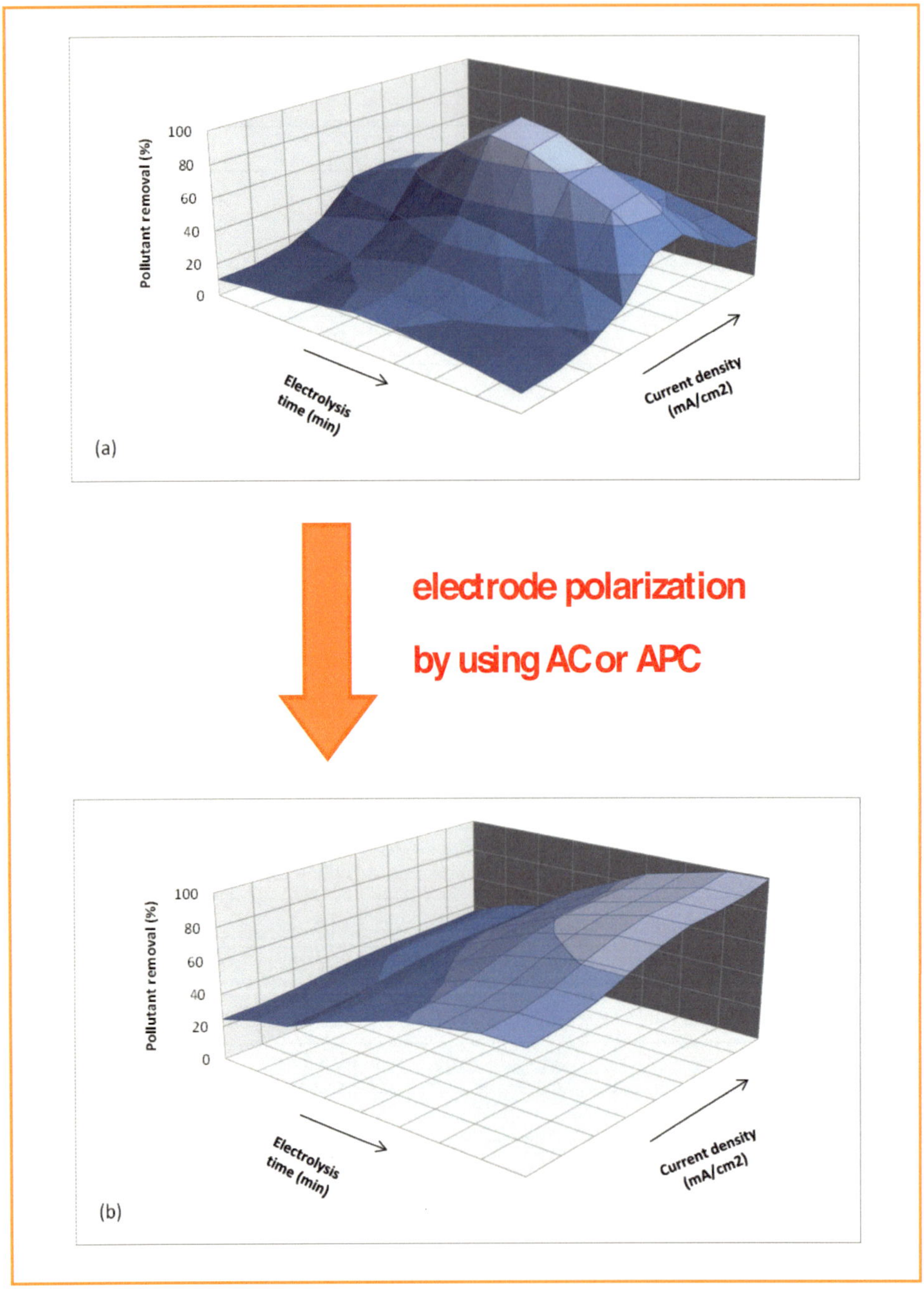

Figure 12. (a) General trend of pollutant removal efficiency changing with respect to current density and operating time in EC using DC power supply (one-way current), (b) Improved removal efficiency behaviour with respect to current density and operating time in EC using AC power supply or APC (two-way current).

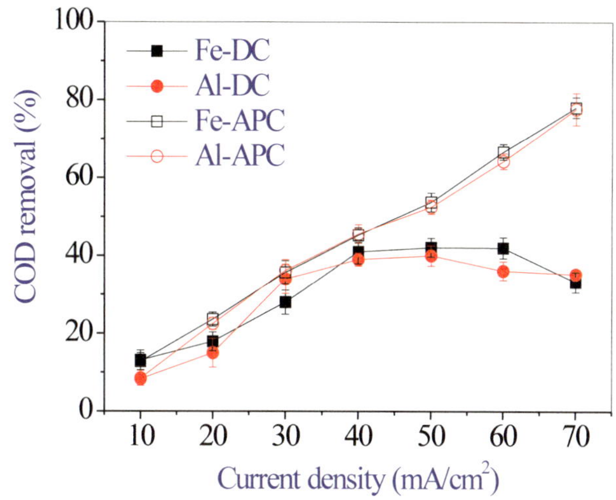

Figure 13. Comparison of DC and APC for two electrode materials in view of COD removal performances at different current densities.

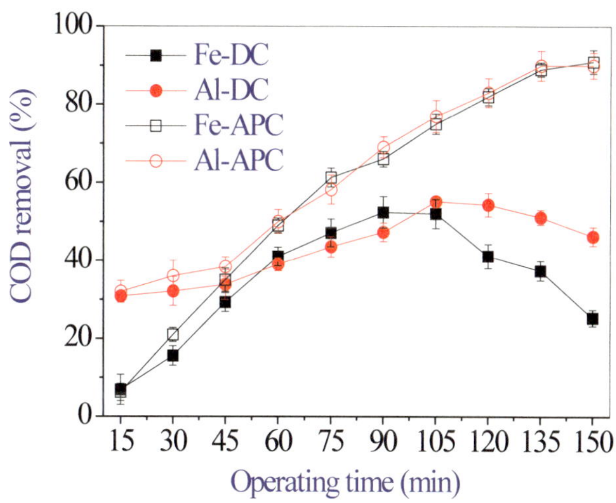

Figure 14. Comparison of DC and APC for two electrode materials in view of COD removal performances at different operating times.

Pollutants	Electrode material	Operational parameters investigated	Summary of the work	Ref.
Cr (VI)	Aluminium, iron	Initial pH, current density, reaction time, initial Cr (IV) concentrations, solution conductivity, electrical energy consumption, type of circuit	The APC mode was found to be more efficient than the DC mode with a lower anode over-voltage, slower anode polarization and passivity, and lower tank voltage. The operating time is 3–25% less when APC mode is used. The APC mode minimizes waste and increases sludge stability. In the APC mode, the "water recovery" was very significant, measuring as high as 0.92 m^3/m^3 wastewater. For DC mode, the water recovery was less than 0.5 m^3/m^3 of wastewater.	[11]
Cadmium	Aluminium	Initial pH, current density, initial Cd concentrations, effect of coexisting ions	The optimized removal efficiency of 97.5% and 96.2% was achieved for AC and DC source at a current density of 0.2 A/dm^2 and pH of 7.0 using aluminum alloy as anode and cathode. For both AC and DC electrolysis the adsorption of cadmium preferably fitting Langmuir adsorption isotherm better than Freundlich isotherm.	[10]
Cadmium	Zinc	Initial cadmium ion concentration, initial pH, current density and temperature.	The optimum removal efficiency of cadmium is 97.8% and 96.9% with the energy consumption of 0.665 and 1.236 kWh/m^3 was achieved for AC and DC source at a current density of 0.2 A/dm^2 and pH of 7.0 For both AC and DC electrolysis the adsorption of cadmium preferably fitting Langmuir adsorption isotherm.	[97]
Synthetic Methyl Orange wastewater	Aluminium	Initial pH, initial MO concentration, solution conductivity	Electrocoagulation with periodic electrode reversal (PREC) can effectively retard cathodic polarization and anodic passivation. Decolorization of MO wastewater is described well by a first-order reaction equation. The rate constant was fitted to be 0.183 min−1 for PREC, an increase of 20% compared to the EC.	[12]
Dianix Yellow CC, Procion Yellow dyes	Aluminium	Initial pH, current density, operating time, frequency of anode-cathode polarization.	Higher removal efficiencies of TOC and dye can be acquired in shorter operation times by using APC system. Removal efficiencies increase in APC system after optimum operation time belongs to DC system as well.	[9]

Table 9. A brief literature review of some studies on power supply effects on EC performance

4. Conclusions

The EC technique has gained a remarkable attention in the wastewater treatment applications due to its benefits including environmental compatibility, versatility, energy efficiency, safety, selectivity, amenability to automation, and cost effectiveness. The EC process contains an in-situ generation of metal hydroxide ions by electrolytic oxidation of the sacrificial anode. These metal hydroxide ions act as coagulant and remove the pollutants from the solution by sedimentation. Majority of the studies reported in the literature have traditionally used DC in the EC process. However, the traditional process also has the serious disadvantages of cathodic passivation which can impede the electrolytic process in a continuous operation. The use of DC leads to the corrosion formation on the anode due to oxidation. An oxidation layer also form on the cathode reducing the flow of current between the cathode and the anode and thereby lowering the pollutant removal efficiency and highering the operational cost. Therefore, in this research an almost new method, APC electrocoagulation was used to overcome the cathode passivation. A real high strength industrial wastewater, winery wastewater, was selected as the model electrolytic solution. In the EC reactor, aluminium and iron were used as sacrificial electrodes separately in three different connection modes namely, monopolar – parallel,,monopolar-serial, and bipolar serial connections. DC was obtained from a DC power supply operated at galvanostatic mode while APC was obtained by a time relay device integrated with the DC power supply. COD, color, and turbidity removal efficiencies were considered when DC and APC were compared technically. Furthermore, various cost items were used to calculate the total operation cost of both DC and APC systems by means of pollutant removals. A comprehensive literature survey from numerous references is also stated on EC and power supply effects. According to the experimental results, the following conclusions may be exposed:

- Higher removal efficiencies can be acquired in both same and shorter operation times by using APC system. In the same operating time, conditions, APC provide 40% more COD removal than DC. Similarly, APC reach 30 % more faster to DC' s COD removal performance. Thus, it can be said that anode–cathode polarization reduces the reaction time which is necessary for metal hydroxides removing the pollutants.

- COD, turbidity and color removal efficiencies increase until a certain current density and operation time and then they decrease so long as DC system goes on working. It may be due to the cathode passivization arisen from accumulation of contaminants on the cathode material. Therefore, electrode surfaces are needed to be cleaned and then put into use again. However, removal efficiencies increase in APC system after optimum operation time belongs to DC system as well. Thus, APC system can prolong the electrode life in each batch round of EC process.

- ACP can be easily obtained by a simple time relay device from the existing DC power supply and can be used in EC applications.

- ACP provides regular polarization to each electrode in the EC reactor, so, the sacrificial electrodes could be consumed in reasonable similar times.

- Fill-and-draw periods of reactor could be easily increased for batch EC processes by using time relay to eliminate cathode passivization. An increasing in fill-and-draw periods is important to decrease operating costs for batch processes.
- According to the results of the study, color and turbidity can be removed successfully from winery wastewaters but remained COD concentration is still too high for discharge. So, EC process should be applied with other treatment technologies such as anaerobic treatment that can remove the high COD concentrations.

Based on the promising results achieved in this research, different electrode materials can be used together by changing the anode-cathode polarization; ACP system can be also evaluated for different wastewater types or electrolytic solutions in further researches.

Author details

Murat Eyvaz*, Ercan Gürbulak, Serdar Kara and Ebubekir Yüksel

*Address all correspondence to: meyvaz@gyte.edu.tr

Gebze Institute of Technology, Environmental Engineering Department, Gebze-Kocaeli, Turkey

References

[1] Khandegar V, Saroha AK Electrochemical Treatment of Distillery Spent Wash Using Aluminum and Iron Electrodes Chinese Journal of Chemical Engineering 2013; 20(3) 439-443, ISSN 1004-9541

[2] Bayramoglu M, Eyvaz M, Kobya M Treatment of the textile wastewater by electrocoagulation: economic evaluation Chemical Engineering Journal 2007; 128 155–161, ISSN 1385-8947

[3] Mollah MYA, Morkovsky P, Gomes JAG, Kesmez M, Parga J, Cocke DL Fundamentals, present and future perspectives of electrocoagulation Journal of Hazardous Material B 2004; 114 199–210, ISSN: 0304-3894

[4] Mollah MYA, Schennach R, Parga JR, Cocke DL Electrocoagulation EC-science and applications Journal of Hazardous Material B 2001; 84 29–41, ISSN: 0304-3894

[5] Rajeshwar K, Ibanez JG, Swain GM Electrochemistry and the environment by regarding some operational parameters on the process efficiency Journal of Applied Electrochemistry 1994; 24 (11) 1077-1091, ISSN 1572-8838

[6] Avsar Y, Kurt U, Gonullu T Comparison of classical chemical and electrochemical processes for treating rose processing wastewater J Hazard Mater 2007; 148 340–345

[7] Parekh BK. The role of hydrolyzed metal ions in charge reversal and flocculation phenomena, PhD Dissertation, Pennsylvania State University, State College, PA. 1979.

[8] Mao X, Hong S, Hua Z Alternating pulse current in electrocoagulation for wastewater treatment to prevent the passivation of Al electrodes, J Wuhan Univ Technol Mater Sci Ed, 2008; 239–241, ISSN 1993-0437

[9] Eyvaz M, Kirlaroglu M, Aktas TS, Yuksel E The effects of alternating current electrocoagulation on dye removal from aqueous solutions Chemical Engineering Journal 2009; 153 (1–3) 16-22 ISSN: 1385-8947

[10] Vasudevan S, Lakshmi J, Sozhan G Effects of alternating and direct current in electrocoagulation process on the removal of cadmium from water Journal of Hazardous Materials 2011; 192 (1) 26–34 ISSN: 0304-3894

[11] Keshmirizadeh E, Yousefi S, Rofouei MK An investigation on the new operational parameter effective in CrVI removal efficiency: A study on electrocoagulation by alternating pulse current Journal of Hazardous Materials 2011; 190 (1–3) 119–124 ISSN: 0304-3894

[12] Pi KW, Xiao Q, Zhang HQ, XiaM, Gerson AR Decolorization of synthetic Methyl Orange wastewater by electrocoagulation with periodic reversal of electrodes and optimization by RSM Process Safety and Environmental Protection 2014; In Press Accepted Manuscript http://dxdoiorg/101016/jpsep201402008 ISSN: 0957-5820

[13] Vepsäläinen M. Electrocoagulation in the treatment of industrial waters and wastewaters. PhD Thesis. Julkaisija – Utgivare. 2012.

[14] Zaroual Z, Azzi M, Saib N, Chainet E Contribution to the study of electrocoagulation mechanism in basic textile effluent Journal of Hazardous Material 2006; 131 73-78 ISSN 0304-3894

[15] Daneshvar N, Oladegaragoze A, Djafarzadeh N Decolorization of basic dye solutions by electrocoagulation: An investigation of the effect of operational parameters Journal of Hazardous Material 2006; 129 116-122 ISSN: 0304-3894

[16] Dubrawski K L, Fauvel M, Mohseni M Metal type and natural organic matter source for direct filtration electrocoagulation of drinking water Journal of Hazardous Materials 2013; 244–245 135-141 ISSN: 0304-3894

[17] Ibanez J G, Singh MM, Szafran Z Laboratory experiments on electrochemical remediation of the environment Part 4 Color removal of simulated wastewater by electrocoagulation–electrocflotation Journal of Chemical Education 1998; 75 1040–1041 ISSN 0021-9584

[18] Daneshvar N, Sorkhabi HA, Tizpar A Decolorization of orange II by electrocoagulation method Seperation and Purification Technology 2003; 31 153–162 ISSN 1383-5866

[19] APHA AWWA and WEF Standard Methods for the Examination of Water and Wastewater 21st ed American Public Health Association Washington DC 2005

[20] Kobya M, Delipinar S Treatment of the baker's yeast wastewater by electrocoagulation Journal of Hazardous Materials 2008; 154 1133–1140 ISSN: 0304-3894

[21] Kobya M, Hiz H, Senturk E, Aydiner C, Demirbas E Treatment of potato chips manufacturing wastewater by electrocoagulation Desalination 2006; 190 201–211 ISSN 0011-9164

[22] Ponselvan F I A, Kumar M, Malviya JR, Srivastava VC, Mall ID Electrocoagulation studies on treatment of biodigester effluent using aluminum electrodes Water Air & Soil Pollution 2009; 199 371–379 ISSN 1573-2932

[23] Cerqueira A A and da Costa Marques M R Electrolytic Treatment of Wastewater in the Oil Industry. New Technologies in the Oil and Gas Industry, 2012. ISBN: 978-953-51-0825-2 InTech

[24] Vasudevan S, Kannan B S, Lakshmi J, Mohanraj S, Sozhan G Effects of alternating and direct current in electrocoagulation process on the removal of fluoride from water Journal of Chemical Technology and Biotechnology 2011; 86 428–436 ISSN 1097-4660

[25] Wang L K, Hung Y, Shammas NK Advanced Physicochemical Treatment Technologies Humana Press USA 2007; 57–106 ISBN 978-1-59745-173-4

[26] Mosteo R, Ormad MP, Ovelleiro J L Photo-Fenton processes assisted by solar light used as preliminary step to biological treatment applied to winery wastewaters Water Science and Technology 2007; 56 89–94 ISSN 0273-1223

[27] Lucas MS, Peres JA, Puma GL Treatment of winery wastewater by ozone-based advanced oxidation processes O_3, O_3 /UV and O_3 /UV/H_2O_2 in a pilot-scale bubble column reactor and process economics Seperation and Purification Technology 2010; 72 235–241 ISSN: 1383-5866

[28] Mulidzi AR Winery and distillery wastewater treatment by constructed wetland with shorter retention time Water Science and Technology 2010; 61 2611–2615 ISSN 0273-1223

[29] Valderrama C, Ribera G, Bahí N, Rovira M, Giménez T, Nomen R, Lluch S, Yuste M, Martinez-Lladó X Winery wastewater treatment for water reuse purpose: Conventional activated sludge versus membrane bioreactor MBR: A comparative case study Desalination 2012; 306 15 1-7 ISSN 0011-9164

[30] Bustamante M, Paredes C, Moral R, Moreno-Caselles J, Pérez-Espinosa A, Pérez-Murcia M D Uses of winery and distillery effluents in agriculture: characterisation of

nutrient and hazardous components Water Science and Technology 2005; 51 (1) 145-51 ISSN 0273-1223

[31] Kumar A, Kookana R. Impact of winery wastewater on ecosystem health – an introductory assessment, Final Report Grape and Wine Research and Development Corporation Adelaide, Australia.2006.

[32] Mosse K, Patti A, Christen E, Cavagnaro T Review: winery wastewater quality and treatment options in Australia Australian Journal of Grape and Wine Research 2011, 17 111–122 ISSN 1755-0238

[33] van Schoor L H Guidelines for the management of wastewater and solid waste at existing wineries Winetech, 2005, 35

[34] Moletta R Biological treatment of wineries and distillery wastewater Proceedings of 5th International Specialized Conference on Sustainable Viticulture Winery Wastes and Ecological Impact Management 2009; 389-398 ISBN 978-88-8443-284-1

[35] Oliveira M, Duarte E Winery Wastewater Treatment-Evaluation of the Air Micro-Bubble Bioreactor Performance Mass Transfer-Advanced Aspects Dr Hironori Nakajima Ed 2011. ISBN: 978-953-307-636-2 http://wwwintechopencom/books/mass-transfer-advanced-aspects/winery-wastewater-treatment-evaluation-of-the-air-micro-bubble-bioreactor-performance

[36] Masi F, Conte G, Martinuzzi N, Pucci, B. Winery high organic contentwastewater treated by constructed wetlands in mediterranean climate in: Conference Proceedings, 2002, IWA 8thInternational, 1 274–282

[37] Rytwo G, Rettig A, Gonen Y Organo-sepiolite particles for efficient pretreatment of organic wastewater: Application to winery effluents Applied Clay Science 2011; 51 (3) 390-394 ISSN 1872-9053

[38] Brito AG, Peixoto J, Oliveira JM, Oliveira JO, Costa C, NogueiraR, Rodrigues A Brewery and winery wastewater treatment: some focal points of design and operation In: Oreopoulou Utilization of By-Products and Treatment of Waste in the Food Industry 2006; 3 109–131

[39] Souza B S, Moreira FC, Dezotti M W C, Vilar V J P, Boaventura R A R Application of biological oxidation and solar driven advanced oxidation processes to remediation of winery wastewater Catalysis Today 2013; 209 201-208 ISSN 0920-5861

[40] Chapman J, Baker P, Wills S Winery Wastewater Handbook: Production, Impacts and Management Winetitles Publishers Adelaide, South Australia 2001

[41] Sheehan GJ, Greenfeld PF Utilization, treatment and disposal of distillery wastewater Water Research 1980; 14 257–277 ISSN 0043-1354

[42] Vlissidis A, Zouboulis AJ Thermophilic anaerobic digestion of alcohol distillery wastewaters Bioresource Technology 1993; 43 131–140 ISSN 0960-8524

[43] Wilkie AC, Riedesel KJ, Owens JM Stillage characterization and anaerobic treatment of ethanol stillage from conventional and cellulosic feedstocks Biomass Bioenergy 2000; 19 63–102 ISSN 0961-9534

[44] Keyser M, Witthuhn RC, Ronquest LC, Britz TJ Treatment of winery ef?uent with upflow anaerobic sludge blanket UASB—granular sludges enriched with enterobacter sakazakii Biotechnology Letters 2003; 25 22 1893–1898 ISSN: 1573-6776

[45] Melamane XL, Strong PJ, Burgess JE Treatment of wine distillery wastewater: a review with emphasis on anaerobicmembrane reactors South African Journal for Enology and Viticulture 2007; 28 1 25–36 ISSN 0253-939X

[46] Kirzhner F, Zimmels Y, Shraiber Y Combined treatment of highly contaminated winery wastewater Separation and Purification Technology 2008; 631-1 38-44 ISSN 1383-5866

[47] Kara S, Gurbulak E, Eyvaz M and Yüksel E Treatment of Winery Wastewater by Electrocoagulation Process Desalination and Water Treatment 2013; 51 28-30 5421-5429 ISSN 1944-3986

[48] Jimenez AM, Rafael B, Martin A Aerobic–anaerobic biodegradation of beet molasses alcoholic fermentation wastewater Process Biochemistry 2003; 38 1275–1284 ISSN 1359-5113

[49] Yavuz Y EC and EF processes for the treatment of alcohol distillery wastewater Seperation and Purification Technology 2007; 53 1 135–140 ISSN 1383-5866

[50] Pena M Coca M Gonzalez G Rioja R Garcia MT Chemical oxidation of wastewater from molasses fermentation with ozone Chemosphere 2003; 51 893–900

[51] Thakur C, Srivastava V C, Mall I D Electrochemical treatment of a distillery wastewater: Parametric and residue disposal study Chemical Engineering Journal 2009; 148 2–3 496-505 ISSN 1385-8947

[52] Chen W J, Su W T, Hsu H Y Continuous flow electrocoagulation for MSG wastewater treatment using polymer coagulants via mixture-process design and response-surface methods Journal of the Taiwan Institute of Chemical Engineers 2012; 43 246–255 ISSN: 1876-1070

[53] Drogui P, Blais J F, Mercier G Review of electrochemical technologies for environmental applications Recent Patents on Engineering 2007; 1 3 257–272 ISSN: 1872-2121

[54] Drogui P, Asselin M, Brar S K, Benmoussa H, Blais J F Electrochemical removal of pollutants from agro-industry wastewaters Seperation and Purification Technology 2008; 61 301–310 ISSN 1383-5866

[55] Asselin M, Drogui P, Brar SK, Benmoussa H, Blais J F Organics removal in oily bilgewater by electrocoagulation process Journal of Hazardous Materials 2008; 151 2–3 446-455 ISSN: 0304-3894

[56] Asaithambi P, Susree M, Saravanathamizhan R, Matheswaran M Ozone assisted electrocoagulation for the treatment of distillery effluent Desalination 2012; 297 1-7 ISSN 0011-9164

[57] Kannan N, Karthikeyan G, Tamilselvan N Comparison of treatment potential of electrocoagulation of distillery effluent with and without activated Areca catechu nut carbon Journal of Hazardous Materials 2006; 137 3 1803-1809 ISSN 0304-3894

[58] Susree M, Asaithambi P, Saravanathamizhan R, Matheswaran M Studies on various mode of electrochemical reactor operation for the treatment of distillery effluent Journal of Environmental Chemical Engineering 2013; 1 3 552-558 ISSN 2213-3437

[59] Tchamango S, Nanseu-Njiki C P, Ngameni E, Hadjiev D, Darchen A Treatment of dairy effluents by electrocoagulation using aluminium electrodes Science of The Total Environment 2010; 4084 947-952 ISSN 0048-9697

[60] Şengil İ A, Özacar M Treatment of dairy wastewaters by electrocoagulation using mild steel electrodes Journal of Hazardous Materials 2006; 137 2 1197-1205 ISSN 0304-3894

[61] Ün U T, Ozel E Electrocoagulation of yogurt industry wastewater and the production of ceramic pigments from the sludge Separation and Purification Technology 2013; 120 386-391 ISSN 1383-5866

[62] Roa-Morales G, Campos-Medina E, Aguilera-Cotero J, Bilyeu B, Barrera-Díaz C Aluminum electrocoagulation with peroxide applied to wastewater from pasta and cookie processing Separation and Purification Technology 2007; 54 1 124-129 ISSN 1383-5866

[63] Bayar S, Yıldız Y Ş, YılmazA E, İrdemez Ş The effect of stirring speed and current density on removal efficiency of poultry slaughterhouse wastewater by electrocoagulation method Desalination 2011; 280 1–3 103-107 ISSN 0011-9164

[64] Bayramoglu M, KobyaM, Eyvaz M, Senturk E Technical and economic analysis of electrocoagulation for the treatment of poultry slaughterhouse wastewater Separation and Purification Technology 2006; 51 3 404-408 ISSN 1383-5866

[65] Kobya M, Senturk E, Bayramoglu M Treatment of poultry slaughterhouse wastewaters by electrocoagulation Journal of Hazardous Materials 2006; 133 1–3 172-176 ISSN 0304-3894

[66] Gengec E, Kobya M, Demirbas E, Akyol A, Oktor K Optimization of baker's yeast wastewater using response surface methodology by electrocoagulation Desalination 2012; 286 200-209 ISSN 0011-9164

[67] Orescanin V, Kollar R, Nad K, Mikelic I L, Gustek S F Treatment of winery wastewater by electrochemical methods and advanced oxidation processes Journal of environmental science and health Part A Toxic/hazardous substances & environmental engineering 2013; 48 12 1543-7 ISSN 1532-4117

[68] Katal R, Pahlavanzadeh V Influence of different combinations of aluminum and iron electrode on electrocoagulation efficiency: Application to the treatment of paper mill wastewater Desalination 2011; 265 1–3 199-205 ISSN 0011-9164

[69] Vepsäläinen M, Kivisaari H, Pulliainen M, Oikari A, SillanpääM Removal of toxic pollutants from pulp mill effluents by electrocoagulation Separation and Purification Technology 2011, 81 2 141-150 ISSN 1383-5866

[70] Zodi S, Louvet J N, Michon C, Potier O, Pons M N, Lapicque F, Leclerc J P Electrocoagulation as a tertiary treatment for paper mill wastewater: Removal of non-biodegradable organic pollution and arsenic Separation and Purification Technology 2011; 81 1 62-68 ISSN 1383-586

[71] Kobya M, Bayramoglu M and Eyvaz M Techno-economical evaluation of electrocoagulation for the textile wastewater using different electrode connections Journal of Hazardous Materials 2007, 148 1-2 311-318 ISSN: 0304-3894

[72] Yuksel E, Eyvaz M, Gurbulak E Electrochemical treatment of colour index reactive orange 84 and textile wastewater by using stainless steel and iron electrodes Environmental Progress & Sustainable Energy 2013; 32 1 60–68 ISSN 1944-7450

[73] Yilmaz A E, Boncukcuoğlu R, Kocakerim M M An empirical model for parameters affecting energy consumption in boron removal from boron-containing wastewaters by electrocoagulation Journal of Hazardous Materials 2007; 144 1–2 101-107 ISSN 0304-3894

[74] Sayiner G, Kandemirli F, Dimoglo A Evaluation of boron removal by electrocoagulation using iron and aluminum electrodes Desalination 2008; 230 1–3 205-212 ISSN 0011-9164

[75] Isa M H, Ezechi E H, Ahmed Z, Magram S F, Kutty S R Boron removal by electrocoagulation and recovery Water Research 2014; 51 113-123, ISSN 0043-1354

[76] Yetilmezsoy K, Ilhan F, Sapci-Zengin Z, Sakar S, Gonullu M T Decolorization and COD reduction of UASB pretreated poultry manure wastewater by electrocoagulation process: A post-treatment study Journal of Hazardous Materials 2009; 162 1 120-132 ISSN 0304-3894

[77] Labanowski J, Pallier V, Feuillade-Cathalifaud G Study of organic matter during coagulation and electrocoagulation processes: Application to a stabilized landfill leachate Journal of Hazardous Materials 2010; 179 1–3 166-172 ISSN: 0304-3894

[78] Li X, Song J, Guo J, Wang Z, Feng Q Landfill leachate treatment using electrocoagulation Procedia Environmental Sciences Part B 2011; 10 1159-1164 ISSN 1878-0296

[79] Ghosh D, Medhi CR, Purkait MK Treatment of fluoride containing drinking water by electrocoagulation using monopolar and bipolar electrode connections Chemosphere 2008; 73 9 1393-1400 ISSN 0045-6535

[80] Mólgora C C, Domínguez A M, Avila E M, Drogui P, Buelna G Removal of arsenic from drinking water: A comparative study between electrocoagulation-microfiltration and chemical coagulation-microfiltration processes Separation and Purification Technology 2013; 118 645-651 ISSN 1383-5866

[81] Ün Ü T, Uğur S, Koparal AS, Öğütveren Ü B Electrocoagulation of olive mill wastewaters Separation and Purification Technology 2006; 52 1 136-141 ISSN 1383-5866

[82] Khoufi S, Feki F, Sayadi S Detoxification of olive mill wastewater by electrocoagulation and sedimentation processes Journal of Hazardous Materials 2007; 142 1–2 58-67 ISSN 0304-3894

[83] Hanafi F, Assobhei O, Mountadar M Detoxification and discoloration of Moroccan olive mill wastewater by electrocoagulation Journal of Hazardous Materials 2010; 174 1–3 807-812 ISSN 0304-3894

[84] Fouad YO Separation of cottonseed oil from oil–water emulsions using electrocoagulation technique Alexandria Engineering Journal In Press 2013; Corrected Proof ISSN 1110-0168

[85] HarizI B, Halleb A, Adhoum N, Monser L Treatment of petroleum refinery sulfidic spent caustic wastes by electrocoagulation Separation and Purification Technology 2013; 107 150-157 ISSN 1383-5866

[86] Akbal F, Camcı S Copper, chromium and nickel removal from metal plating wastewater by electrocoagulation Desalination 2011; 269 1–3 214-222 ISSN 0011-9164

[87] Al Aji B, Yavuz Y, Koparal A S Electrocoagulation of heavy metals containing model wastewater using monopolar iron electrodes Separation and Purification Technology 2012; 86 248-254 ISSN 1383-5866

[88] Tong X, Xu R Removal of CuII from acidic electroplating effluent by biochars generated from crop straws Journal of Environmental Sciences 2013; 25 4 652-658 ISSN 1001-0742

[89] Eyvaz M. Treatment of textile wastewaters by using electrocoagulation with different electrode connections Fe and Al. MSc Thesis. In Turkish Kocaeli Turkey. 2006.

[90] Rebhun M, Lurie M Control of organic matter by coagulation and floc separation Water Science and Technology 1993; 27 1–20 ISSN 0273-1223

[91] Pykhteev O Y, Ofimov A A, Moskvin L N Hydrolysis of iron III aqua complexes Russian Journal of Applied Chemistry 1999; 72 9–20 ISSN: 1070-4272

[92] Pouet M F, and Grasmick A Urban wastewater treatment by electrocoagulation and flotation Water Science and Technology 1995; 31 275–283 ISSN 0273-1223

[93] Ün Ü T, Koparal A S, Öğütveren Ü B Hybrid processes for the treatment of cattle-slaughterhouse wastewater using aluminum and iron electrodes Journal of Hazardous Materials 2009; 164 580–586 ISSN: 0304-3894

[94] Chen G Electrochemical technologies in wastewater treatment Separation and Purification Technology 2004; 38 (1) 11-41 ISSN 1383-5866

[95] Bellebia S, Kacha S, Bouyakoub A Z, Derriche Z Experimental ınvestigation of chemical oxygen demand and turbidity removal from cardboard paper mill effluents using combined lectrocoagulation and adsorption processesi Environmental Progress & Sustainable Energy 2011; 31 (3) 361–370 ISSN 1944-7450

[96] Martínez-Huitle C A, Brillas E Decontamination of wastewaters containing synthetic organic dyes by electrochemical methods: a general review Applied Catalysis B: Environmental 2009; 87 105–145 ISSN 0926-3373

[97] Vasudevan S & Lakshmi J Effects of alternating and direct current in electrocoagulation process on the removal of cadmium from water – A novel approach Separation and Purification Technology 2011; 80 3 18 643–651 ISSN 1383-5866

[98] Dubrawski K L, Mohseni M Standardizing electrocoagulation reactor design: Iron electrodes for NOM removal Chemosphere 2013; 91 55–60 ISSN 0045-6535

Chapter 9

Role of Modern Localised Electrochemical Techniques to Evaluate the Corrosion on Heterogeneous Surfaces

R. Leiva-García, R. Sánchez-Tovar,
C. Escrivà-Cerdán and J. García-Antón

Additional information is available at the end of the chapter

http://dx.doi.org/10.5772/57204

1. Introduction

Corrosion is a great problem in a lot of chemical industries and in the systems of energy production. When metals are exposed to an aggressive environment or to atmosphere effects tend to reverse to the lesser energy state of ore. All these processes are enhanced by the aggressive conditions that sometimes can be found in the industry, high temperature and pressure or corrosive environments. Therefore, corrosion may imply a lot of economic costs in the productive system. According to NACE [1] in 2002 the cost of corrosion in USA could be established in 276 million of dollars per year. Therefore, it is quite important to develop and use techniques that can allow determining and monitoring the corrosion of the metals.

Since 1903, when the electrochemical character of corrosion was widely accepted after the publication of the paper of Whitney [2], several electrochemical techniques has been developed and used in the study of the corrosion processes. Some of the main milestones developed in the electrochemistry science during the 20th century can be summarised as follows:

1. Alexander Naumovich Frumkin (1895-1976). made vital contributions to our knowledge of the fundamentals of electrode reactions — particularly the influence of the electrode-electrolyte interface on the rate of electron transfer across it [3].

2. Veniamin Grigorevich Levich (1917-1987), an associate of theoretical physicist Lev Davidovich Landau (1908-1968), helped Frumkin in relating his experimental results to theory. The collaboration led to the development of the rotating disc electrode [4].

3. The quantitative measurement of electrochemical corrosion got established with a 1932 publication of Thomas Percy Hoare (1907-1978) and Ulick Richardson Evans (1889-1980).

INTECH
open science | open minds

© 2014 The Author(s). Licensee InTech. This chapter is distributed under the terms of the Creative Commons Attribution License (http://creativecommons.org/licenses/by/3.0), which permits unrestricted use, distribution, and reproduction in any medium, provided the original work is properly cited.

His 1937 book "Metallic Corrosion, Passivity, and Protection" is probably one of the clearest books written on corrosion science.

4. 1938, the Belgian electrochemist Marcel Pourbaix (1904-1998) constructed his famous potential–pH diagrams "Pourbaix diagrams".

5. Much of the theory behind cyclic voltammetry and electrochemical impedance spectroscopy came from the work of the English electrochemist John Edward Brough Randles (1912-1998).

6. One important development in electrochemical instrumentation is the invention of the potentiostat by the German engineer-physicist Hans Wenking (1923).

Nowadays, electrochemistry and corrosion science looks interconnect with parallel advancements in materials science and characterization techniques. For example, there is increasing thrust toward exploiting nanoscopic materials and architectures. Expectations in this direction are high because surface plays a key role in electrochemical processes. Continual developments in the synthesis and characterization of materials have also led to welcome changes in electrochemical research.

Therefore, the use of the electrochemical techniques has demonstrated being quite helpful in the study and control of corrosion in several environments. In this chapter a review of the conventional and localised electrochemical techniques applied to the analysis of corrosion processes has been made. Additionally, some experimental results of the authors, which are a good example of the use of localised electrochemical techniques in the study of heterogeneous materials, are presented.

2. Stainless steels

Stainless steel is a generic name commonly used for that entire group of iron-based metal which are the most commonly used metallic materials. The main characteristics are their resistance to corrosion in many environments, their good mechanical properties over an extremely wide range of temperatures and their resistance to oxidation and scaling at very high temperatures [5, 6]. These compounds contain at least a 12% of chromium, low carbon content and they might also have percentages of other elements such as nickel, molybdenum, vanadium, titanium, aluminium and so on. The corrosion resistance of stainless steels is provided by a very thin and protective surface oxide film, known as passive film. It is generally accepted that passive films formed on stainless steels have a duplex structure which consists of an inner region rich in chromium and an outer region rich in iron [7-10]. Nickel and molybdenum are commonly added in the stainless steel composition in order to increase its corrosion resistance.

There are five main classes of stainless steel: austenitic stainless steels, martensitic stainless steels, ferritic stainless steels, duplex stainless steels and precipitation hardening stainless steels.

2.1. Austenitic stainless steels

Austenitic stainless steels are the most common and familiar types of stainless steel. They are not magnetic. They have a face-centred-cubic lattice structure of austenite over the whole temperature range from room temperature (and below) to the melting point. The austenitic stainless steels, because of their high chromium and nickel content, are the most corrosion resistant of the stainless group. Moreover, they are weldable and can be divided into three groups: common chromium-nickel (300 series), manganese-chromium-nickel-nitrogen (200 series) and specialty alloys. The commonly used austenitic alloys contain chromium (17-27 wt. %), nickel (6-32 wt. %) and low carbon amounts (0.02-0.1 wt. %) in an iron matrix. Austenitic stainless steels are used for numerous industrial and consumer applications, such as in chemical plants, power plants, food processing equipment [11].

The most common austenitic grades are the 304 and the 316 ones [12]. Austenitic 304 grade contains approximately 18 wt. % of chromium and 8 wt. % of nickel contents. This kind of stainless steel is widely used for chemical processing equipment, for food, dairy, and beverage industries, for heat exchangers and for the milder chemicals. Type 316 is an austenitic chromium nickel stainless steel containing molybdenum. This addition increases general corrosion resistance, improves resistance to pitting from chloride ion solutions and provides increased strength at elevated temperatures [13]. Properties are similar to those of Type 304 except that this alloy is somewhat stronger at elevated temperatures. Corrosion resistance is improved, particularly against sulphuric, hydrochloric, acetic, formic and tartaric acids, acid sulphates and alkaline chlorides. Table 1 shows the typical composition of an austenitic 304 and 316 grade, respectively [14].

	Type 304	**Type 316**
Carbon	0.08 max.	0.08 max.
Manganese	2.00 max.	2.00 max.
Phosphorus	0.045 max.	0.045 max.
Sulfur	0.03 max.	0.03 max.
Silicon	0.75 max.	0.75 max.
Chromium	18.00-20.00	16.00-18.00
Nickel	8.00-12.00	10.00-14.00
Molybdenum	-	2.00-3.00
Nitrogen	0.10 max.	0.10 max.
Iron	Balance	Balance

Table 1. Composition of type 304 and 316 stainless steels, values are given in wt. %

2.2. Martensitic stainless steels

Martensitic stainless steels are a chromium-iron alloy containing 10.5 to 17 wt %. chromium and controlled amounts of carbon. They are ferromagnetic, hardenable by heat treating, and generally resistant to corrosion only in relatively mild environments. This kind of alloy is extremely strong and tough, as well as highly machinable. The martensitic grades are mainly used where hardness, strength, and wear resistance are required, i.e. they are specified when the application requires good tensile strength, creep and fatigue strength properties, in combination with moderate corrosion resistance and heat resistance up to approximately 650 °C [15].

The most commonly used alloy within the martensitic stainless steel family is type 410, which contains approximately 12 wt% Cr and 0.1 wt% C to provide strength. It is widely used where corrosion is not severe (air, water, some chemicals, and food acids). Typical applications include highly stressed parts needing the combination of strength and corrosion resistance such as fasteners.

2.3. Ferritic stainless steels

Ferritic stainless steel grades contain between 10.5 wt % and 27 wt % chromium and very little nickel, if any, but some types can contain lead. Most compositions include molybdenum; some, aluminium or titanium. They were developed to provide a group of stainless steel to resist corrosion and oxidation, while being highly resistant to stress corrosion cracking [16, 17]. These kind of alloys are magnetic but cannot be hardened or strengthened by heat treatment, as martensitic grades. They can be cold worked and softened by annealing. As a group, they are more corrosive resistant than the martensitic grades, but generally inferior to the austenitic grades. The basic ferritic grade is type 430, with a little less corrosion resistance than type 304. Type 430 combines high resistance to such corrosives as nitric acid, sulfur gases and many organic and food acids.

2.4. Duplex stainless steels

Duplex stainless steels have a mixed microstructure of austenite and ferrite and roughly twice the strength compared to austenitic stainless steels. They also have improved resistance to localized corrosion, particularly pitting, crevice corrosion and stress corrosion cracking. They are characterized by high chromium (19–32 wt %) and molybdenum (up to 5 wt %) and lower nickel contents than austenitic stainless steels. That is why they have lower cost [18, 19]. Duplex stainless steels are considered a very attractive structural material in the field of energy/environmental systems due to their good mechanical and corrosion resistance properties. However, the higher Cr and Mo content in duplex stainless steels can promote the precipitation of secondary phases, such as sigma phase [20, 21]. These phases, which are rich in alloying elements, appear when the steel is heated in a specific range of temperatures. During these heat treatments, the microstructure of duplex stainless steels undergoes morphological changes. The ratio of the ferrite and the austenite phase, that in the beginning is 1/1, changes with the heat treatment temperature. Duplex stainless steels are typically used in the main following areas: chemical processing, transport and storage; oil and gas exploration and offshore rigs; oil and gas refining; marine environments; pollution control equipment; pulp and paper manufacturing and chemical process plants.

2.5. Precipitation hardening stainless steels

Precipitation hardening grades are iron-nickel-chromium alloys containing one or more precipitation hardening elements such as aluminum, titanium, copper, niobium, and molybdenum. The two main characteristics of all precipitation-hardening stainless steels are high strength (at the expense of toughness) and high corrosion resistance (comparable to that of the standard 304 and 316 austenitic alloys) [22]. The aging treatments are designed to optimize strength, corrosion resistance, and toughness. On the other hand, to improve toughness, the amount of carbon is kept low.

3. Generation of heterogeneous surfaces in stainless steels

Welding is widely used in the fabrication, maintenance, and repair of many industrial parts and structures, especially for the joint of stainless steels. However, welding introduces residual stresses into the material and can cause cracking as the weld cools and contracts. Minor differences in composition and microstructure between a weld and a base zone can create an electrochemical potential between the zones and may cause galvanic corrosion [23]. In fact, corrosion in welds has always been a problem due to local variations in material composition and structure. When austenitic stainless steels are incorrectly heat treated, in the temperature range between 500 and 900 °C, chromium and carbon combine at the grain boundaries to form chromium carbides (typically $Cr_{23}C_6$), whilst simultaneously as these carbides form, chromium depletion occurs at the adjacent zones. This process is called sensitisation and leads to a decrease in the corrosion resistance of stainless steels, notably resistance to intergranular corrosion, being due to the depleted regions becoming anodic in the presence of an electrolyte [24-30]. These heat treatments generate heterogeneous surfaces on the stainless steels. The "L" grades are used to provide extra corrosion resistance after welding. The letter "L" after a stainless steel type indicates low carbon (as in 304L). The carbon is kept to 0.03 wt. % or under to avoid carbide precipitation [31]. The most common grade L stainless steel types are 304L and 316L. In this way, the welding procedure itself or the generation of sensitization zones in stainless steels due to welding or heat treatments create heterogeneities in the alloy, which might lead to corrosion problems or contribute to aggravate them. For this reason, the electrochemical study of these materials is a very important issue. All these corrosion mechanisms have mainly been studied using 'macro' experiments (working electrode in the $mm^2 - cm^2$ range) [32, 33]. However, such processes are due to mechanisms on a smaller scale. Therefore, it is beneficial to develop appropriate 'micro' experimental devices that require a reduced working electrode surface area [34]. In this context, microelectrochemical methods are powerful techniques to study localised corrosion processes on small areas of passive metals, microgalvanic cells, etc [35-37].

4. Conventional techniques in the study of stainless steels

Corrosion science and engineering have benefit tremendously from the use of electrochemical methods that can probe the thermodynamic and kinetic aspects of corrosion, including the rate of corrosion. These methods have proved of great utility to corrosion engineers and scientists in predicting the performance of materials and devising corrosion mitigation strategies, understanding the effects of changes in process and environment conditions, and assessing the accuracy of corrosion monitoring techniques [38]. Some of the most common techniques used in research studies are described in the following sections.

DC electrochemistry, and in particular, the potentiodynamic polarisation scan, is a rapid method of obtaining an insight into the corrosivity of a particular environment for a steel. Through the DC polarisation technique, information on the corrosion rate, pitting susceptibility, passivity, as well as the cathodic behaviour of an electrochemical system may be obtained.

In a potentiodynamic experiment, the current represents the rate with which the anodic and cathodic reactions are taking place on the working electrode. Typically, the current is expressed in terms of the current per unit area of the working electrode, or the current density. Numerous variables will influence the rate of a given electrochemical reaction, including the temperature, the surface condition of the surface being interrogated, as well as the chemical environment in which the experiment is performed. In this experiment, the driving force (i.e., the potential) for anodic and cathodic reactions is controlled, and the net change in the reaction rate (i.e., current) is observed.

The reaction rate may be controlled by two different kinetic phenomena. The first is charge transfer or activation control. In this case, the reaction is controlled by the size of the driving force available (e.g., hydrogen evolution reaction or water reduction reaction). As the driving force increases, so does the reaction rate. The other mechanism which may control the reaction rate is mass transfer. In this case, the reaction rate is controlled by mass transfer through the electrolyte to the electrode surface (e.g., oxygen reduction reaction). Since the reaction rate is controlled by diffusion, it cannot increase indefinitely as the driving force increases. Instead, the current reaches a maximum, or limiting, current density which is itself a function of the concentration of the species of interest in the solution as well as its diffusivity. Once the rate for a particular reaction has reached its limiting value, further increases in driving force will not result in any additional increase of the reaction rate [39, 40].

During the potentiodynamic polarisation scan, the potential of the electrode is controlled, while the current is determined as a function of time. The opposite case, where the current is controlled (frequently held constant), and the potential becomes the dependent variable, which is determined as a function of time. The experiment is carried out by applying the controlled current between the working and auxiliary electrodes with a current source (called a *galvanostat*) and recording the potential between the working and reference electrodes (e.g., with a recorder, oscilloscope, or other data acquisition device) (Figure 1). These techniques are generally called *chronopotentiometric techniques*, because E is determined as a function of time, or galvanostatic techniques, because a small constant current is applied to the working electrode [39].

Another DC technique is the ZRA (Zero resistance ammeter), this device is a current to voltage converter that produces a voltage output proportional to the current flowing between its to

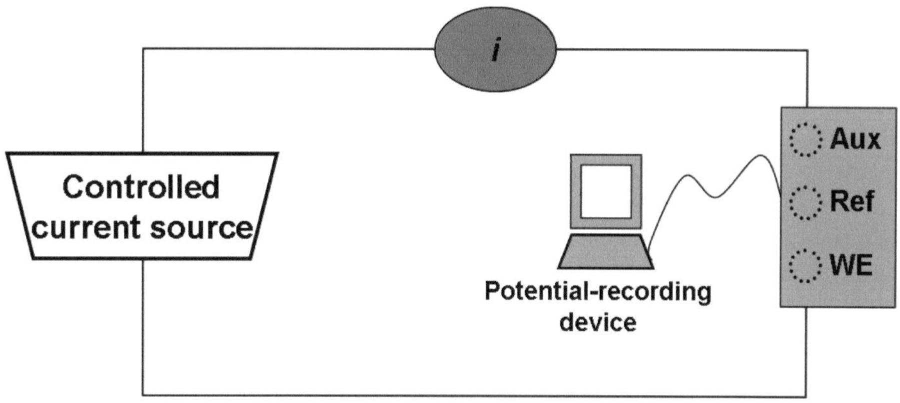

Figure 1. Simplified block diagram of the setup for chronopotentiometric measurements.

input terminals while imposing a 'zero' voltage drop to the external circuit. In corrosion test a ZRA is typically used to measure the galvanic coupling current between two dissimilar electrodes. An interesting application is when the coupling current between two nominally identical electrodes is measured. If both electrodes were identical then very little coupling current would flow. In real situations these electrodes will be slightly different, one being more anodic or cathodic than the other and a small coupling current will exist [41, 42].

On the other hand, *AC impedance methods* are widely used for the characterisation of electrode processes and complex interfaces. These methods are commonly called Electrochemical Impedance Spectroscopy (EIS) and study the system response to the application of a periodic small amplitude AC signal. These measurements are carried out at different AC frequencies and, thus, the name impedance spectroscopy was later adopted. Analysis of the system response contains information about the interface, its structure and reactions taking place there. EIS is now described in the general books on electrochemistry and specific books [39], and there are also numerous articles and reviews [43-46]. It became very popular in the research and applied chemistry [47].

5. Localised techniques in the study of stainless steels

Once the generalised techniques have been presented, a brief summary of the available localised electrochemical techniques is presented, indicating the principles and main applications of each of them.

5.1. Scanning electrochemical microscopy (SECM)

Scanning electrochemical microscopy has become a powerful technique for quantitative investigations of interfacial physicochemical processes, in a wide variety of areas, as considered in several reviews [48-50]. In the simplest terms, SECM involves the use of a mobile ultramicroelectrode (UME) probe, either amperometric or potentiometric, to investigate the activity and/or topography of an interface on a localized scale [51].

Therefore, SECM is based on the possibility of precisely positioning a probe close to the object under investigation (substrate). In the case of SECM, the probe is an ultramicroelectrode, which is an electrode of nanometer to micrometer dimension. Ultramicroelectrode and substrate are both immersed into an electrolyte solution containing either an oxidizable (or reducible) chemical species, which is called mediator. The ultramicroelectrode is electrically biased so that a redox current, the tip-current, is generated. When the ultramicroelectrode is brought near the substrate the tip-current changes and information about the surface activity of the substrate can be extracted. By scanning the ultramicroelectrode laterally above the substrate one can acquire an image of its topography and/or its surface reactivity [51]. Several modes of SECM have been developed to allow the local chemical properties of interfaces to be investigated [51], two of them are the *Feedback Mode* and the *Generation-Collection Mode.*

In the first one (*Feedback Mode*), the species are reduced or oxidised at the tip, producing a steady-state current that is limited by hemispherical diffusion. As the tip approaches to a conductive substrate in the solution, the reduced species formed at the tip is oxidized at the conductive surface, yielding an increase in the tip current and creating a regenerative "positive" feedback. The opposite effect is observed when insulating surfaces are probed, as the oxidized species cannot be regenerated and diffusion to the electrode is inhibited as a result of physical obstruction as the tip approaches the substrate, creating a "negative" feedback loop and decreasing the tip current [52]. One scheme of this effect is presented in Figure 2. However, sometimes the negative feedback is also observed when there is a competitive reaction between the tip and the substrate; this is the case when the oxygen reduction is the followed reaction and the substrate is a stainless steel [53].

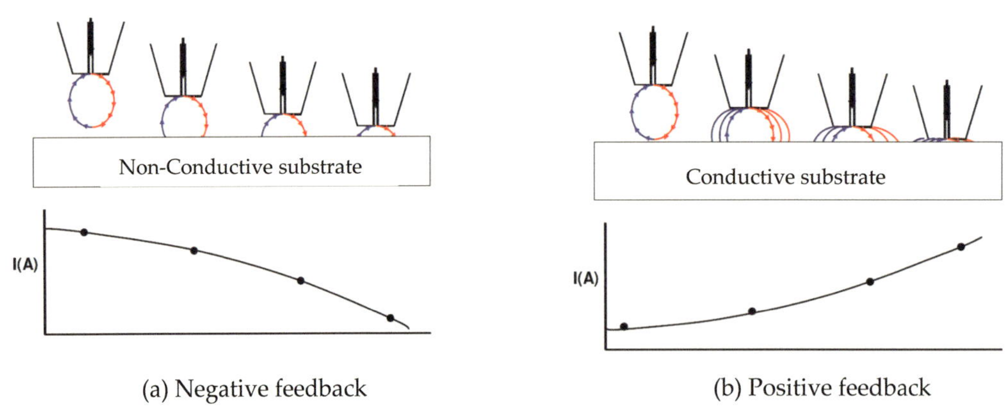

Figure 2. Scheme of the evolution of the current registered on the tip during the feedback mode.

On the other hand, the other modes of operation are the Generation Collection modes. In the Tip generation/substrate collection (TG/SC) mode, the tip is held at a potential sufficient for an electrode reaction to occur and "generate" a product while the substrate is held at a potential sufficient for the product to react with or be "collected" by the substrate [53]. The reciprocal to this method is the substrate generation/tip collection (SG/TC), where the substrate acts to generate species that are measured at the tip. Both TG/SC and SG/TC variations are also categorized as "direct" modes [51].

Among the main applications of the SECM technique, some of them are: corrosive activity of the tested materials, topographic maps of the samples, measurements of heterogeneous of homogeneous kinetics of reactions, study of the electrochemical behaviour of membranes or surface reactions.

5.2. Scanning droplet cell (SDC)

A new electrochemical device, the capillary-based droplet cell [54, 55] is a technique which confines a liquid in contact with a sample surface in order to measure electrochemical and corrosion reactions over a limited region where the droplet is actually in contact with the sample. This Scanning Droplet Technique provides facilities for micro-electrochemical investigations at high resolution, in which small electrolyte droplets are positioned on the sample surface, enabling a spatially resolved surface analysis or modification. The small area of the working electrode determined by the tip size of a capillary with a diameter of 20–600 μm enables the investigation of localized corrosion or passivation of small areas within a single grain or phase. In addition, the device has a smaller resolution than some special probe techniques (e.g. STM), but enables a complete range of potentiostatic (or dynamic) and galvanostatic (or dynamic) techniques including impedance spectroscopy.

This is accomplished by providing a convenient three-electrode system consisting of the working electrode of interest (the wetted surface area under investigation) and the capillary contains the counter and reference electrodes which are electrically connected to the surface through the drop. The wetted area is approximately determined by the capillary radius. The small distance between the counter electrode and the sample allows high current densities due to the small ohmic resistance. The electrolyte drop is then scanned at high resolution across the surface of the sample.

The scanning droplet system allows a spatially resolved, in-situ investigation by the standard electrochemical techniques such as line scans, area maps, potentiostatic pulse steps and open circuit measurements.

5.3. Scanning Kelvin Probe (SKP)

The Kelvin Probe is a non-contact, non-destructive measurement device used to investigate properties of materials [56;57]. It is a tool with wide applications for both surface science and industrial uses. It is based on a vibrating capacitor and measures the work function difference or, for non-metals, the surface potential, between a conducting specimen and a vibrating tip [58-60]. This method exploits the well-established principles of direct correlation between work function and surface condition. The work function is an extremely sensitive indicator of surface condition and is affected by adsorbed or evaporated layers, surface reconstruction, surface charging, oxide layer imperfections, surface and bulk contamination, etc. [61].

Regarding the operation mode, a metal microprobe is positioned close to the surface of the sample (on the order of 100-microns). If the microprobe and sample are of different metals, there is an energy difference between their electrons. The microprobe is then electrically shorted to the sample, through internal electronics of the system. As a consequence, one metal forms a positive charge on its surface and the other metal forms a negative charge on its surface. The probe and sample are separated by a dielectric (air), so a capacitor is formed. The probe is then vibrated and "backing potential" or "nulling potential" is then applied sufficient to minimize this capacitance. At the applied voltage that causes the capacitance to go to zero, the original state is achieved. This value is recorded and charted. Experiments are typically performed in ambient gaseous conditions, but several published examples use humidified environments.

One improvement of this technique is the Scanning Kelvin Force Microscopy (SKPM) [62-64], this technique allows to measure the local potential between a conducting atomic force microscopy (AFM) tip and the sample. Therefore, with this method it is possible to characterise the nano-scale electronic/electrical properties of metal/semiconductor surfaces and semiconductor devices. Among the main applications of the SKP technique, some of the most important are: study of coatings interfaces, study of galvanic pairs or the oxide formation effect.

5.4. Scanning vibrating electrode technique (SVET)

The scanning vibrating probe electrode measures the potential drop generated by current flow in a small volume of solution by vibrating Pt-Ir tipped microelectrodes between two virtual points in the electrolyte at a defined distance from the surface [65]. In corrosion of metals, this will reflect the distribution of current associated with anodic and cathodic current at the metal surface but with decreasing resolution of local anodic and cathodic processes with distance from the surface and with increasing conductivity of the solution. For the latter reason, relatively low conductivity solutions may be necessary for some systems, sufficient to generate distinguishable potential gradients in solution. Then, the scanning Vibrating Electrode Technique (SVET) uses a scanning microelectrode to measure these gradients in situ, thereby, locating and quantifiying the corrosion activity at specific points on the sample surface. Measurements are made by vibrating a fine tipped microelectrode a few hundred microns about the sample in a perpendicular plane to the surface. The electrochemical potential of the microelectrode is recorded at the extremes of vibration amplitude, resulting in the generation of a sinusoidal AC signal. This signal is then measured using a lock-in amplifier, which is tunned to the frequency of the probe vibration. Furthermore, the resulting signal, which is the effect in effect a measure of the DC potential gradients in solution, can be converted to current density by a simple calibration procedure [66]. This technique is, therefore, able to make in-situ measurements of the localised corrosion activity occurring at the sample surface.

This method was originally devised by biologists for the measurement of extracellular currents near to living cells. Isaacs, later developed the technique to study various localised corrosion processes, including stress corrosion cracking of AISI 304 stainless steels and corrosion inhibition by cerium salts [65]. The main applications of the SVET technique to the material science are: study of defect evolution in coatings, galvanic study during a corrosion process, evolution of artificial defects or determination of precursor sites for corrosion.

5.5. Localised electrochemical impedance spectroscopy (LEIS)

As previously mentioned, electrochemical impedance spectroscopy (EIS) is an in-situ non-destructive technique, which has been widely employed in the study of corrosion processes, and in particular is now widely accepted as a standard technique for the investigation of corrosion an coatings. However, difficulties arise in EIS, when attempting to study localised electrochemical processes, as impedance is calculated from the bulk current and voltage data, and it is, therefore, averaged across the entire surface area of the sample. In order to make impedance measurements on a local scale, work by a number of authors has been conducted to combine established DC scanning techniques with AC impedance techniques [67, 68], resulting in the evolution of LEIS.

The principles of LEIS are similar to those used in the conventional bulk EIS; a small sinusoidal voltage perturbation is applied to a working electrode sample and the resulting current is measured to allow the calculation of the impedance. However, rather than measure the bulk current from the whole surface area, an electrochemical probe is scanning close to the surface in order to measure the local current distributions in the electrolyte. This probe incorporates two platinum electrodes; one constitutes the tip of the probe and is electrochemically sharpened to approximately 5 mm of diameter and the other is a ring positioned at a distance of 3 mm over the tip. Each of these electrodes is platinised in order to increase the active area and therefore, to reduce the interfacial area between the electrode and the electrolyte. The potential difference between both platinum electrodes is measured via an electrometer and, for known solution conductivity, the local current is calculated. Then the ratio between the AC perturbation applied to the sample and the local current density determine the local impedance. Therefore, this approach gives spatial resolution to the measurement of the electrochemical impedance, overcoming the limitations of the surface averaging encountered in the conventional EIS measurements [69, 70].

5.6. Scanning refererence electrode technique (SRET)

Scanning reference electrode technique (SRET) is an in situ technique used to study the electrochemical process during localized corrosion without interrupting the process taking place. Within the electrolyte directly above a localized electrochemical active site, there exists an electromagnetic field. A probe consisted of two platinum wires can sense the potential difference between two different places where the two platinum wires are located. This potential difference then is processed by a differential amplifier. By scanning the SRET probe over the surface of the test material immersed in an electrolyte, these fields may be mapped out as a function of x and y or monitored with respect to time. Resulting data can be instantly display, stored and manipulated in a number of formats including individual linescans, area maps and sequential time related images.

The potential field distribution over the surface of the electrochemically active site is representative of the localized current flow. By using a calibration procedure the SRET can also be used to make discrete current measurements.

5.7. Summary

Once the main principles of the localised electrochemical techniques have been presented in the previous point, an overview of the current application of these techniques to the study of heterogeneous surfaces will be commented. Three different kind of studies will be taken into consideration:

5.7.1. Coatings and protective films

One way to protect metals from corrosion is the use of protective films such as coatings. These films can undergo defects that can decrease their functionality. The use of the localised electrochemical techniques can be quite helpful to study the behaviour of these defects. Kinlen [71] used the SRET technique in order to analyse the effect of conductive polymers such as polyaniline (PANI) in the use of coatings formulations and which was the its toleration to minor scratches. This technique has been also used by several authors to study different sort of coatings, such as PVD ceramic-coated steels [72], or physical vapour deposition (PVD) coated samples [73]. Other way to study the local defects on coatings is the SECM technique, with this technique is possible to evaluate the electrochemical activity that surrounds a defect or the initial points where corrosion begins. Examples of that are: the studies of the cathodic protection of aluminium by means of a magnesium coating conducted by Simoes [74, 75] or the studies of damage to paint coatings carried out by Souto [76, 77]. On the other hand, using the SVET or the LEIS devices it is also possible to evaluate the cathodic and anodic sites that appears in the coatings when samples are immersed in an electrolyte, some conducted studies are: the study of microdefects on a coated AZ31 magnesium alloy [78], the observation of self-healing functions of galvanised steel [79] or the evaluation of the PEO-coating/substrate interface [80]. Finally, also the Scanning Kelvin probe can be used in order to study electrochemical differences between coated and uncoated metals[81].

5.7.2. Welds and sensitised samples

As it has been explained previously, when the microstructure on an alloy changes as a consequence of a heat treatment, the corrosion properties change too. Therefore, it is very useful the use of the localised electrochemical techniques in order to evaluate the local electrochemical behaviour. Microcell techniques, with a similar configuration to the Droplet cell have been widely used to determine the properties of the different parts of a weld specimen. One example of this is the cell developed by García in order to study the weld behaviour in AISI 304 and 316L [34]. On the other hand, other authors prefer a different approach to the problem using the SECM system. Some of the most relevant studies carried out are: visualization of local electrochemical activity and local nickel ion in weld Ni/Ti steels [82], characterisation of the electrochemical activity at the interface of a dissimilar explosive joint of stainless steel [83], evaluation of the effects of microplasma arc AISI 316L welds [84] or the study of the sensitisation of a highly alloyed austenitic stainless steel [53]. In the same way, the SVET technique can be used in the study of welds on order to determine the anodic and the cathodic sites of the heated samples. One example of that is the investigation into the effect of a spot weld electrode using 3-D SVET [85] conducted by Benjamin in 2012.

5.7.3. *Identification of anodic and cathodic sites*

Finally, all the localised electrochemical techniques can be used to determine the anodic and cathodic distribution on the surface of the tested samples or the areas that are more susceptible to corrosion, showing higher current density or lower potential. An example of this kind of work is the visualisation of pits by means of the SECM technique [86-88]. In this way it is possible to evaluate the points where the localised corrosion will begin. On the other hand, these cathodic and anodic areas or their activity can be determined using the Scanning Vibrating Electrode or Localised Impedances, the first one allows to establish the anodic and cathodic areas or study inclusions [89, 90]; the second one permits analysing the admittance of the metal/electrolyte interphase in the different areas [91]. Furthermore, when metals with different characteristics are coupled, the galvanic activity can be determined. Simoes used the SECM and the SVET system to determine the galvanic properties of an iron-zinc cell, observing that both techniques has a comparable sensibility [92].

Table 2 presents different papers and communications where the localised techniques are used to study corrosion in metallic materials:

Authors	Journal	Research topic	Technique
Jean-Baptiste et al.	Corrosion Science 48 (2006) 1779–1790	Delaminated areas beneath organic coating: A local electrochemical impedance approach [93]	LEIS
Z.Y. Liu et al.	Electrochimica Acta 60 (2012) 259–263	Understand the occurrence of pitting corrosion of pipeline carbon steel under cathodic polarization [94]	LEIS
Darya Snihirova et al.	Electrochimica Acta, In Press	"SMART" protective ability of water based epoxy coatings loaded with CaCO3 microbeads impregnated with corrosion inhibitors applied on AA2024 substrates [95]	LEIS
Halina Krawiec et al.	Electrochimica Acta 53 (2008) 5252-5259	Numerical modelling of the electrochemical behaviour of 316 stainless steel based upon static and dynamic experimental micro-capillary based techniques: effects of electrolyte low and capillary size [96]	SDC
S. Gnefid, R. Akid	The European Corrosion Congress 2009	The Effects of Flow Rate on Pitting Corrosion of DSS2205 [97]	SDC
Leiva García et al.	Electrochimica Acta, 70, 2012, pp 105-111	Study of the sensitisation of a highly alloyed austenitic stainless steel, Alloy 926 (UNS N08926), by means of scanning electrochemical microscopy [53]	SECM
Yawei Shao et al.	Corrosion Science 51 (2009) 371–379	The role of a zinc phosphate pigment in the corrosion of scratched epoxy-coated steel [98]	SECM

Authors	Journal	Research topic	Technique
Yuehua Yin et al.	Applied Surface Science 255 (2009) 9193–9199	In situ characterization of localized corrosion of stainless steel by scanning electrochemical microscope [99]	SECM
Xiaolan Liu et al.	Corrosion Science 51 (2009) 1772–1779	Effect of alternating voltage treatment on the corrosion resistance of pure magnesium [100]	SECM
A.Q. Fu et al.	Corrosion Science 51 (2009) 914–920	Characterization of corrosion of X65 pipeline steel under disbonded coating by scanning Kelvin probe [101]	SKP
A.Q. Fu et al.	Corrosion Science 51 (2009) 186–190	Characterization of corrosion of X70 pipeline steel in thin electrolyte layer under disbonded coating by scanning Kelvin probe [102]	SKP
Cavalcolid et al.	Journal of the Electrochemical Society. Vol. 150, G456-G460 2003	Surface contaminant detection in semiconductors using noncontacting techniques. [103]	SKP
K.Borgwarth et al.	Electrochimica Acta, Jul 1995	Applications of scanning ultra micro electrodes for studies on surface conductivity [104]	SRET
I.M. Zin et al.	Progress in organic coatings 2005, 52, 126-135	Under-film corrosion of epoxy-coated galvanised steel an EIS and SVET study of the effect of inhibition at defects [105]	SVET
Kiran B	Electrochimica Acta 56 (2011) 1737–1745	Numerical modeling of micro-galvanic corrosion [106]	SVET
Francois Berger et al.	Electrochimica Acta, volume 53, 2010, 2852-2861	Hybrid coating on steel: ZnNi electrodeposition and surface modification with organothiols and diazonium salts [107]	SVET
G.A. Zhang, Y.F. Cheng	Corrosion Science 51 (2009) 1714–1724	Micro-electrochemical characterization of corrosion of welded X70 pipeline steel in near-neutral pH solution [108]	SVET & LEIS
R. Akid et al.	Local Probe Techniques for corrosion research. 978-1-4200-5405-7 (CRC Press)	Application of scanning vibrating electrode technique (SVET) and scanning droplet cell (SDC) techniques to the study of weld corrosion [109]	SVET & SDC

Table 2. Research studies where localised electrochemical techniques are applied to the study of materials.

6. Applications of localised techniques in the study of heterogeneities in stainless steels

After presenting the different localised electrochemical techniques and the main application of then to different heterogeneous surface, two different applications of localised techniques conducted in the "Ingeniería Electroquímica y Corrosión" group from the Universitat Politècnica the València will be explained. The first one corresponds to some experiments carried out with a duplex stainless steel under different sensitisation conditions. The second one, correspond to the evaluation of the local electrochemical properties of a weld in AISI 316L.

6.1. Sensitisation process in a duplex stainless steel

Conventional electrochemical techniques have been widely used to study sensitisation and provide important information about the sensitisation processes on steels [25, 27, 29, 30]. In contrast to pure metals, alloys have rarely been investigated using the micro-droplet cell because of their complexity in structure and chemical composition. Duplex stainless steel (Alloy 900) is potentially a good candidate for micro-electrochemical investigations with the droplet cell, because this alloy is composed of an approximately equal volume fraction of ferrite (α) and austenite (γ) phase, and, sometimes, secondary phases such as sigma (σ) or alpha prime (α') phase [20, 110, 111]. Figure 3 shows the SEM micrographs of Alloy 900 after different heat treatments in an inert atmosphere; these images were obtained using backscattered electrons. In the images it can be observed the change in the percentage of the present phases and the formation of new intermetallic phases such as sigma and chi.

The advances in the field of localised microelectrochemical techniques have provided the facility to acquire spatially resolved information about the corrosion processes occurring on the surface of the metal. The Scanning Droplet cell and the Scanning Electrochemical Microscopy have been used to evaluate the effect of the sensitisation on the electrochemical behaviour of the samples.

In order to conduct the tests with different samples of Alloy 900 at the same time, individual unsensitised and sensitised samples were mounted in epoxy resin as shown in Figure 4.

The linear sweep voltammetry in Figure 5, which has been obtained using the droplet cell, has been used to present the electrochemical behaviour of the different zones in the sensitised and unsensitised Alloy 900. These results show that the electrochemical activity in the sensitised sample at 850 °C during 2 hours is higher than in the unsensitised sample, indicating the detrimental effect of the intermetallic phase precipitation. In the potential range where oxygen reduction occurs, the oxide film is partly reduced but an oxide film remains on the steel surface. According to the literature [112], oxygen reduction on polished surfaces is limited by the mass transport in the solution; however, on passivated surfaces, the oxygen reduction is limited by access of the oxygen to the metal surface and the electronic conductivity of the oxide film [112, 113]. Therefore, this result is indicative of the formation of a different passive film on the sensitised sample as opposed to that on the unsensitised sample.

Furthermore, according to the linear sweep voltammetry, the passivation current is higher in the case of the sensitised sample (Figure 5). A higher passivation current indicates a 'less-protective' passive film with higher electronic conductivity, this being due to the presence of areas depleted in alloying elements, notably chromium [53].

The Scanning Droplet cell also allows performing area scans in the samples investigated. As an example, Figure 6 shows the area scans carried out in an unsensitised and sensitised sample of Alloy 900.

These results show that the free potential in the sensitised sample (Figure 6b) is more negative than that obtained in the unsensitised sample (Figure 6a). This fact indicates that the free potentials of the unsensitised sample are more noble and, as a result the passive film on the

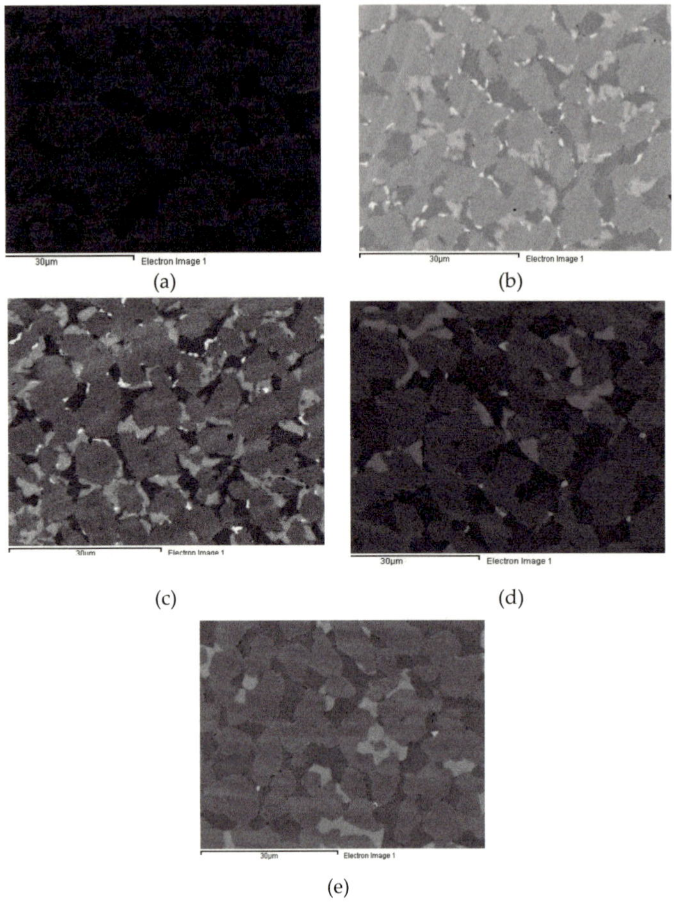

Figure 3. SEM images of the different Alloy 900 samples: a) unsensitised Alloy 900, b) sensitised Alloy 900 at 850 °C during 1 hour, c) sensitised Alloy 900 at 850 °C during 2 hours, d) sensitised Alloy 900 at 950 °C during 1 hour and e) sensitised Alloy 900 at 950 °C during 2 hours. Images were obtained with backscattered electrons.

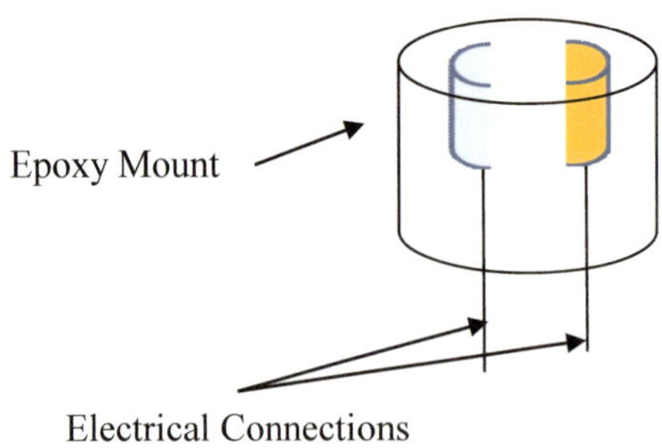

Figure 4. Schematic diagram of the working electrode tested by the different electrochemical techniques. Unsensitised sample on the rigth side.

http://dx.doi.org/10.5772/57204

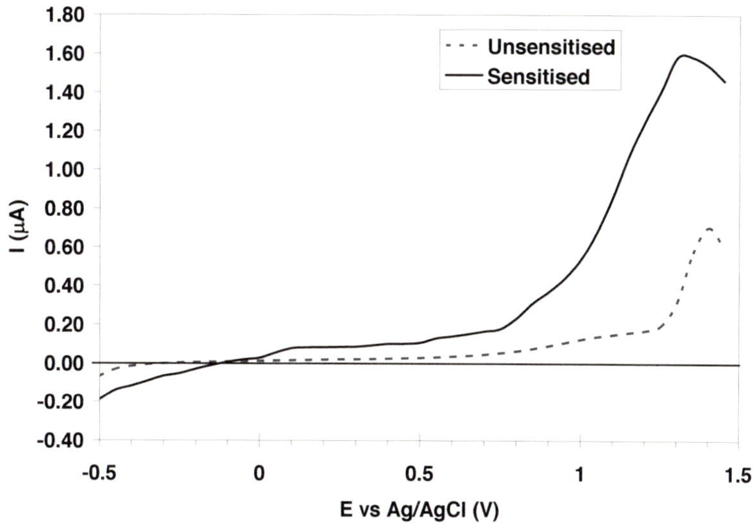

Figure 5. Linear sweep voltammetry of the sensitised and unsensitised samples of Alloy 900 carried out in the Scanning Droplet Cell.

sensitised sample presents a higher electrochemical activity, indicating a greater electronic conductivity of the passive film. Several authors [112, 114-116], proposed that the reduction pathway is influenced by the surface composition of the electrode and oxides have an important role to play in the oxygen reduction kinetics. A homogeneous mixture of chromium oxide and hydroxide constitutes a barrier to oxygen reduction, whereas no diffusion barrier is observed when the surface is only partially covered with a non-reducible chromium oxide [117]. Therefore, any chromium depleted areas formed in the sensitised alloy can lead to the formation of a more conductive passive film that promotes higher oxygen reduction on the electrode surface.

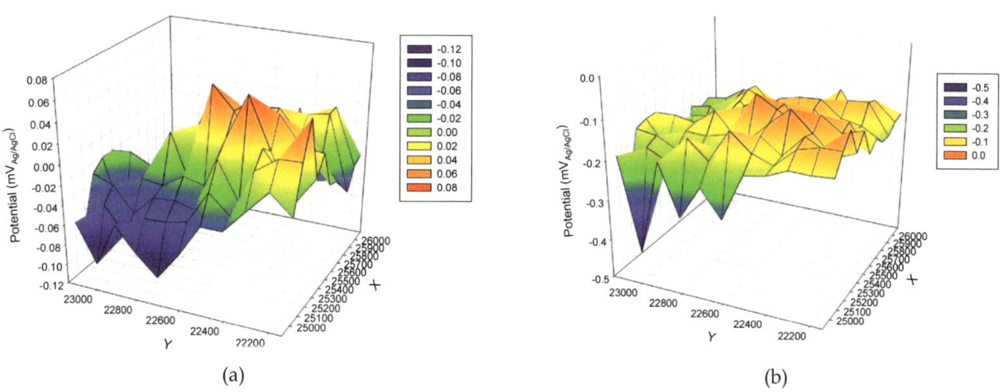

Figure 6. SDC area scans of the free potential in the (a) unsensitised and (b) sensitised sample of Alloy 900.

On the other hand, a similar behaviour can be observed if the different samples of Alloy 900 are analysed by means of the SECM technique. Figure 7 shows different line scans carried out on the samples of Alloy 900 after different states of sensitisation, notably, heated at 950 °C during 1 and 2 hours. It can be observed that the current registered with the tip over the sensitised sample is lower than over the unsensitised one. These differences in the oxygen reduction on one and another sample confirm the different electrochemical behaviour as a consequence of the heat treatment, which was also observed using the scanning droplet cell.

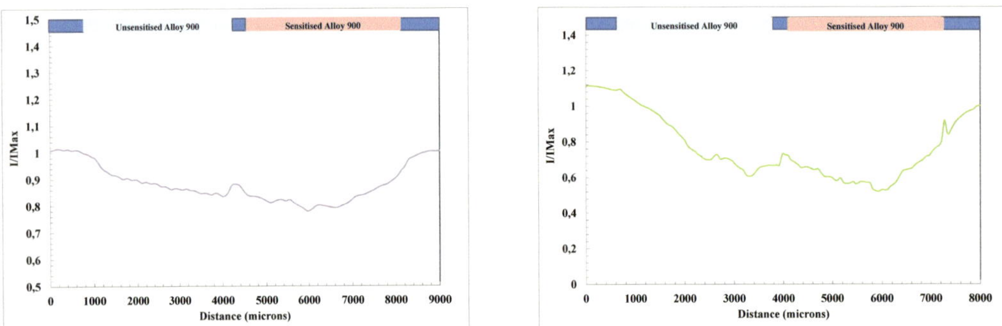

(a) 950 °C during 1 hour

(b) 950 °C during 2 hours

Figure 7. Line scans carried out over the Alloy 900 specimens in the 1 mM NaCl solution at 25 °C. The sensitised sample is on the right.

6.2. Welds in an austenitic stainless steel

Since approximately 20% of the pipeline construction costs are due to welding, coating, and subsequent maintenance to confirm their integrity [118], in the present study the application of the scanning electrochemical microscope to a welded type 316L tube will be shown. The most popular method for welding pipes is the shielded metal-arc process. However, the plasma arc welding technique is an alternative welding procedure which, thanks to its greater arc stability and energy concentration together with its deeper and narrower penetration, produces minimal detrimental residual stresses and distortion at the weld joints. This technique is an adequate welding method because its energy density is high. Microplasma arc welding (MPAW) is a variation of the plasma arc process with amperages between 0.05 and 15 A, used to weld thin pipelines [119].

The materials used to show the power of the SECM technique in order to analyse heterogeneous surfaces were tubes of type 316L SS and microplasma arc welded 316L type welded with filler alloy. Filler alloy was also 316L SS type. The tubes were 20 mm in length and 14 and 16 mm in inner and external diameter, respectively.

The MPAW process was performed manually using argon as backing gas, parameters of the welding process are shown in Table 3.

Welding process		Micro-plasma arc welding (MPAW)
Process type		Manual
Backing gas		Argon (99.9 %)
Flow rate		2.5 L/min
Plasma gas flow rate		6.5 L/min
Number of passes		2
Step 1	Current	11.3 A
	Voltage	20 V
	Welding speed	2.6 mm/s
Step 2	Current	13 A
	Voltage	20 V
	Welding speed	2.6 mm/s

Table 3. Welding parameters for the microplasma arc welding of the 316L SS type.

The materials were examined by light microscopy (LM) in order to estimate possible microstructural variations in them during the MPAW procedure. For this purpose, each material was cut lengthwise and covered in resin; then, the samples were wet abraded from 220 silicon carbide (SiC) grit to 4000 SiC grit. Then, they were polished with 1 and 0.3 micron alumina and were rinsed with distilled water and ethanol. Once the samples were polished, metallographic etching was carried out according to ASM International [120]. The etchant composition consisted of 10 mL of nitric acid, 10 mL of acetic acid, 15 mL of hydrochloric acid and 5 mL of glycerine. The samples were immersed in the etching solution during 90 seconds and then rinsed with distilled water and ethanol.

Figure 8 shows, the microstructure of the welded 316L SS. The microstructure of type 316L SS is a single-phase austenitic microstructure with equiaxed grains. A heat affected zone (HAZ) can be observed in Figure 8. The microstructure of this zone is characterized by an increase in grain size due to the fact that there is no transformation point at temperatures higher than room temperature [121]. Figure 8 also shows the microstructure of the weld zone (WZ). Usually, the WZ of austenitic stainless steels has a cast structure with the presence of 2-10 wt. % delta-ferrite in the austenite matrix [122]. Delta-ferrite presents two opposite phenomena: increasing its content causes a higher resistance to hot cracking [123, 124], but also reduces corrosion resistance due to the formation of a less stable passive film [125]. In Figure 8 it can be observed that the weld zone has columnar grains where delta-ferrite can be found. Alloys with especially low carbon contents (such as 316L SS type), to minimize susceptibility to sensitization during welding, present a greater tendency towards delta-ferrite stabilization [126]. The microstructure of delta-ferrite occurs in vermicular form. Vermicular morphology is typical of welds that solidify ferrite, followed by a subsequent formation of austenite involving the grains of ferrite until complete solidification. According to this fact, it is clearly observed that the microplasma arc welding process affects on the microstructure of the SS.

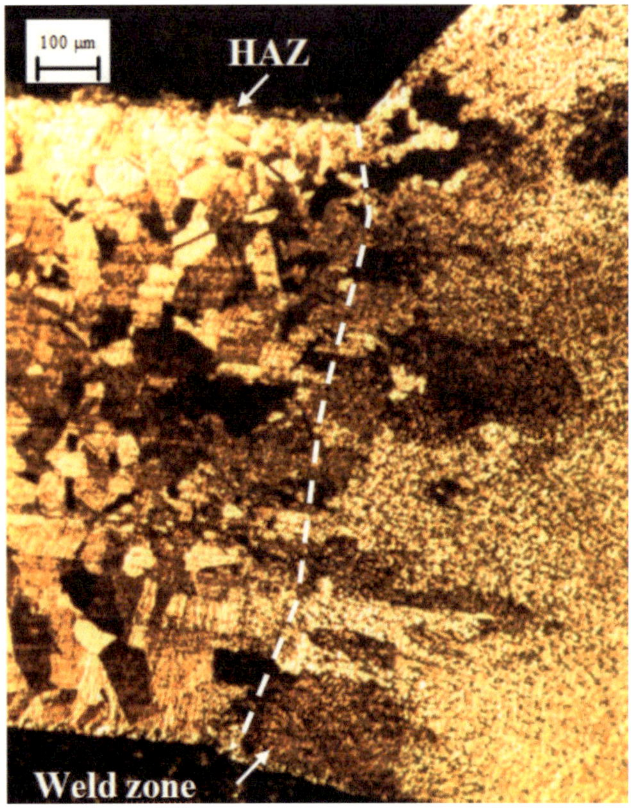

Figure 8. Microstructure of the microplasma arc welded 316L SS type [27].

Additionally, in order to know the influence of the microplasma arc welding technique on the mechanical properties of the stainless steel, Vickers microhardness measurements were carried out using a microhardness tester (Struers Duramin) with a diamond pyramid indenter at a load of 300 g and duration of 15 s [127]. Figure 9 shows the microhardness profiles across the base (non-welded) 316L SS and the microplasma arc welded 316L SS. As it is shown in the figure, the weld and the heat affected zones are softer than the base one. Since there is a positive correlation between hardness and strength; the higher the hardness values, the higher the strength. This fact indicates that the strength of the join decrease in the weld and heat affected zones [128].

In order to know how this microstructure and microhardness variations affect on the electrochemical behaviour of the stainless steel, SECM tests were performed. To carry out the scanning electrochemical measurements, the tubes were cut lengthwise and covered with resin; then, the samples were wet abraded from 220 silicon carbide (SiC) grit to 4000 SiC grit. Finally, they were rinsed with distilled water, ethanol and dried.

SECM tests were performed using a Sensolytics device connected to a bipotentiostat. The SECM was operated in "feedback mode". This technique measures a faradic current at the

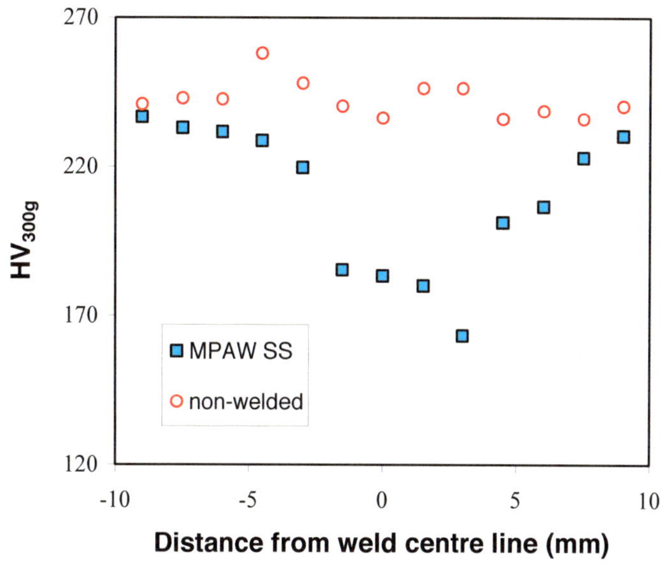

Figure 9. Microhardness profiles of the base and welded 316L SS.

microdisk, while the tip is scanned over the specimen surface. A platinum microelectrode tip of 25 μm in diameter was used as counter-electrode, and a silver/silver chloride (Ag/AgCl) 3M potassium chloride (KCl) microelectrode was used as reference electrode. The samples were immersed in a 35 g/L sodium chloride (NaCl) solution at 25 °C. All the tests were carried out in natural aerated solutions. Oxygen was used as the electrochemical mediator at the tip. The reduction of oxygen on the microelectrode was used to establish the height of the tip over the sample, and also to know the reactivity of the surface during the test. The cyclic voltammogram was recorded at a scan rate of 0.05 V/s from 0 $V_{Ag/AgCl}$ to -1.2 $V_{Ag/AgCl}$. The analysis of the data was used to choose the polarisation potential of the tip; in this study the experiments were performed at -0.6 $V_{Ag/AgCl}$.

Then approach curves (intensity vs distance from the surface of the sample) were performed in welded and base 316L SS. The progress of the dissolved oxygen reduction reaction was followed by setting the tip at -0.6 $V_{Ag/AgCl}$. Figure 10a shows the cyclic voltammetry of the welded alloy; where it can be clearly observed that at a potential of -0.6 $V_{Ag/AgCl}$ the detection of oxygen is under limiting current of oxygen reduction.

Figure 10b shows the approach curves of the base and welded SS. They show that the proximity of the tip to the specimen affects the diffusion of oxygen to the tip and hence the magnitude of the current. The drop in the current measured at the tip occurs at a shorter distance working with the welded 316L SS type than with the base one, indicating that the oxygen reduction reaction is higher over the SS surface in its welded state. Thus, there is a higher electrochemical activity on the welded alloy. This is due to the presence of chromium depleted regions and delta-ferrite originated in weldments [129, 130].

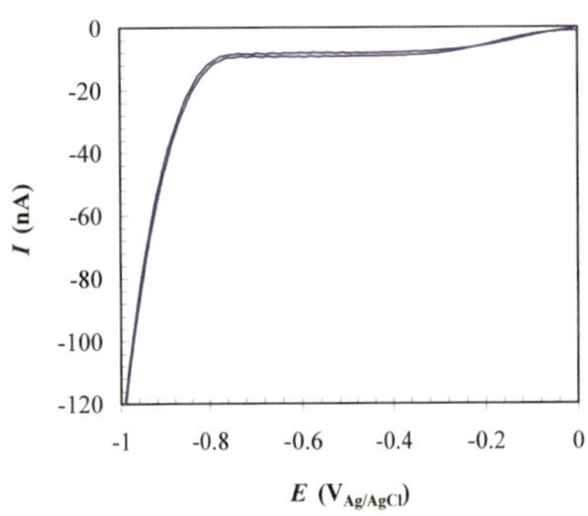

(a)

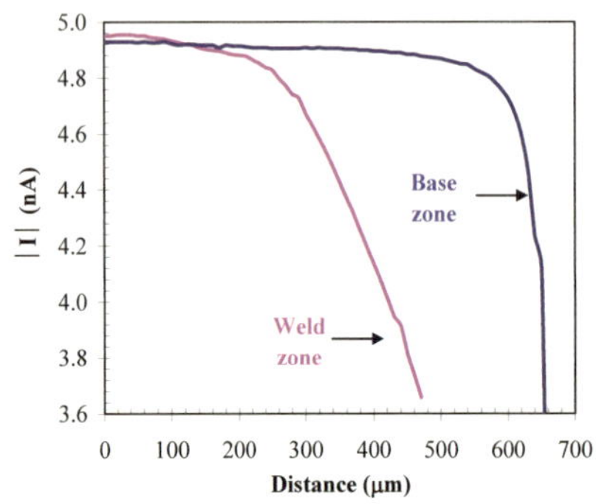

(b)

Figure 10. a) Cyclic voltammogram of the welded sample measured at the SECM tip, at the beginning of the test and (b) approach curves in the 35 g/L NaCl solution at 25 °C [84].

7. Conclusions

It can be concluded that SECM and SDC are powerful techniques that reveals substantial differences in the electrochemical activity of base and welded samples. These techniques can also contribute to the study of the galvanic effect in different materials, helping to the elucidation of the precursor sites for the pit formation. The use of localised techniques can complement the results obtained with conventional electrochemical techniques, leading to a better comprehension of the corrosion processes.

Acknowledgements

We wish to express our gratitude to MICINN (CTQ2009-07518) and to Universitat Politècnica de València (CEI-01-11).

Author details

R. Leiva-García, R. Sánchez-Tovar, C. Escrivà-Cerdán and J. García-Antón*

*Address all correspondence to: jgarciaa@iqn.upv.es

Univ. Politécnica de Valencia, Ingeniería Electroquímica y Corrosión (IEC), Dep. Ingeniería Química y Nuclear, Valencia, Spain

References

[1] Corrosion Costs and Preventive Strategies in the United States NACE (2002).

[2] Whitney, W. R. THE CORROSION OF IRON. Journal of the American Chemical Society 1903; 25(4), 394-406.

[3] Frumkin A., Damaskin B., Grigoryev N., Bagotskaya I. Electrochimica Acta 1974; 19 p. 69.

[4] Levich V. G., "Physico-Chemical Hydrodynamics," Academy of Sciences of the USSR Press (1959).

[5] Cobb HM. Steel Products Manual: Stainless Steels. Warrendale: Iron & Steel Society; 1999.

[6] Sedriks AJ. Corrosion Resistance of Stainless Steels and Nickel Alloys. In: Cramer SD, Covino BS Jr. (Eds.), Corrosion: Fundamentals, Testing and Protection. Vol. 13A. USA: ASM Handbook; 2003. 697-702.

[7] Hakiki N. E., Boudin S., Rondot B., Da Cunha Belo M. The electronic structure of passive films formed on stainless steels, Corrosion Science 1995; 37, 1809-1822.

[8] Wijesinghe T. L. S., Blackwood D. J.. Photocurrent and capacitance investigations into the nature of the passive films on austenitic stainless steels, Corrosion Science 2008; 50, 23-34.

[9] Lothongkum G., Chaikittisilp S., Lothongkum A. W.. XPS investigation of surface films on high Cr-Ni ferritic and austenitic stainless steels, Applied Surface Science 2003; 218, 203-210.

[10] Freire L., Carmezim M. J., Ferreira M. G. S., Montemor M. F.. The passive behaviour of AISI 316 in alkaline media and the effect of pH: A combined electrochemical and analytical study, Electrochimica Acta 2010; 55, 6174-6181.

[11] Almarshad A.I., Jamal D.. Electrochemical investigations of pitting corrosion behaviour of type UNS S31603 stainless steel in thiosulfate-chloride environment, Journal of Applied Electrochemistry 2004; 34, 67-70.

[12] Montañés M.T., Sánchez-Tovar R., García-Antón J., Pérez-Herranz V. The influence of Reynolds number on the galvanic corrosion of the copper/AISI 304 pair in aqueous LiBr solutions, Corrosion Science 2009; 51, 2733-2742.

[13] Honeycombe RWK, HKDH. Bhadeshia. Steels, Microstructure and Properties. Oxford: Butterworth Heinemann; 1995.

[14] Brinkman CR, Garvin HW. Properties of Austenitic Stainless Steels and Their Weld Metals: Influence of Slight Chemistry Variations. Vol. 679 Baltimore: ASTM special technical publication; 1979.

[15] Krauss J. Steels: Processing, Structure, and Performance. USA: ASM international; 2005.

[16] L. Antoni and B. Baroux: "Cyclic oxidation behaviour of stainless steels - appliation to the automotive exhaust lines", La Revue de métallurgie-CIT Février 2002; 178-188.

[17] Santacreu P.O., Sassoulas, H., Moser, F., Cleizergues, O., Lovato, G. Study of the thermal fatigue of stainless steels and its application to the life prediction of automotive exhaust line components. In: Skrzypek JJ., Hetnarski RB. (eds.). 3rd Int. Congress on Thermal Stresses: Thermal Stresses 1999, June 13-17, Cracow, Poland.

[18] Gunn RN. Duplex Stainless Steels: Microstructure, Properties and Applications. England: Abington publishing, Woodhead publishing Ltd in association with The Welding Institute; 1997.

[19] Alvarez-Armas I., Degallaix-Moreuil S. Duplex stainless steels. Great Britain; John Wiley and Sons Inc; 2009.

[20] Park C.J., Shankar Rao V., Kwon H.S. Effects of sigma phase on the initiation and propagation of pitting corrosion of duplex stainless steel, Corrosion 2009; 61, 76-83.

[21] Iacoviello F., Casari F., Gialanella S. Effect of "475°C embrittlement" on duplex stainless steel localized corrosion resistance, Corrosion Science 2005; 47, 909-922.

[22] Davis JR. Stainless Steels. Ohio: ASM International; 1994.

[23] Hemmingsen T., Hovdan H., Sanni P., Aagotnes N.O. The influence of electrolyte reduction potential on weld corrosion, Electrochimica Acta 2002; 47, 3949-3955.

[24] Shaikh H., Rao B.P.C., Gupta S., George R.P., Venugopal S., Sasi B., Jayakumar T., Khatak H.S. Assessment of intergranular corrosion in AISI Type 316L stainless steel weldments, British Corrosion Journal 2002; 37, 129-140.

[25] Lima A.S., Nascimento A.M., Abreu H.F.G., De Lima-Neto P. Sensitization evaluation of the austenitic stainless steel AISI 304L, 316L, 321 and 347, Journal of Materials Science 2005; 40, 139-144.

[26] Parvathavarthini N., Dayal R.K., Khatak H.S., Shankar V., Shanmungan V. Sensitization behaviour of modified 316N and 316L stainless steel weld metals after complex annealing and stress relieving cycles, Journal of Nuclear Materials 2006; 355, 68-82.

[27] Rao R.V.S., Parvathavarthini N., Pujar M.G., Dayal R.K., Khatak H.S., Kaul R., Ganesh P., Nath A.K. Improved pitting corrosion resistance of cold worked and thermally aged AISI type 316 L(N) SS by laser surface modification, Surface Engineering 2007; 23, 83-92.

[28] Stella J., Cerezo J., Rodríguez E. Characterization of the sensitization degree in the AISI 304 stainless steel using spectral analysis and conventional ultrasonic techniques, NDT & E International 2009; 42, 267-274.

[29] Leiva-García R., Muñoz-Portero M.J., García-Antón J. Evaluation of Alloy 146, 279, 900, and 926 sensitization to intergranular corrosion by means of electrochemical methods and image analysis, Corrosion Science 2009; 51, 2080-2091.

[30] Li S., Li L., Yu S.R., Akid R., Xia H.B. Investigation of intergranular corrosion of 316L stainless steel diffusion bonded joint by electrochemical potentiokinetic reactivation, Corrosion Science 2011; 53, 99-104.

[31] Sánchez-Tovar R., Montañés M.T., García-Antón J. Effect of different micro-plasma arc welding (MPAW) processes on the corrosion of AISI 316L SS tubes in LiBr and H_3PO_4 solutions under flowing conditions, Corrosion Science 2010; 52, 1508-1519.

[32] ASM Metals Handbook, Corrosion vol. 13, Corrosion of Weldments, ASM International, OH, 1992, pp. 771-838.

[33] ASM Specialty Handbook, Stainless Steels, Corrosion of Weldments, ASM International, OH, 1994, pp. 238-257.

[34] Garcia C., de Tiedra M.P., Blanco Y., Martin O., Martin F.. Intergranular corrosion of welded joints of austenitic stainless steels studied by using an electrochemical minicell. Corrosion Science 2008; 50, 2390-2397.

[35] Reclaru L., Lerf R., Eschler P.Y., Meyer J.M.. Corrosion behavior of a welded stainless-steel orthopedic implant Biomaterials 2001; 22, 269-279.

[36] Perren R.A., Suter T.A., Uggowitzer P.J., Weber L., Magdowski R., Böhni H., Speidel M.O.. Corrosion resistance of super duplex stainless steels in chloride ion containing environments: investigations by means of a new microelectrochemical method - I. Precipitation-free states. Corrosion Science 2001; 43, 707-726.

[37] Wloka J., Laukant H., Glatzel U., Virtanen S., Corrosion Properties of Laser Beam Joints of Aluminium with Zinc- Coated Steel. Corrosion Science 2007; 49, 4243-4258.

[38] Robert G. Kelly, John R. Scully, David Shoesmith, Rudolph G. Buchheit. Electrochemical Techniques in Corrosion Science and Engineering. Marcel Dekker: New York; 2003.

[39] Bard A. J., Faulkner L. R.. Electrochemical Methods. Fundamentals and Applications. John Wiley & Sons. Inc. New York; 2001.

[40] Enos D.G., Scribner L.L. The Potentiodynamic Polarization Scan. Technical Report 33. Solartron Instruments: UK; 1997.

[41] Baboian R., Electrochemical Techniques for Predicting Galvanic Corrosion, Galvanic and Pitting Corrosion - Field and Laboratory Studies, ASTM STP 576, American Society for Testing and Materials, 1976.

[42] Wolstenholme J., Inexpensive Zero-Resistance Ammeter for Galvanic Studies, British Corrosion Journal 1974; 9, No.2, 116-117.

[43] Macdonald D.D. Techniques for Characterization of Electrodes and Electrochemical Processes, Ed. by R. Varma and J.R. Selman, J. Wiley & Sons, New York, 1991, p. 515.

[44] Mansfeld F., Lorenz W.J. Techniques for Characterization of Electrodes and Electrochemical Processes, Ed. by R. Varma and J.R. Selman, J. Wiley & Sons, New York, 1991, p. 581.

[45] Szpak S. Techniques for Characterization of Electrodes and Electrochemical Processes, Ed. by R. Varma and J.R. Selman, J. Wiley & Sons, New York, 1991, p. 677.

[46] Electrochemical Impedance: Analysis and Interpretation, Ed. by J.R. Scully, D.C. Silverman and M.W. Kendig, ASTM, Philadelphia, 1993.

[47] Lasia A., Electrochemical Impedance Spectroscopy and Its Applications, Modern Aspects of Electrochemistry, B. E. Conway, J. Bockris, and R.E. White, Edts., Kluwer Academic/Plenum Publishers, New York, 1999, Vol. 32, p. 143-248

[48] Barker A L, Gonsalves M, Macpherson JV, Slevin CJ, Unwin P.R.. Scanning electrochemical microscopy: beyond the solid/liquid interface Analitical Chimica Acta 1999; 385, 223-40.

[49] Barker AL, Slevin CJ, Unwin PR, Zhang J 2001 Liquid Interfaces in Chemical, Biological, and Pharmaceutical Applications ed A G Volkov (New York: Marcel Dekker) p 283.

[50] Pu G, LongoML and BordenMA. Effect of microstructure on molecular oxygen permeation through condensed phospholipid monolayers Journal of American Chemical Society 2005; 127, 6524-5.

[51] Bard A J and Mirkin M V 2001 Scanning Electrochemical Microscopy (New York: Marcel Dekker).

[52] Bard, Allen; Fan, Kwak, Lev (1989). "Scanning Electrochemical Microscopy. An Introduction and Principles". Analytical Chemistry 61 (2): 132-138.

[53] Leiva-García R., Akid R., Greenfield D., Gittens J., Munoz-Portero M.J., García-Antón J. Study of the sensitisation of a highly alloyed austenitic stainless steel, Alloy 926 (UNS N08926), by means of scanning electrochemical microscopy Electrochimica Acta 2012; 70, 105-111.

[54] Arjmand F., Adriaens A.. Investigation of 304L stainless steel in a NaCl solution using a microcapillary electrochemical droplet cell: Comparison with conventional electrochemical techniques Electrochimica Acta 2012; 59, 222-227.

[55] Rees N.V., Compton R.G.. Hydrodynamic microelectrode voltammetry Russian Journal of Electrochemistry 2006; 44, 368-389.

[56] Bozek R.. Application of Kelvin Probe Microscopy for Nitride Heterostructures Acta Physica Polonica A 2005; 108, 541-554.

[57] Zisman W A 1932 A new method of measuring contact potential differences in metals Rev. Sci. Instrum. 3 367.

[58] Baikie et al, 'Work Function study of rhenium oxidation using an ultra-high vacuum scanning Kelvin probe, Journal of Applied Physics 2000; 88, 4371.

[59] Kelvin L, Fitzgerald G and Francis W Contact electricity of metals Phil. Mag. 1898; 46, 82.

[60] Christmann K, Ertl G and Pignet T Adsorption of hydrogen on a Pt(111) surface Surface Science 1976; 54, 365.

[61] Moores B., Simons J., Xu S., Leonenko S. Kelvin probe force microscopy in application to biomolecular films: Frequency modulation, amplitude modulation, and lift mode. Ultramicroscopy 2010; 110, 708-711.

[62] Nonnenmacher M, Boyle OMP, Wickramasinghe H.K. Kelvin probe force microscopy Applied Physic Letters 1991; 58, 2921.

[63] Weaver JMR, Abraham DW High resolution atomic force microscopy potentiometry Journal of Vacuum Science and Technology B 1991; 9, 1559-1561.

[64] Barth C., Hynninen T., Bieletzki T., Henry C.R., Foster A.S., Esch F., Heiz U. AFM tip characterization by Kelvin probe force microscopy. New Journal of Physics 2010; 12, 093024.

[65] Isaacs, H. S., Applications of current measurement over corroding metallic surfaces. Progress in Clinic Research 1986; 210, 37-44.

[66] SVP100 Manual, Version 1.23, Uniscan Instruments Ltd., Sigma House, Buxton, UK, SK17 9JB.

[67] Hughes M.C., Parks J.M. "An AC Impedance Probe as an indicator of corrosion and defects in polymer/metal substrate system", in Corrosion Control by Organic Coatings, Henry Leidheiser Jr., Editor, p. 45-50, NACE, Houston, Texas (1981).

[68] Lillard R.S., Moran P.J., Isaacs H.S., Method for Determining the AC Current Distribution at an Electrochemical Interface and Generating Local Impedance Data. Journal of Electrochemical Society 1992; 139, 1007-1012.

[69] Mierisch A.M., Yuan J., Kelly R.G., Taylor S.R., 'Probing Coating Degradation on AA2024-T3 Using Local Electrochemical and Chemical Techniques', Journal of Electrochemical Society, 1999; 146, 4449.

[70] Annergren I., Thierry D., F. Zou, Localized Electrochemical Impedance Spectroscopy for Studying Piting Corrosion on Stainless Steels" Journal of Electrochemical Society 1997; 144, 1208.

[71] Kinlen P.J., Menon V., Ding Y.. A Mechanistic Investigation of Polyaniline Corrosion Protection Using the Scanning Reference Electrode Technique. Journal of The Electrochemical Society 1999; 146, 3690-3695.

[72] Liu C., Bi Q., Leyland A., Matthews A.. An electrochemical impedance spectroscopy study of the corrosion behaviour of PVD coated steels in 0.5 N NaCl aqueous solution: Part II. EIS interpretation of corrosion behaviour. Corrosion Science 2003; 45 1257-1273.

[73] Akid R., Hovsepian P., Kok, Y.N. Tribocorrosion testing of stainless steel (SS) and PVD coated SS using a modified scanning reference electrode technique. Wear 2005; 259, 1472-1481.

[74] Simoes A.M., Battocchi D., Tallman D.E., Bierwagen G.P.. SVET and SECM imaging of cathodic protection of aluminium by a Mg-rich coating.Corrosion Science 2007; 49, 3838-3849.

[75] Simoes A.M., Battocchi D., Tallman D.E., Bierwagen G.P. Assessment of the corrosion protection of aluminium substrates by a Mg-rich primer: EIS, SVET and SECM study. Progress in Organic Coatings 2008; 63, 260-266.

[76] Souto R.M., Gonzalez-Garcia Y., Gonzalez S., Burstein G.T.. Damage to paint coatings caused by electrolyte immersion as observed in situ by scanning electrochemical microscopy. Corrosion Science 2004; 46, 2621-2628.

[77] Souto R.M., Gonzalez-Garcia Y., Gonzalez S., Burstein G.T.. Imaging the Origins of Coating Degradation and Blistering Caused by Electrolyte Immersion Assisted by SECM. Electroanalysis 2009; 21, 2569-2574.

[78] Karavai O.V., Bastos A.C., Zheludkevich M.L., Taryba M.G.,. Lamaka S.V, Ferreira M.G.S.. Localized electrochemical study of corrosion inhibition in microdefects on coated AZ31 magnesium alloy. Electrochimica Acta 2010; 55, 5401-5406.

[79] Oglea K., Morel S., Jacquet D.. Observation of Self-Healing Functions on the Cut Edge of Galvanized Steel Using SVET and pH Microscopy. Journal of The Electrochemical Society 2006; 153 1, B1-B5.

[80] Sinebryukhov S.L., Gnedenkov A.S., Mashtalyar D.V., Gnedenkov S.V.. PEO-coating/substrate interface investigation by localised electrochemical impedance spectroscopy. Surface & Coatings Technology 2010; 205, 1697-1701.

[81] McMurray H.N., Williams G., O'Driscoll S.. Chromate Inhibition of Filiform Corrosion on Organic Coated AA2024-T3 Studied Using the Scanning Kelvin Probe. Journal of The Electrochemical Society 2004; 151-7, B406-B414.

[82] Ruhlig D., Gugel H., Schulte A., Theisen W., Schuhmann W.. Visualization of local electrochemical activity and local nickel ion release on laser-welded NiTi/steel joints using combined alternating current mode and stripping mode SECM. Analyst, 2008; 133, 1700-1706.

[83] Jayaraj, J.; Mudali, U. Kamachi. Electrochemical Activity at the Interface of Dissimilar Explosive Joint of Stainless Steel with Zircaloy by Scanning Electrochemical Microscopy. Journal of Advanced Microscopy Research 2012; 7, 214-217.

[84] Sánchez-Tovar R., Montañés M.T., García-Antón J. Effects of microplasma arc AISI 316L welds on the corrosion behaviour of pipelines in LiBr cooling systems, Corrosion Science 2013; 73, 365-374.

[85] Benjamin P. Wilson, Justin R. Searle, Kirsi Yliniemi, T. Bryan Jones, David A. Worsley, H. Neil. McMurray. Investigation into the Effect of Spot Weld Electrode Life and Quality on the Corrosion Behavior of Galvanized Automotive Steel Using the Three-dimensional Scanning Vibrating Technique. ECS Transactions 2013; 50, 53-64.

[86] Eckhard K., Etienne M., Schulte A., Schuhmann W.. Constant-distance mode AC-SECM for the visualisation of corrosion pits. Electrochemistry Communications 2007; 9, 1793-1797.

[87] Gabrielli C., Joiret S., Keddam M., Portail N., Rousseau P., Vivier V.. Single pit on iron generated by SECM. An electrochemical impedance spectroscopy investigation. Electrochimica Acta 2008; 53, 7539-7548.

[88] Gonzalez-Garcia Y., Burstein G.T., Gonzalez S., Souto R.M.. Imaging metastable pits on austenitic stainless steel in situ at the open-circuit corrosion potential. Electrochemistry Communications 2004; 6, 637-642.

[89] Vuillemin B., Philippe X., Oltra R., Vignal V., Coudreuse L., Dufour L.C., Finot E.. SVET, AFM and AES study of pitting corrosion initiated on MnS inclusions by microinjection. Corrosion Science 2003; 45, 1143-1159.

[90] Krawiec H., Vignal V., Oltra R.. Use of the electrochemical microcell technique and the SVET for monitoring pitting corrosion at MnS inclusions. Electrochemistry Communications 2004; 6, 655-660.

[91] Ramana M. Pidaparti, Ronak R. Patel. Correlation between corrosion pits and stresses in Al alloys. Materials Letters 2008; 62, 4497-4499.

[92] Simoes A.M., Bastos A.C., Ferreira M.G., González-García Y., González S., Souto R.M.. Use of SVET and SECM to study the galvanic corrosion of an iron-zinc cell. Corrosion Science 2007; 49, 726-739.

[93] Jorcin J.B., Aragon E., Merlatti C., Pebere N. Delaminated areas beneath organic coating: A local electrochemical impedance approach. Corrosion Science 2006; 48, 2006 1779-1790.

[94] Liu Z.Y., Li X.G., Cheng Y.F. Understand the occurrence of pitting corrosion of pipeline carbon steel under cathodic polarization. Electrochimica Acta 2012; 60 259- 263.

[95] Snihirova D., Sviatlana V. Lamaka, M.F. Montemor, "SMART" protective ability of water based epoxy coatings loaded with CaCO3 microbeads impregnated with cor-

rosion inhibitors applied on AA2024 substrates, Electrochimica Acta 2012; 83, 439-447.

[96] Krawiec Halina, Vignal Vincent, Akid Robert. Numerical modelling of the electrochemical behaviour of 316 stainless steel based upon static and dynamic experimental micro-capillary based techniques: effects of electrolyte low and capillary sizeElectrochimica Acta 2008; 53 5252-5259.

[97] Gnefid S., Akid R. The Effects of Flow Rate on Pitting Corrosion of DSS2205. The European Corrosion Congress 2009, Nice (France).

[98] Shao Y., Jia C., Meng G., Zhang T., Wang F. The role of a zinc phosphate pigment in the corrosion of scratched epoxy-coated steel. Corrosion Science 2009; 51, 371-379.

[99] Yin Y., Niu L., Lu M., Guo W., Chen S. In situ characterization of localized corrosion of stainless steel by scanning electrochemical microscope. Applied Surface Science 2009; 255, 9193-9199.

[100] Liu X., Zhang T., Shao Y., Meng G., Wang F. Effect of alternating voltage treatment on the corrosion resistance of pure magnesium. Corrosion Science 2009; 51, 1772-1779.

[101] Fu A.Q., Cheng Y.F.. Characterization of corrosion of X65 pipeline steel under disbonded coating by scanning Kelvin probe. Corrosion Science 2009; 51 914-920.

[102] Fu A.Q., Cheng Y.F.. Characterization of corrosion of X70 pipeline steel in thin electrolyte layer under disbonded coating by scanning Kelvin probe. Corrosion Science 2009; 51, 186-190.

[103] Cavalcoli D, Cavallini A, Rossi M. Surface contaminant detection in semiconductors using noncontacting techniques. Journal of the Electrochemical Society 2003; 150, G456-G460.

[104] Borgwarth K., Ebling D., Heinze J.. Applications of scanning ultra micro electrodes for studies on surface conductivity. Electrochimica Acta 1995; 40, 1455-1460.

[105] Zin I.M., Lyon S.B., Hussain A.. Under-film corrosion of epoxy-coated galvanised steel an EIS and SVET study of the effect of inhibition at defects. Progress in organic coatings 2005; 52, 126-135.

[106] Kiran B. Deshpande. Numerical modeling of micro-galvanic corrosion. Electrochimica Acta 2011; 56, 1737-1745.

[107] Berger F., Delhalle J., Mekhalif Z.. Hybrid coating on steel: ZnNi electrodeposition and surface modification with organothiols and diazonium salts. Electrochimica Acta 2010; 53, 2852-2861.

[108] Zhang G.A., Cheng Y.F.. Micro-electrochemical characterization of corrosion of welded X70 pipeline steel in near-neutral pH solution. Corrosion Science 2009; 51, 1714-1724.

[109] Akid R., Roffey P, Greenfield D., Guillen D.. Application of scanning vibrating electrode technique (SVET) and scanning droplet cell (SDC) techniques to the study of weld corrosion. Local Probe Techniques for corrosion research. 978-1-4200-5405-7 (CRC Press): 2013.

[110] Kim J.S., Kwon H.S. Effects of tungsten on corrosion and kinetics of sigma phase formation of 25% chromium duplex stainless steels. Corrosion 1999; 55 (5) 512-521.

[111] Park C.J., Kwon H.S., Lohrengel M.M. Micro-electrochemical polarization study on 25% Cr duplex stainless steel. Materials Science and Engineering A, 2004; 372 180-185.

[112] Le Bozec N., Compère C., L'Her M., Laoueman A., Costa D., Marcus P.. Influence of stainless steel surface treatment on the oxygen reduction reaction in seawater. Corrosion Science 2001, 43(4) 765-786.

[113] Okuyama M., Haruyama S., The cathodic reduction of oxygen on stainless steels in a neutral solution. Corrosion Science 1990; 31, 521-526.

[114] Gojkovic S.L.J., Zecevic S.K., Obradovic M.D., Drazic D.M.. Oxygen reduction on a duplex stainless steel. Corrosion Science 1998; 40(6) 849-860.: 2013.

[115] Calvo E.J., Schiffrin D.J.. The reduction of hydrogen peroxide on passive iron in alkaline medium. Journal of Electroanalytical Chemistry 1984; 163, 257-275.

[116] Zecevic S., Drazic D.M., Gojkovic S.L. Journal of Electroanalytical Chemistry 1989; 265, 179.

[117] Ramasubramanian N., Preocanin N., Davidson R.D.. Analysis of Passive Films on Stainless Steel by Cyclic Voltammetry and Auger Spectroscopy Journal of Electrochemical Society 1985; 132, 793-798.

[118] Mohammadi F., Eliyan F.F., Alfantazi A. Corrosion of simulated weld HAZ of API X-80 pipeline steel, Corrosion Science 2012; 63, 323-333.

[119] Jiang Y., Xu B., Lu Y., Xiang Y., Liu C., Xia D. Application prospects of plasma welding within remanufacturing engineering. 4th World Congress on Maintenance 2008: conference proceedings, November 24-26, Haikou, Hainan, China.

[120] Vander Voort G.F. Metallography: Principles and Practice. New York: ASM International; 1999.

[121] Conde y Santiago G. Aceros inoxidables, refractarios y criogénicos (Stainless Steels, refractory and cryogenic). Madrid: INTERCIENCIA; 1971.

[122] Lu B.T., Chen Z.K., Luo J.L., Patchett B.M., Xu Z.H. Pitting and stress corrosion cracking behavior in welded austenitic stainless steel, Electrochimica Acta 2005; 50, 1391-1403.

[123] Patchett BM., Bringas J. The Metals Blue Book, Filler metals. USA: CASTI Publishing Inc. and American Welding Society (AWS); 1998.

[124] Lo I.-H., Tsai W.-T. Effect of heat treatment on the precipitation and pitting corrosion behavior of 347 SS weld overlay, Materials Science Engineering A 2003; 355, 137-143.

[125] Pujar M.G., Dayal R.K., Gill T.P.S., Malhortra S.N., Role of delta-ferrite in the dissolution of passive films on the austenitic stainless-steel weld metals, Journal of Materials Science Letters 1999; 18, 823-826.

[126] Vander Voort GF., James HM., Wrought Stainless Steels, Metallography and Microstructure. Vol. 9. USA: ASM International; 1992.

[127] ASTM E-384, Standard Test Method for Microindentation Hardness of Materials. ASTM International; 1999.

[128] De Lima-Neto P., Farias J.P., Herculano L.F.G., de Miranda H.C., Araújo W.S., Jorcin J.B., Pébère N. Determination of the sensitized zone extension in welded AISI 304 stainless steel using non-destructive electrochemical techniques, Corrosion Science 2008; 50, 1149-1155.

[129] Olsson C.-O. A., Landolt D. Passive films on stainless steels: chemistry, structure and growth, Electrochimica Acta 2003; 48, 1093-1104.

[130] White W. E. Observations of the influence of microstructure on corrosion of welded conventional and stainless steels, Materials Characterization 1992; 28, 349-358.

Chapter 10

Enhancement of Photoelectrocatalysis Efficiency by Using Nanostructured Electrodes

Guilherme Garcia Bessegato,
Thaís Tasso Guaraldo and
Maria Valnice Boldrin Zanoni

Additional information is available at the end of the chapter

http://dx.doi.org/10.5772/58333

1. Introduction

This chapter describes some fundamental features of photoelectrocatalytic processes, including the basic concepts of the technique, the phenomena at the electrode/electrolyte interface and the development of new materials employed in the last few years related to the specific applications. The nanostructured materials used in the photoelectrochemical field can be called photoanodes (n-type) when oxidation reactions take place at the interface, and photocathodes (p-type) when the reduction is the main process [1, 2]. This chapter focuses on photoanode materials and how their surface influences the applications of this technique.

Photoelectrocatalysis could be described as a multidisciplinary field, involving surface science, electrochemistry, solid-state physics and optics. The basic concept is that when a semiconductor surface is irradiated by light ($h\nu \geq E_g$) there is generation of electron/hole pairs (e^-/h^+) by the promotion of an electron from the valence band (lower energy level) to the conduction band (higher energy level). The electrons are forwarded to the counter electrode under positive bias potential (n-type) in order to minimize the recombination of these pairs due to the short life-time. When immersed in electrolyte the adsorbed water molecules and/or hydroxyl ions react with the holes on the valence band to generate hydroxyl radicals ($^{\bullet}OH$), which are a powerful oxidizing agent (+2.80 V) [3-5].

The first findings, from 1839, found that the photoelectrochemistry field was stimulated by the Becquerel effect [6]. They observed a photocurrent flow of electrons due to illumination of a material connected by two electrodes immersed in solution. In 1972, the work of Fujishima and Honda had a huge impact on this field. They studied the use of a TiO_2 semiconductor on

INTECH
open science | open minds

© 2014 The Author(s). Licensee InTech. This chapter is distributed under the terms of the Creative Commons Attribution License (http://creativecommons.org/licenses/by/3.0), which permits unrestricted use, distribution, and reproduction in any medium, provided the original work is properly cited.

the photoelectrolysis of water (water splitting) under anodic bias potential in a photoelectrochemical (PEC) cell [7, 8]. Nowadays, photoelectrocatalysis is an emerging field with many applications, such as organic compounds oxidation [9-11], inorganic ions reduction [12, 13], disinfection [14, 15] and production of electricity and hydrogen [16-18].

The development of this technique is intimately related to a better understanding of materials' surfaces and properties. Highly ordered nanomaterial arrays have promoted a revolution in applications of these materials as nanotubes, nanowires, nanofibres, nanorods, nanowalls, etc. [19]. The main applications of the technique include the degradation of unwanted environmental pollutants (organic and inorganic compounds) and converting sunlight directly into an energy carrier [4, 19, 20].

This work presents an overview of the fundamentals of photoelectrocatalysis and the huge contribution made by nanostructured architectures, as well as explaining the efficiency of the technique as a treatment method for organic and inorganic compounds and for water splitting.

2. Photoelectrocatalysis: Basic concepts

Advanced oxidation processes (AOPs) have been proposed as alternative methods for the degradation of recalcitrant organic compounds in water [21], air [22] and soil [23] in recent years [4]. AOPs are based on the generation of hydroxyl radicals ($^{\bullet}OH$) as highly oxidant species, which are responsible for the oxidation of the major pollutants [4, 21]. Among the AOPs, heterogeneous photocatalysis deserves particular attention [5]. The method is based on the use of a semiconductor (mostly TiO_2) irradiated with light energy equal to or greater than its band-gap energy. Since 1972 it has been known that is possible to promote photoelectrolysis of water (water splitting) under anodic bias potential [8]. Since then, photocatalysis has been explored to promote organics oxidation [9-11], inorganics reduction [12, 13], disinfection of water containing biological materials [14, 15] and production of electricity and hydrogen [16-18].

A semiconductor material is characterized by two energy bands separated by the band-gap energy, E_g. A semiconductor at absolute zero is insulating, because the valence band (lower energy level) is completely occupied and the conduction band (higher energy level) totally empty (Figure 1). To become conductive, charge carriers need to be created, usually by photoexcitation. The basic concept is that when a semiconductor surface is irradiated by light ($h\nu \geq E_g$) there is generation of an electron/hole pair (e^-/h^+) by promotion of an electron from the valence band (VB) to the conduction band (CB) (Equation 1) [5, 24].

The oxidizing nature of the holes (h^+) in the valence band means they generate $^{\bullet}OH$ radicals by the oxidation of H_2O molecules or OH^- ions adsorbed on the semiconductor surface, and are also able to oxidize organic molecules directly. The photoexcitation of TiO_2 and possible oxidation of an organic compound (RX) are represented in Equations 1–4 [21, 25].

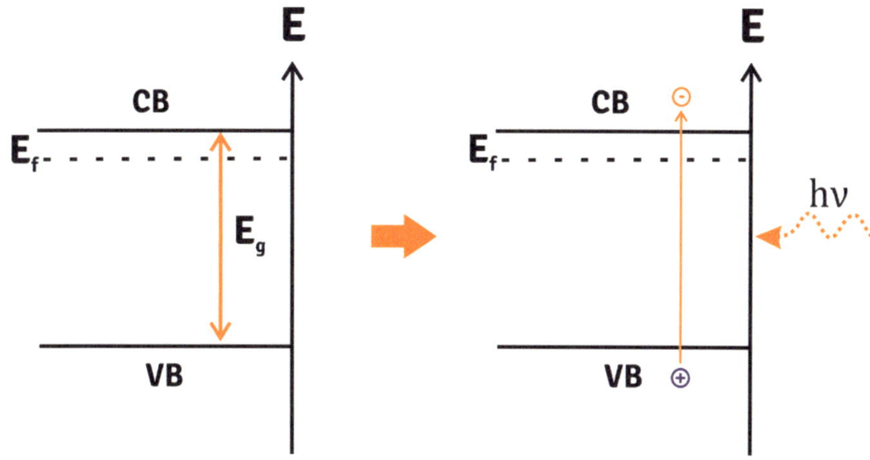

Figure 1. Schematic representation of the energy band diagram in a semiconductor and the mechanism of charge carrier generation by photoexcitation

$$TiO_2 \xrightarrow{hv} TiO_2 - e^-_{CB} + TiO_2 - h^+_{VB} \tag{1}$$

$$TiO_2 - h^+_{VB} + H_2O_{ads} \rightarrow TiO_2 - HO^{\bullet}_{ads} + H^+ \tag{2}$$

$$TiO_2 - h^+_{VB} + HO^-_{ads} \rightarrow TiO_2 - HO^{\bullet}_{ads} \tag{3}$$

$$TiO_2 - h^+_{VB} + RX_{ads} \rightarrow TiO_2 + RX^{\bullet +}_{ads} \tag{4}$$

Although heterogeneous photocatalysis is a well understood process, and despite its promising results in water decontamination, its practical exploitation has been restricted by its low photonic efficiency, which is mainly due to recombination of the e^-/h^+ pair, as shown in Equation 5 [25, 26].

$$TiO_2 - e^-_{CB} + TiO_2 - h^+_{VB} \rightarrow TiO_2 + heat \tag{5}$$

Therefore, there are considerable efforts being made to obtain new processes able to separate charge carriers and minimize their recombination rate [26, 27]. The combination of electrochemical and photocatalysis processes (photoelectrocatalysis) offers the opportunity to separate photo-generated e^-/h^+ pairs by gradient potential [28, 29]. Specifically, when the photocatalyst is attached to a conductive substrate (photoanode), there is the possibility to apply an anodic bias potential to the semiconductor and to modify the substrate/electrolyte interface. This alternative improves the efficiency of charge separation by driving the photo-generated electrons via the external circuit to the counter electrode [26, 28-30]. Figure 2 illustrates the mechanism of photoelectrocatalysis.

Furthermore, the great goal is to avoid the removal of photocatalyst suspensions. The immobilization of the photocatalyst particles on a solid substrate is usually applied on photoelectrocatalysis and therefore the process dispense next filtration step [28, 29].

It is interesting to understand why photoelectrocatalysis is efficient in charge separation. When a semiconductor is in contact with an electrolyte there is formation of a junction semiconductor/electrolyte interface, which determines the electron hole separation kinetics. The junction in a redox electrolyte causes a change in the electrochemical potential (Fermi level) due to discrepant potentials at the interface [19]. Thus, the equilibration of this interface needs the flow of charge from one phase to another, and a band-bending is created within the semiconductor phase. The amount of band-bending in this Schottky junction will depend on the difference of the Fermi levels of semiconductor and electrolyte. The region where there is bending is called the space charge layer (SCL), which is characterized by the accumulation of electrons or holes at the surface [5, 19, 24, 31]. Figure 3 shows the behaviour of these charges in the semiconductor before and after this equilibration when it is in contact with an electrolyte.

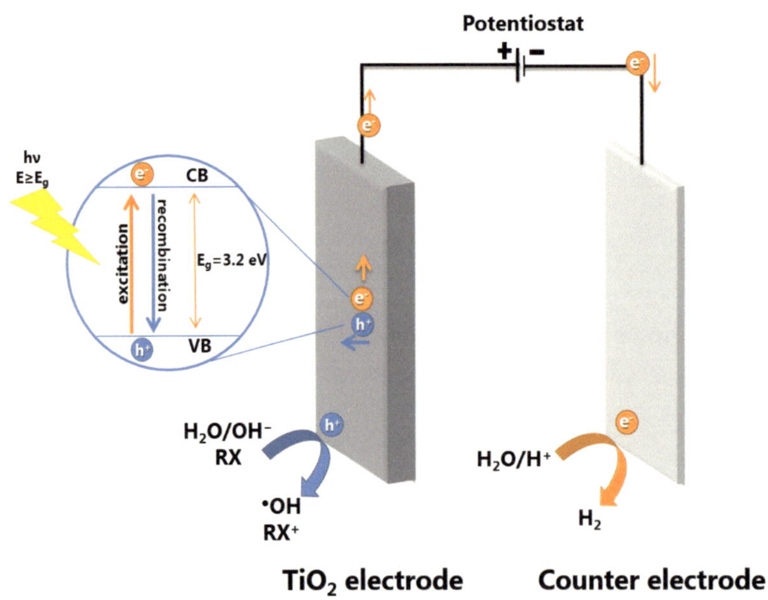

Figure 2. Schematic representation of the mechanism of separation and recombination of charges in the photocatalysis or photoelectrocatalysis and mechanism of charge separation in a photoelectrochemical system, where a gradient of potential is created

http://dx.doi.org/10.5772/58333

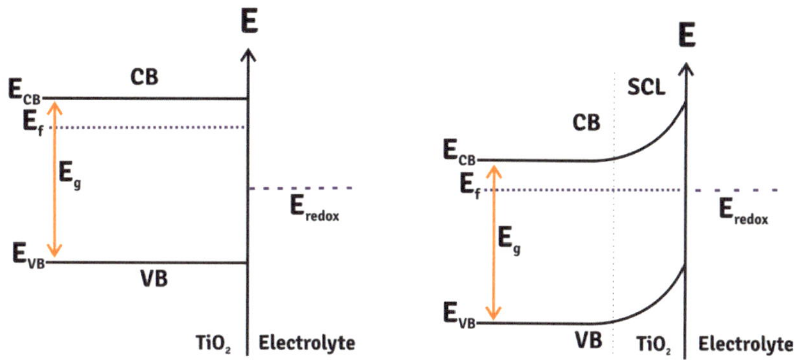

Figure 3. Energy band diagram for an n-type semiconductor before and after the equilibration of Fermi levels at the interface semiconductor/electrolyte, and the appearance of band-bending and the space charge layer (SCL)

Another method to control the Fermi level (and therefore the band-bending) is by applying a bias potential [19]. For any given semiconductor and electrolyte, there is an exact potential for which the potential drops between the surface and the bulk of the electrode is zero; in other words, there is no space charge layer [31]. Because the band edges are flat, this potential is called flat-band potential, V_{fb} (Figure 4). The application of any potential greater than the flat-band potential will increase the band-bending at the n-type semiconductor electrode, such as TiO_2. In this case electrons are depleted and holes enriched at the surface, as we can see in Figure 4. When TiO_2 is irradiated, it is observed that the photogenerated holes have an oxidizing power equivalent to the potential of the valence band edge, and are able to oxidize an RED molecule, whose formal potential is more negative than the valence band. In the case of TiO_2, the H_2O can be oxidized producing •OH radicals. The electron in the conduction band flows via an external circuit to the counter electrode, where reduction reactions may occur, such as the reduction of H^+ ions to H_2 (Figure 2). It is important to note that in photo(electro)catalysis, the greater the band-bending (and therefore the SCL) the faster the electron/hole separation occurs, and then the recombination of charges is minimized [5, 19, 24, 31].

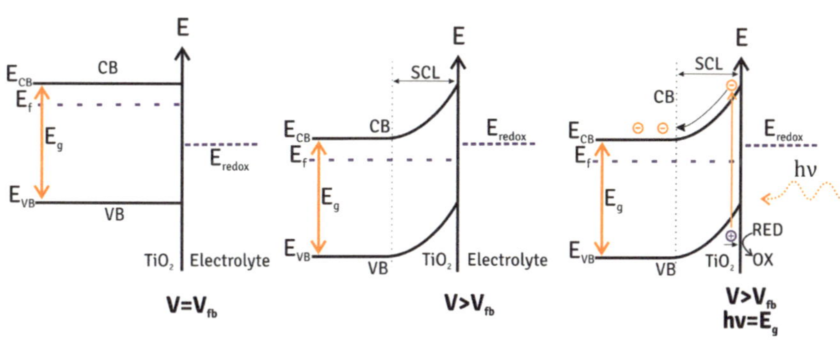

Figure 4. Energy band diagram for a n-type semiconductor when the applied potential (V) is equal to flat-band potential (V_{fb}) and when the applied potential (V) is greater than V_{fb}. The last schematic shows the mechanism of charge separation when the electrode is submitted for a potential higher than the V_{fb} and irradiated with $\lambda \geq E_g$.

Thus, considering the high oxidative power of •OH that is easily generated by irradiation of the TiO_2 surface, an increased number of applications of photoelectrocatalysis has developed with the aim of promoting the degradation of organic pollutants to CO_2 and minerals.

3. The degradation of organic compounds on thin films

The presence of recalcitrant organic pollutants such as pesticides, hormones, pharmaceuticals, phenols, surfactants and dyes in water and wastewater has been described in the literature as one of the most serious problems for human beings and the environment [32, 33]. The great concern is mainly that the genotoxic and mutagenic properties of these pollutants can cause bioaccumulation problems and transportation that is magnified in the food chain [34]. They have therefore received great attention since they are released into the environment through a variety of human and industrial activities. Conventional techniques such as adsorption, precipitation, flocculation and reverse osmosis simply transfer organic pollutants from different phases or concentrate them in one phase, without actually removing them [33].

Different methodologies have been proposed to promote the complete degradation of organic matter. Among them, the use of advanced oxidative processes (AOPs) has been seen as an efficient alternative for pollutant degradation and has received a great deal of attention from several researchers. The *in situ* generation of hydroxyl radicals (HO•) has proved effective in the oxidation of most organic substances because it is both a non-selective reagent and a highly oxidizing agent [21]. However, the complete mineralization which is the conversion of organic molecules into CO_2, H_2O and other small molecules, the reaction mechanisms and the characterization of secondary products and intermediates have not been frequently investigated [35].

Over the past decades, electrochemical methods such as electrocoagulation, electrocatalysis oxidation and reduction, electro-Fenton, photoelectro-Fenton, photocatalysis and photoelectrocatalysis (Figure 5) have been pointed out as good alternatives to promote the degradation and mineralization of organic pollutants, since they combine the advantages of hydroxyl radicals formation and the efficiency of electrochemistry [21, 36].

In Electrochemically Mediated Oxidative Advanced Processes (EOAPs), hydroxyl radicals can be generated by direct electrochemistry (anodic oxidation) or indirectly through electrochemical generation of Fenton's reagent. In photoelectrocatalytic oxidation the •OH is generated heterogeneously by direct water discharge on specific anodes such as DSA and BDD electrodes [36]. During the electro-Fenton reaction the hydroxyl radicals are generated homogenously via Fenton's reaction [37].

Photoelectrochemical methods have been intensively investigated as promising alternative methods not only to remove organic pollutants but also to decrease toxicity, since they degrade substances in a short period of time. The degradation mechanism of photocatalysis can be classified into five steps: (1) transfer of reactants in the fluid phase to the surface; (2) adsorption of the reactants; (3) reaction in the adsorbed phase; (4) desorption of the products; and (5) removal of products from the interface region [38].

http://dx.doi.org/10.5772/58333

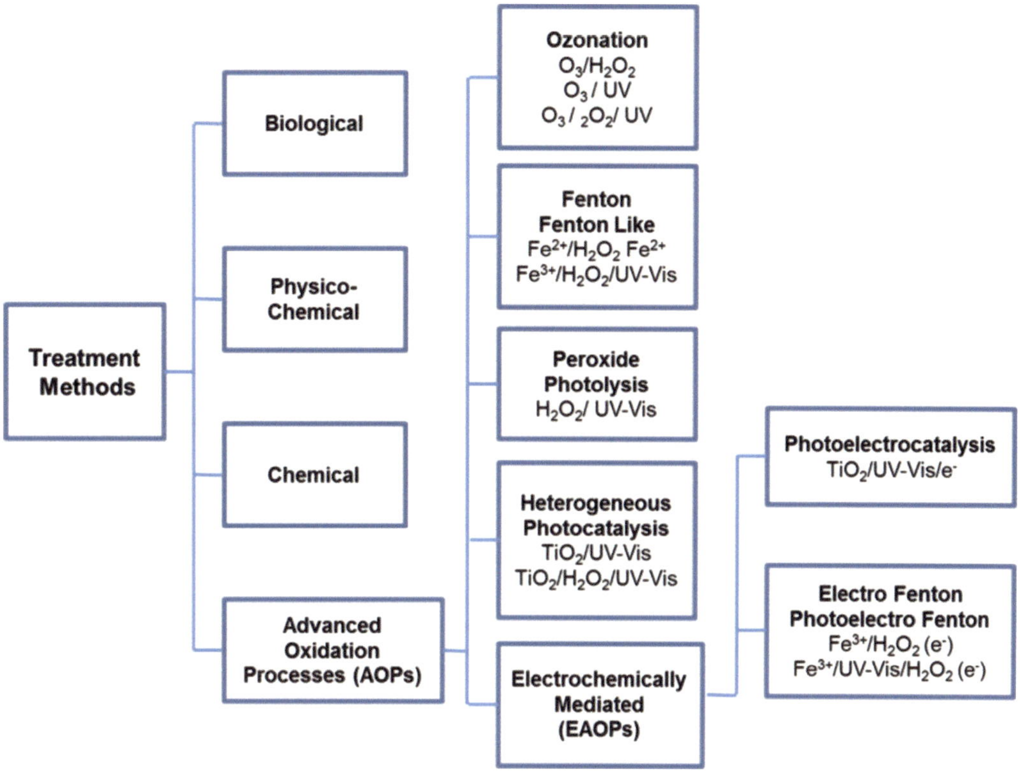

Figure 5. Treatment methods described for the degradation of organic pollutants, including conventional techniques and advanced oxidation processes

The key to obtaining success with photocatalytic and photoelectrocatalytic methods is the development of novel efficient materials as working electrodes, which present good optical, mechanical, electronic, electrochemical and catalytical properties [39]. The choice of the synthesis method to produce the semiconductor material is of fundamental relevance, as it will determine the efficacy of the PEC treatment. All factors related to the surface material will influence the success of photoelectrochemical processes as morphological and structural features (particle size, surface area), good charge separation (e^-/h^+), suitable photonic efficiency and band-gap energy level [40].

3.1. Synthesis of thin film semiconductor materials

Emerging technologies providing feasible alternatives for the development of new materials have been the subject of several studies. Titanium dioxide is the most used material and can be prepared in the form of powder, crystals or thin films. To obtain good-quality materials there are many methods described in the literature, based on precipitation and co-precipitation [41, 42], solvothermal [5], sol-gel [43], microemulsion [44], electrochemical [40] and gas-phase methods [40].

Heterogeneous photocatalysis started with the use of TiO_2 semiconductors in a slurry system (suspension of fine powder). The most efficient powder reported in the literature is the Degussa P25, which is a combination of rutile and anatase allotropic phases in the ratio 3:1. There are many advantages of using this powder: it provides high surface area showing excellent photocatalytic activity because of the adsorptive affinity of organic compounds on the surface of anatase [45]. However, a post-treatment filtration step is required to separate it from the solution, which limits practical application as this is a time-consuming and costly process. Moreover, the suspended particles tend to aggregate, especially at high concentrations, which makes the separation more complicated and limits application in continuous flow systems [46].

Since 1993, the immobilization of TiO_2 on a substrate has offered an alternative way of using powder and started a search for thin films [28, 47]. Several researchers have anchored photocatalysts onto a variety of surfaces, such as glass (ITO and FTO), silica gel, metal, ceramics, polymer, thin films, fibres, zeolite, alumina clays, activated carbon, cellulose, reactor walls and others [33]. To support TiO_2 there must be four main criteria: strong adherence, stability of the catalyst, high specific surface area to promote strong adsorption of the pollutant on the electrode surface [38]. The substrate material has a great influence on the electron transfer along the film. It is reported that conducting glasses have a relatively poor connection within the film; on the other hand, metal substrates present a lower impedance because there is a reduction of charge transfer resistance leading to better PEC activity [33].

The photocatalytic activity of a TiO_2 system mainly depends on its intrinsic properties, such as particle size, surface area, film thickness, crystallinity and crystal phase [33, 48]. For this purpose, many different techniques emerge from the need for immobilization, since the photocatalytic activity of the film is highly dependent on the preparation method [46]. For instance, the most reported preparation routes are sol-gel [43], chemical vapour deposition [49], electrodeposition [50], sol-spray [51], and hydrothermal [38]. Besides the preparation routes, the coating techniques also influence the resulting material properties. Deposition methods such as dip-coating [52], spin coating [53] and even the development of new coating methods based on conventional dip and spin coating [54] have been shown to be simple and able to produce stable materials.

When compared to other methods, the advantages of the sol-gel technique are easy control of deposits, reliability and reproducibility, resulting in good-quality nanostructured thin films [55]. In fact, successful formation of the desired crystal phase is directly related to the starting material, composition, and deposition, as well as the annealing temperature. The crystal morphology has a direct relation to the light absorption as incident light affects photoelectrocatalytic efficiency. Film thickness can affect the efficiency of both light energy conversion and electron transfer; thick films may lower efficiency as these processes have a higher resistance [33]. It has been also shown that the pH of the original solution can influence particle size [56]. It is known [56] that acidic conditions favour the formation of smaller particles, while at higher pH values larger particle size is observed. The use of sol-gel methods has inspired a great number of studies on the development of new semiconductors for the suppression of electron/hole recombination and enhancement of the photosensitivity of titania for successful application [57]. Therefore, the use of nanoporous thin films for photoelectrochemical purposes has been widely described in studies on the removal of organic matter such as dyes [58], phenol [59], tetracycline [60], toxic metals [61] and microorganisms [62]. Annealing temperature has been intimately related to the crystal structure formation because phase transfer is temperature dependent. For many uses, including photoelectrocatalysis and solar cells, the most desired crystal structure is anatase, because this structure shows a higher charge carrier mobility than rutile [19, 63]. However, in many cases of photocatalysis, combinations of anatase and rutile have been used due to the higher photocatalytic activity that these display compared to pure anatase (probably due to the smaller band-gap energy of rutile (E_g=3.0 eV vs. anatase E_g=3.2 eV) absorbing more visible light radiation).

The use of mesoporous TiO_2 thin films has also been studied. According to the definition of IUPAC, porous solids can be classified into three groups based on their pore diameter, namely microporous (5–20 Å), mesoporous (20–500 Å), and macroporous (>500 Å) materials [64]. The success of mesoporous materials depends on the availability of precursor materials and the precision of control over the hydrolysis reaction, as well as the choice of an appropriate surfactant. All these parameters interfere with the obtaining of highly organized materials. In order to obtain mesoporous materials with good photocatalytic features it is necessary to use an appropriate method to produce films with a large surface area, pore-wall structure and crystallinity [65].

Other thin-film semiconductors have been used in the degradation of such organic compounds as WO_3 [66], ZnO [67] and Fe_2O_3 [68, 69]. The anodic growing of tungsten trioxide thin film has been described as a good alternative to TiO_2, mainly because of its intrinsic characteristics like lower band-gap energy of E_g=2.8–3.0 eV and higher photoactivity [70]. Iron oxide (α-Fe_2O_3) has the desirable property of narrowing the band gap (E_g=2.2 eV), as well as low cost, electrochemical stability and low toxicity [68]. ZnO (E_g=3.2 eV) has good properties for use as a photocatalyst, such as high photocatalytic efficiency, low cost and environmental friendliness [71]. It can also be used for degradation and disinfection purposes, as it can degrade dirt and inhibit the growth of microorganisms [67].

3.2. Operational characteristics on the PEC systems

The basic photoelectrochemical reactor setup consists of three conventional electrodes (working, reference and counter electrode) immersed in an aqueous electrolyte contained within a vessel for the potentiostatic mode. A two-electrode system (working and counter) can also be used when current density is used to supply the system. The vessel containing the aqueous electrolyte is transparent to light or fitted with an optical window, usually quartz, that allows light to reach the photoactive electrode [72].

Besides material properties, some operational parameters such as pH, biased potential, initial concentration of analyte and electrolyte composition have a direct influence on the degradation of organic pollutants. The point of zero surface charge (pzc) of the TiO_2 at the electrode/electrolyte interface will determine the adsorption of the pollutant in relation to the pH and pKa of the pollutant. In acidic conditions TiO_2 is positively charged, while in basic conditions it is negatively charged, according to the equations below [25, 33, 73]:

$$TiO_2 + H^+ \rightarrow TiOH_2^+ \quad (6)$$

$$TiO_2 + OH^- \rightarrow TiO^- + H_2O \quad (7)$$

The influence of biased potential on the degradation rate must be optimized as a function of the flat band potential. Generally, when the potential is increased, the degradation rate increases as well until no more gain is observed because electrons and holes have a good separation and recombination rate is minimized [33, 74]. Current density can be applied instead of potential, as it requires a much simpler arrangement of two electrodes, lowering costs and favouring the photoelectrocatalytic application on large-scale reactors [4, 43].

The initial pollutant concentration, especially for wastewaters and coloured solution, will limit the photoanode activation by light [61]. Moreover, at high concentration the photoelectrochemical efficiency is decreased and longer treatment periods will be required to achieve complete pollutant removal. Depending on the pollutant, it is possible to promote the degradation at high concentrations [4, 25].

Recent investigations prove that light intensity and lamp irradiance are critical factors in photoelectrochemical systems. It has been reported in the literature that the higher light intensity achieved, the faster the degradation rate will be [33]. Zainal and colleagues [75] demonstrated that a 100 W UV lamp was almost equivalent to a 300 W halogen lamp, probably due to the higher intensity of the halogen lamp.

When the degradation is conducted in the presence of different electrolytes, there will be significant change in the degradation rate. In the presence of chloride, the degradation is improved because there will be generation of chlorine radicals, with a high oxidizing power which is not observed in sulphate and nitrate mediums [58].

The PEC reactor also plays an important role in the efficiency of photoelectrochemical methods. Different materials (glass, quartz and Teflon) and shapes are employed on these systems. The photoanode irradiation can be used either externally or internally [4]. The reactor could be rectangular or cylindrical, although the latter makes greater use of light and hence better performance. There are single chamber reactors and double-vessel reactors, also known as H-type [72].

4. Strategies to enhance the PEC efficiency

Several photocatalysts have been applied in photoelectrocatalysis, among them TiO_2, WO_3 [66], ZnO [67], CdS, Fe_2O_3 [68, 69] and SnO_2. Over the years considerable effort has been devoted to the improvement of the materials used in photocatalysis. TiO_2 has become one of the most common materials used in materials science [20] as it is environmentally friendly, low cost, has a long lifetime of electron/hole pairs, presents a compatible energy position of BV and BC, and has good chemical and thermal stability and superior catalytic stability [20, 76]. Among these features, the band edge positions relative to H_2O oxidation represent a very important characteristic that improves the applicability of TiO_2 in photo(electro)catalysis to decompose H_2O to H_2 and O_2 and also to create •OH radicals [19]. There are many transition metal oxides with semiconductor properties, but many of them do not have suitable electronic properties (energy position of bands edges) for useful electron transfer reactions.

Some of the main applications of TiO_2 photoelectrocatalysis have involved water-splitting [16, 77, 78] inactivation of microorganisms [14, 79] and degradation of contaminants in water [10, 33, 39, 78, 80]. Although it is the most suitable material for such applications, titanium dioxide has some limitations that hinder its use in technological applications. For example, it is activated only under ultraviolet irradiation ($\lambda \leq 387$ nm), and thus the use of sunlight is limited because it provides up to 5% of UV light; it also presents recombination of electron/hole pairs. In order to obtain a better utilization of the photocatalytic properties of TiO_2 and to achieve more responsiveness to the visible wavelengths, the preparation of nanostructured materials and their surface modification or doping (band-gap engineering) has emerged as a potential method.

Thus, in order to increase the efficiency of photoelectrocatalysis, organized nanostructured materials, especially those involving electrochemical methods of preparation, have attracted attention. The main advantages are discussed below.

4.1. Nanostructured morphologies

Nanostructured materials represent an important challenge of current science, and the new materials have presented special physical and chemical properties. Recently, one-dimensional (1D) nanostructures such as rods, belts, wires and tubes have become a focus of intensive research, mainly due to their high surface area (ideal for catalysis as it facilitates reaction/ interaction between the devices and the interacting media) and other exceptional properties such as electrical properties: charge carrier transfer is mainly governed by the quantum confinement phenomenon [81].

The discovery of carbon nanotubes by Iijima in 1991 [82], with their variety of interesting properties, boosted research focused on the synthesis of tubular nanostructures of other materials. Among the various nanotube materials, titanium dioxide nanotube arrays are of particular interest because of their many applications, for example in photo(electro)catalysis [10, 78, 83-87], sensors [88, 89], biosensors [90], dye-sensitized solar cells [91, 92], hydrogen generation by water photoelectrolysis [77, 78, 93], photocatalytic reduction of CO_2 [94, 95] and biomedical-related applications [96, 97].

In recent years, a great number of investigations have focused on the photocatalytic activity of TiO_2 nanomaterials and effective ways to improve their photocatalytic efficiency. Various nanostructures have been reported, such as nanowires [98], nanofibres [99], nanorods [100, 101], and nanowalls [101], but TiO_2 nanotubes are certainly the most promising and explored architecture.

4.1.1. TiO_2 nanotube arrays

As previously mentioned, TiO_2 is a widely studied materialdue to its versatility, and in photoelectrocatalysis it is undoubtedly the most explored semiconductor. The use of the TiO_2 nanotubes morphology has allowed advances in photo(electro)catalysis due to specific improvement of properties and will be further discussed.

For photoelectrocatalytic applications TiO_2 nanotubes (TiO_2 NTs) present interesting properties, such as large internal surface area, which can be easily filled with liquid enabling intimate contact with electrolytes and excellent charge transport [39, 94]. Due to its high structural organization, the nanotubes architecture exhibits excellent electron percolation pathways for vectorial charge transfer between interfaces, thereby minimizing the recombination of charges. Figure 6 illustrates an image of scanning electron microscopy (SEM) of TiO_2 NTs prepared under electrochemical anodization. As the TiO_2 film grows on the metal surface (is not deposited) there is a good electrical connection between the oxide and the metal. Zhu and colleagues [102] found charge carrier recombination much slower in the TiO_2 NTs films than in the nanoparticulate TiO_2 films in dye-sensitized solar cells.

Additionally, the morphological parameters of the architecture can be precisely controlled when the material is prepared by electrochemical anodization. The control of the nanotube dimensions is important because each application may require morphological surfaces with particular characteristics. For example, Liu et al. [103] found that the photoelectrocatalytic activity shows a dependence on the length of the nanotube arrays. They studied the degradation of phenol at TiO_2 NTs electrodes with different tube lengths under UV irradiation and applied potential. It was verified that a short nanotube array shows better photoelectrocatalytic activity than a long nanotube array, which can be explained by the reduced recombination effects. However, the photocatalytic degradation (no applying potential) showed that longer nanotubes were more efficient because they favour light trapping.

More information can be obtained in some excellent reviews found in the literature, dealing with preparation, properties, strategies to increase the photoactivity and applications of TiO_2 NTs [19, 20, 39, 81, 94, 104-107]. Titania nanotubes can be synthesized in two forms: powder form and self-organized nanotube arrays grown on a substrate of metallic titanium. Several techniques for the preparation of TiO_2 NTs have been reported, such as hydro/solvothermal methods [108], sol-gel [109], template-assisted methods [110] and electrochemical anodization [39, 105, 106]. The growth of TiO_2 NTs by electrochemical anodization in a fluorinated-based electrolyte is less expensive and simpler that most of these methods and allows precise control of dimensions, presenting a more orderly arrangement of nanotubes [105].

The first self-organized oxide obtained by anodization in electrolytes containing hydrofluoric acid was reported by Zwilling and colleagues in 1999, where a nanoporous structure was achieved [111]. In 2001, Gong and colleagues [112] developed the first generation of highly ordered and vertically oriented nanotube arrays of 500 nm length. The structure was obtained by electrochemical oxidation of titanium in a HF aqueous electrolyte. The fabrication of TiO_2 NTs films was performed in a two-electrode electrochemical cell using aqueous electrolytes

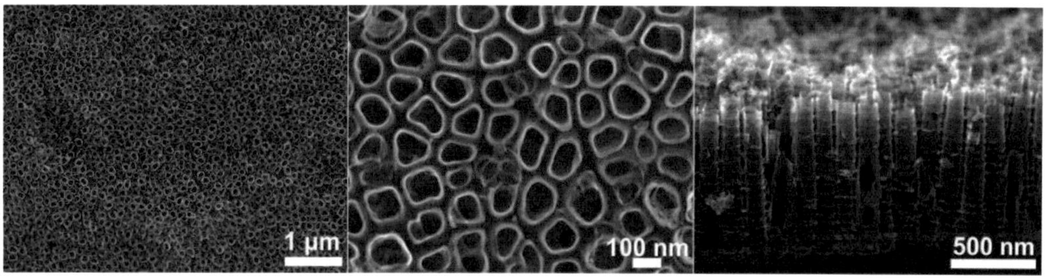

Figure 6. TiO_2 nanotubes scanning electronic microscopy (SEM) images, top view (in different magnifications) and cross section. The TiO_2 NTs were grown by electrochemical anodization of Ti foil in 1 M NaH_2PO_3+0.3 wt.% HF. The TiO_2 NTs presented a diameter of 110 nm, wall thickness of 13 nm and length of 900 nm on average

containing 0.5-3.5 wt. % HF and voltages varying from 3 to 23 V. They found that at low voltage (3 V), porous films are obtained and at higher voltage (23 V) the nanotube structure was destroyed. The ideal conditions were 0.5 wt. % HF electrolyte applying 20 V for 20 min.

In 2005, Cai and colleagues [113] developed the second synthesis generation of titania nanotubes. They found that adequate control of the electrolyte pH can decrease the oxide chemical dissolution rate; thus, the tube length is enhanced using aqueous buffer electrolyte. The pH of a KF-containing electrolyte is adjusted to 4.5 using additives such as sulphuric acid, sodium hydroxide, sodium hydrogen sulphate, and/or citric acid. This usually obtains TiO_2 NTs of 4.4 μm in length.

The third synthesis generation of titania nanotube arrays, initially reported by Ruan and colleagues [114] in 2005, involves improvements in nanotube-array length using non-aqueous electrolytes or polar organic solvents such as formamide, N-methylformamide, dimethyl sulphoxide, and ethylene glycol mixed with HF, NH_4F or KF to provide fluoride ions [112, 115-117]. Ruan and colleagues [114] also studied the anodization of titanium in polar organic solvent using mixtures of dimethyl sulphoxide (DMSO) and hydrofluoric acid. TiO_2 nanotube arrays of 2.3 μm length were obtained in DMSO+4.0% HF electrolyte applying 20 V for 70 h.

The fourth synthesis generation of TiO_2 NTs was developed by Richter and colleagues [118] and Allam et al. [119], and is characterized by the fabrication of nanotube arrays by Ti anodization using fluoride-free HCl aqueous electrolytes. The mechanism of TiO_2 NTs formation on Ti substrate is well studied in the literature [94, 105, 106].

4.1.1.1. Mechanism of formation of nanotubes by electrochemical anodization

The production of oxide films on metal surfaces by oxidation in an electrolytic process can be called electrochemical anodization. In practice, a metallic electrode compatible with oxide growth is connected to the positive pole (anode) of a dc power supply and the cathode, usually a platinum piece (or another material, such as carbon for example) is connected to the negative pole (Figure 7). The electrodes are placed in an electrolytic solution and when a potential is applied in the system the metal reacts with oxygen ions from the electrolyte, growing an oxide film on the surface. The electrons resulting from the oxidation travel through the external circuit to reach the cathode, where they can react with H^+ ions and generate bubbles of H_2 [94]. The key point that determines the form of the oxide is the composition of the electrolyte. The TiO_2 NTs, in this case, can be achieved in electrolytes containing fluoride ions, with adjustments of applied potential and anodization time.

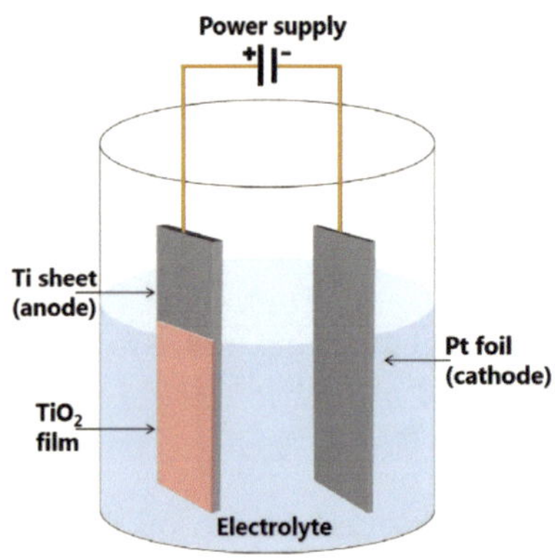

Figure 7. Scheme representing an electrochemical cell used to produce TiO_2 films by anodization of Ti

The growth of self-organized TiO_2 NTs (as well as porous structures of other metals such as Zr, Nb, W, Ta, and Hf) by electrochemical anodization in fluoride-containing electrolyte is governed by competition between steps that occur simultaneously.

First, there is the formation of oxide on the metal surface. In this step, there is a field-assisted oxidation of Ti metal to form Ti^{4+} species which will react with O_2^- (from H_2O). After the formation of an initial oxide layer, further oxide growth is controlled by field-assisted ion transport, where O_2^- anions migrate through the oxide layer until they reach the metal/oxide interface, where they react with the metal [94, 106, 120].

$$Ti + 2H_2O \rightarrow \mathrm{TiO}_2 + 4H^+ + 4e^- \tag{8}$$

In another step, Ti^{4+} ions migrate from the metal through the oxide by field-assisted transport until they reach the oxide/electrolyte interface. Then, small pits are formed due to the localized dissolution of the oxide by the high electrical field, which act as pore-forming centres.

The key step is the chemical dissolution of oxide by fluoride ions at the as-formed pits, forming soluble fluoride complexes. The Ti^{4+} ions field transported at the oxide/electrolyte interface are also complexed [94, 106, 120].

$$\mathrm{TiO}_2 + 6F^- \xrightarrow{H^+} [TiF_6]^{2-} + H_2O \tag{9}$$

$$Ti^{4+} + 6F^- \rightarrow [TiF_6]^{2-} \tag{10}$$

If the chemical dissolution is too high or too low, there is no formation of nanotubes. The dissolution rate can be adjusted by varying the concentration of F^- and pH (more acidic pH and higher concentrations of F^- increases the chemical dissolution) [94]. This was the principle used to obtain longer and smoother nanotubes, leading to the second and third generations of TiO_2 NTs.

When the rate of pore growth at the metal–oxide interface becomes identical to the rate of oxide dissolution at the pore–bottom–electrolyte interface, the thickness of the barrier layer remains unchanged, although it moves further into the metal, making the pore deeper [94, 106, 120]. Commonly, the wall thickness of TiO_2 NTs varies from 5 to 30 nm and the pore size from 20 to 350 nm (tube diameter is reported to be linearly dependent on the applied anodic potential during growth [106, 121]). The length often varies from 0.2 to 1000 μm; the aspect ratio, defined as the ratio between length and diameter of the tube, can be controlled from about 10 to approximately 20,000 by selection of appropriate anodization variables [94].

4.1.2. Nanostructured arrays of other semiconductors

Nanostructured architectures are also fabricated by electrochemical anodization for other semiconductors of interest in photoelectrocatalysis, such as ZnO, WO_3 and Fe_2O_3.

Prakasam and colleagues [69] prepared nanoporous film of Fe_2O_3 by submitting a Fe foil to electrochemical anodization in electrolyte composed of 1% HF+0.5% ammonium fluoride +0.2% 0.1 M nitric acid (HNO_3) in glycerol (pH 3) at 10°C. LaTempa and colleagues [122] produced α-Fe_2O_3 (hematite) nanotubes by potentiostatic anodization of iron foil in an ethylene glycol electrolyte containing NH_4F and deionized water. Hematite has a band gap of ≈2.2 eV (indirect) and can absorb light at $\lambda \leq 560$ nm; it can therefore be activated in a large part of the solar spectrum.

Lai et al. [123] prepared WO_3 nanotubes by electrochemical anodization of W foil in electrolyte composed of 1 M of sodium sulphate+0.5 wt.% of ammonium fluoride at 40 V. The WO_3 is photoactive when irradiated by visible light due to its small band-gap energy (2.4 eV to 2.8 eV) and has attracted scientific interest in photo(electro)catalysis. Some reviews [29, 70] have explored the use of WO_3 photoanodes mainly in photoelectrochemical water splitting.

Park and colleagues [124] reported a synthesis of ZnO nanowires by electrochemical anodization on a Zn foil using as electrolyte 5 mM $KHCO_3$ aqueous solution. ZnO has a similar band gap and band positions of TiO_2 (E_g about 3.2 eV), but higher quantum efficiency than TiO_2. On the other hand it has limited applications due to its photocorrosion in acidic medium [71].

4.2. Band-gap engineering

Despite all the improvements made to TiO_2 as a photoactive catalyst, the material still presents problems, such as activation with UV irradiation ($\lambda \leq 387$ nm), due to its wide band gap (E_g=3.2 eV). Thus, the use of solar energy is limited since the activation of TiO_2 occurs only from UV light, which corresponds to a small fraction (≈5%) of the sun's energy compared to visible light (45%) [39]. In this sense, efforts have been directed at shifting the optical response of titanium dioxide from the UV to the visible spectral range, which would be of great utility in photo(electro)catalysis and other applications of TiO_2. This modification of optical properties of semiconductors has been called band-gap engineering [19, 39, 94, 107].

Modification of TiO_2 properties has been achieved mainly by (i) doping with different transition metal ions (such as Cr [125], Co [126], W [127], Zr [128] and Fe [129]) and with different anions (such as N [130], F [131], S [132], B [133], C [93]) that replace oxygen in the crystal lattice, and (ii) by surface decoration, which includes coupling with other semiconductors and deposition of particles of noble metals [14, 84, 134-137].

However, these arrangements frequently increase only the absorption and do not properly improve material properties such as the stability of the semiconductor under illumination, efficiency of the photocatalytic process, and the wavelength range response. One example is the CdS, which absorbs a good portion of the visible radiation but is usually unstable and photodegrades with time [138].

Dopant/Modifier	Strategies	References
N	Anodization of Ti–N alloy* Anodization in nitrogen-containing electrolyte* Electrodeposition in nitrogen-containing electrolyte	[130], [139], [140], [141], [142], [143], [144], [145]
C	Anodization in carbon containing electrolyte*	[146], [147]
F	Anodization in containing electrolytes*	[143]
B	Anodization in boron-containing electrolyte* Electrodeposition in boron-containing electrolyte	[133], [148], [149] [150]
W	Anodization of Ti–W alloy* Anodization in tungsten-containing electrolyte*	[127], [151], [152]
Zr	Anodization in zirconium-containing electrolytes* Electrodeposition in zirconium-containing electrolyte	[128], [153], [145]
La	Electrodeposition in lanthanum-containing electrolyte	[154]
Si	Anodization in silicon-containing electrolyte*	[152]
Nb	Anodization of Ti–Nb alloy*	[144]
Ag	Electrodeposition in silver-containing electrolyte	[155], [156]
Pt	Electrodeposition in platinum-containing electrolyte	[157], [158]
Pd	Electrodeposition in palladium-containing electrolyte	[159], [160]
CdS	Electrodeposition in Cd and S-containing electrolyte	[135], [161]
CdTe	Electrodeposition in Cd and Te-containing electrolyte	[162]
Cu_2O	Electrodeposition in Cu-containing electrolyte	[163]

*one-step synthesis

Table 1. Electrochemically doped/surface modified TiO_2 nanotube arrays

In order to make materials more photoactive under visible light and more stable under certain conditions, and to have lower band-gap energy, the doping of TiO_2 with several metals and non-metal compounds has also been explored: Table 1 shows a summary of the electrochemical methods adopted to promote doping/surface modification of TiO_2 nanotubes,with the related references.

4.2.1. *Doped TiO_2 nanomaterials*

Asahi et al. [164], in a 2001 study, developed a method for TiO_2 visible light activation through doping of C, N, F, P, or S for O in the anatase TiO_2 crystal using calculated densities of states (DOSs). They found that the substitutional doping of N was the most effective method because nitrogen p states contribute to band gap narrowing by mixing with O 2p states. Nitrogen can be easily introduced into the TiO_2 structure, due to its comparable atomic size with oxygen, small ionization energy and high stability.

There are two main ways to perform anion doping in TiO_2 by electrochemical techniques: (i) electrodeposition and (ii) adding a precursor of the element into the electrolyte during electrochemical anodization to oxide formation. It should be noted that for this the TiO_2 film must be immobilized on a conductive substrate, as in the case of TiO_2 NTs grown on metallic titanium.

In 2006, Shankar and colleagues [139] described a simple way to introduce N atoms into TiO_2. N-doped thin films were fabricated by anodic oxidation of a pure titanium sheet in electrolyte composed of 0.07 M HF, NH_4NO_3 (from 0.2 to 2.5 M) and NH_4OH to adjust the pH to 3.5. The material showed optical absorption in the visible wavelength range from 400 to 530 nm. The XPS data confirmed that all the incorporated nitrogen is substitutional on the oxygen site, and the proportions of N atoms in $TiO_{2-x}N_x$ were x=0.23, x=0.09 and x=0.02. The N-doped samples exhibited a shift in absorption toward the visible spectra from 400 to 510 nm. Antony and colleagues [140] prepared N-doped TiO_2 NTs by anodizing Ti foils in ethylene glycol+NH_4F +water mixture containing urea as a nitrogen source. They used various concentrations of urea and achieved different N concentrations in TiO_2 film, determined by X-ray photoelectron spectroscopy (XPS). There was nitrogen incorporation in TiO_2lattice mainly in substitutional form (substitution of O_2^- ions by N^{3-} ions). The doped samples showed visible light response, and the calculated optical band gaps were 3.27, 3.21, 2.75 and 2.77 eV for pristine TiO_2, $TiO_{1.85}N_{0.115}$, $TiO_{1.813}N_{0.14}$and $TiO_{1.84}N_{0.121}$, respectively. Zhou et al. [141] fabricated N-doped using the same methodology, via anodic oxidation of Ti in electrolyte composed of ammonium fluoride (NH_4F) and triethylamine ($C_6H_{15}N$). Nitrogen was successfully introduced into the TiO_2 lattice replacing oxygen atoms, and as a result there was a shift of TiO_2 band edge from 380 nm to 405 nm in N-doped TiO_2.

Kim et al. [142] produced N-doped TiO_2NTs by anodization of a high-purity TiN alloy with approximately 5 at.% of N in a glycerol+water (50:50 vol%)+0.27 M NH_4F electrolyte. XPS data of the sample surfaces indicated 2–3 at.% of N atoms present as Ti–O–N in the nanotubes. They found that the nanostructured layer grown on TiN alloy showed decreased UV response compared with pure TiO_2 NTs film, but showed a strongly increased photoresponse in visible light spectra.

Li and colleagues [165] used an electrochemical technique to dope TiO_2 with nitrogen atoms, in two steps. N-doped TiO_2 NTs were prepared by electrochemical anodization in glycerol electrolyte, followed by electrochemical deposition in NH_4Cl solution. The optimal conditions in electrodeposition were: voltage of 3 V, reaction time 2 h, and NH_4Cl concentration of 0.5 M. Both the photoelectrochemical properties and photocatalytic activity under visible light irradiation were enhanced after N doping into TiO_2 nanotube arrays.

By using the aforementioned electrochemical techniques for the nitrogen, it is also possible to perform doping with other non-metals, such as C and B, for example. Milad and colleagues [146] achieved carbon-doped titanium oxide nanotubular arrays via anodic oxidation of titanium foil at 20 V in acidic (0.5 M H_3PO_4+0.14 M NaF) and organic media (ethylene glycol +0.3 wt% NH_4F) with 0.5 and 1 wt% carbon source (polyvinyl alcohol). Approximately 2.75% and 8.45% carbon was incorporated into the TNT in the acidic and organic electrolyte, respectively. The highest photocurrent density was observed for the sample with the higher amount of carbon atoms incorporated. Krengvirat et al. [147] produced carbon-incorporated TiO_2 by anodic oxidation in EG containing 0.5 wt% NH_4F+1 wt% water. The interstitial carbon arising from the pyrogenation of ethylene glycol electrolytes induced a new C *2p* occupied state at the bottom of the TiO_2 conduction band, decreasing band-gap energy to 2.3 eV and consequently making the material visible-light active. Lu and colleagues [133] fabricated boron-doped TiO_2 NTs by electrochemical anodization in an electrolyte containing different concentrations of $NaBF_4$ as a boron source. XPS data showed that the boron atoms were incorporated into the TiO_2 lattice, forming a Ti–B–O bond. All the samples presented red shift (photoresponse under visible light) and higher photocurrents under visible light than the bare TiO_2 NTs. Li and colleagues [148] fabricated TiO_2 NTs by electrochemical anodization of Ti in 1 M $(NH_4)_2SO_4$ + 0.5 wt% NH_4F electrolyte, and accomplished boron doping by electrodeposition in 0.1 M H_3BO_3 electrolyte (using current densities of 10 $\mu A/cm^2$ for 27 min). Using XPS data, B atoms were incorporated into TiO_2 matrix, and the B-doped samples exhibited red shift in absorption (380–510 nm) due to the excitation of electrons from the impurity energy levels located above the valence-band edge (provided by the B atoms), to the conduction band edge. The proposed mechanism is consistent with those reported for doping with carbon and nitrogen.

Besides anion doping, there are numerous papers that investigate the effect of doping with metal ions in the TiO_2 lattice. The metal ions can occupy two different positions in the TiO_2 matrix, which are substitutional and interstitial, depending on the ionic radius of the metal. The dopant occupies the interstitial sites if the dopant radius is much smaller than the matrix cation, in this case, titanium. If the dopant has similar ionic radius of Ti, the substitutional mode is adopted [7]. In metal-doped TiO_2, new energy states can be formed either within or beyond the VB and CB, decreasing band-gap energy. However, transition metals may also act as recombination sites and may cause thermal instability in the anatase phase of TiO_2 [7, 27].

Tungsten-doped TiO_2 NTs were prepared by Gong et al. [127] in glycerol/fluoride electrolyte containing sodium tungstate via the electrochemical oxidation of a Ti substrate. XPS data showed that the W^{6+} ions were loaded into TiO_2 lattice by displacing Ti^{4+} ions and forming W–O–Ti bonding. Thus, the UV–Vis spectra of W-doped samples show red shift and decrease the band-gap energy from 3.18 eV (bare TiO_2 NTs) to 2.97 eV (W-doped TiO_2 NTs). These findings can be attributed to the fact that the conduction band of the W-doped samples was reformed in the presence of W^{6+} ions. Das and colleagues [151] prepared tungsten-doped TiO_2 NTs by electrochemical anodization of Ti–W alloys. The sample containing 9% W presented band-gap energy of 2.83 eV and higher visible photocurrents than undoped samples.

Liu and colleagues [153] produced Zr-doped TiO_2 NTs. They prepared TiO_2 NTs by electrochemical anodization in 0.14 M NaF and 0.5 M H_3PO_4 electrolyte, and made the zirconium doping by electrodeposition in 0.1 M $Zr(NO_3)_4$ electrolyte, varying the applied potential. When the amount of zirconium in TiO_2 was small (lower potentials of deposition) zirconium entered into the lattice of TiO_2, acting as defect positions, improving separation of charges. At higher Zr amounts, zirconium atoms were partially unable to enter into the TiO_2 lattice, acting as recombination sites on the TiO_2 surface, decreasing the photocatalytic efficiency. Using a similar approach, Nie and colleagues [154] produced lanthanum-doped (La-doped) TiO_2 NTs. After the preparation of TiO_2 NTs, they executed a cathodic electrochemical process using lanthanum nitrate solution as the La source. The material became visible photoactive, and the band gap was decreased from 3.32 eV (undoped) to 3.03 eV (La-doped NTs).

Another approach reported in the literature focuses on the incorporation of more than one anion (or an anion and a cation) in the structure of TiO_2, which is called codoping. Su et al. [143] prepared N-F-codoped TiO_2 NTs by electrochemical anodization of Ti in oxalic acid +NH_4F electrolyte. N-doping into TiO_2 resulted in the creation of surface oxygen vacancies, and F-doping produced several beneficial effects, such as the creation of surface oxygen vacancies, which enhance the surface acidity, and creation of Ti^{3+} ions, which reduce electron/hole recombination. Zhou and colleagues [149] produced B,N-codoped TiO_2 nanotube arrays. Sun et al. [152] produced Si–W codoped TiO_2 NTs using a one-step anodization process with the presence of silicotungstic acid in the electrolyte, and the doped samples presented visible photocurrent 2.5 times larger than bare TiO_2 NTs. Xua and colleagues [144] produced passivated n–p co-doping of niobium and nitrogen into TiO_2 lattice by anodizing Ti–Nb alloys and posterior N-doping. Liu et al. [145] produced N/Zr-codoped TiO_2nanotube arrays in a two-step process. Firstly they prepared the TiO_2 NTs by electrochemical anodization and then accomplished doping using electrochemical deposition in $Zr(NO_3)_4$ and NH_4Cl electrolyte. The doped materials presented increased photoactivity under UV and visible light; the visible light sensitivity was caused by N-doping, and Zr-doping was responsible for enhancing the charge separation.

Although several mechanisms have been proposed for doping from experimental and theoretical data, it is not possible to clearly understand the role of dopants and therefore there is no consensus in the scientific community [7, 166]. Table 1 shows a summary of the electrochemical strategies fordopingTiO_2 nanotubes, with the related references.

4.2.2. *Composite semiconductor as photocatalysts*

The coupling of two semiconductors with appropriate energy CB and CV can reduce the recombination of e^-/h^+ pairs due to the transfer of carriers from one semiconductor to the other, as can be seen in Figure 8. Furthermore, depending on the band-gap energy of the semiconductor used, the composite can be activated in the visible region [7].

There are few papers that report on the coupling of semiconductors by electrochemical techniques. In this case, the composite is produced by a two-step process. CdS is the most used semiconductor to coupling with TiO_2 due to its small band-gap energy (E_g=2.4 eV). Li and colleagues [135] produced CdS nanoparticles-modified TiO_2 nanotube arrays by electrodeposition via direct current. In the electrodeposition step, they used as electrolyte a mixed solution of 0.01 M $CdCl_2$ in dimethyl sulphoxide (DMSO) with saturated elemental sulphur. CdS was cathodically electrodeposited at the optimum constant DC density of 0.5 mA cm^{-2} for 5–15 min. They found that the photocurrents of CdS/TiO_2 NTs were much larger than those of pure TiO_2 NTs. Under UV−Vis irradiation, both semiconductors are excited and as the conduction band of TiO_2 is more anodic than that of the CdS there is efficient electron transfer between the CdS and TiO_2. Thus, the photogenerated electrons are injected from the conduction band (CB) of CdS to the CB of TiO_2; at the same time, the holes transfer from the valence band (VB) of TiO_2 to the VB of CdS. In heterojunctions such as CdS/TiO_2 there are less electron/hole recombinations and enhanced light absorption, both UV and visible. Zhang et al. [161] prepared water-soluble CdS quantum dots (QDs) and deposited on highly ordered TiO_2NTs by various methods, including cyclic voltammetric (CV) electrodeposition. The QDs were prepared using 0.01 mol L^{-1} cadmium nitrate and 0.01 mol L^{-1} sodium sulphide dissolved in 6 × 10^{-5} mol L^{-1} *N*-cetyl-*N*,*N*,*N*-trimethyl ammonium bromide aqueous solution. The CV electrodeposition was carried out in a conventional three-electrode system with TiO_2 NTs as the working electrode under applied voltage sweeps from −0.8 to 0.2 V versus SCE and a scan rate of 30 mV s^{-1}. The yielding composites of CdS/TiO_2 NTs prepared by CV showed excellent photoelectrical behaviour and superior visible-light photocatalytic activity due to the solid binding and effective coupling between the QDs and the TiO_2 NTs.

Feng and colleagues [162] prepared a heterojunction of CdTe/TiO_2 NTs. CdTe is a direct band-gap semiconductor with E_g=1.5 eV, absorbing almost across the visible spectrum. After the preparation of TiO_2 NTs, CdTe nanoparticles were pulse electrodeposited in a conventional three-electrode system (with the TiO_2 NTs as working electrode) in electrolyte solution containing 0.08 mol L^{-1} $CdSO_4$ and 0.05 mol L^{-1} $NaTeO_3$. The pulse on–off time ratio was 0.2:1, with a running voltage of −1 V. A red shift of 50 nm was observed in CdS/TiO_2 NTs composite and the calculated optical band gap was 1.5 eV. The positions of CB and VB in relation to the TiO_2were similar to the CdS; there was electron injection from the photoexcited CdTe to TiO_2 CB, and the photogenerated holes moved from the TiO_2 VB to the CdTe VB, preventing the recombination of charges.

Tsui and colleagues [163] studied the modification of TiO_2 NTs with Cu_2O by electrodeposition. Cu_2O is a p-type semiconductor with a direct band gap of 1.95–2.2 eV. The junction between p-type Cu_2O and n-type TiO_2 in principle enhances the separation of electron/hole pairs; the Cu_2O is also visible-light responsive. Electrodeposition of Cu_2O was performed using the as-prepared TiO_2 NTs with working electrode using a three-step pulse plating method (−0.5 V for 5 ms, −0.3 V for 0.5 ms, and 0 V for 5 s) from a solution containing 0.02 M $Cu(CH_3COO)_2$ and 0.1 M $NaCH_3COO$ (pH 5.7). The Cu_2O/TiO_2 composite presented visible light absorption and

http://dx.doi.org/10.5772/58333

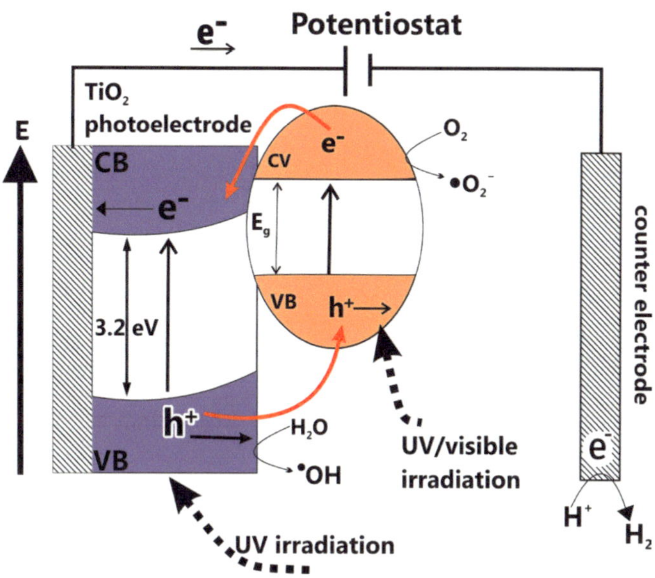

Figure 8. Schematic representation of the mechanism of charges separation in a photoelectrochemical system operatedby coupling a visible active semiconductor to a TiO_2 electrode

the band gap values obtained were 3.27 eV for TiO_2 and 2.21 eV for Cu_2O/TiO_2 heterojunction. However, Cu_2O on TiO_2 NTs dissolves under intense light, limiting the use of Cu_2O in photoelectrochemical devices.

4.2.3. Metal deposition

The decoration of TiO_2 by dopants of nanoparticles of noble metals (such as Ag, Au, Pt, and Pd) has attracted attention in order to enhance the photoactivity of the material. Due to different Fermi levels of TiO_2 and the metal nanoparticles, a Schottky barrier can be formed in the new material. Therefore, there is a rectification of the charge carrier transfer where the energetic difference at the semiconductor/metal interface drives the e^- from the CB of the TiO_2 into the metal nanoparticles. In other words, the metal acts as an electron trap, promoting interfacial charge transfer and therefore minimizing recombination of the e^-/h^+ pairs, as shown in Figure 9 [7].

Xie and colleagues [155] produced Ag-loaded TiO_2 NTs using pulse current deposition technique in 0.01 M $AgNO_3$ and 0.1 M $NaNO_3$ electrolyte, using the as-prepared TiO_2 NTs as working electrode. They applied −15 mA cm^{-2} of pulse current with 0.1 s on-time and 0.3 s off-time. Highly dispersed Ag nanoparticles of 10–40 nm were deposited on TiO_2. TiO_2 NTs and Ag/TiO_2 NTs showed a similar maximum photocurrent density λ (i_{max} 330 nm), but Ag/TiO_2 NTs displayed much more intensive photocurrent response, which can be explained by the Schottky barrier formation separating the charge carriers more efficiently. Zhang and colleagues [156] prepared N-doped TiO_2 NTs and loaded Ag nanoparticles on the TiO_2 surface by

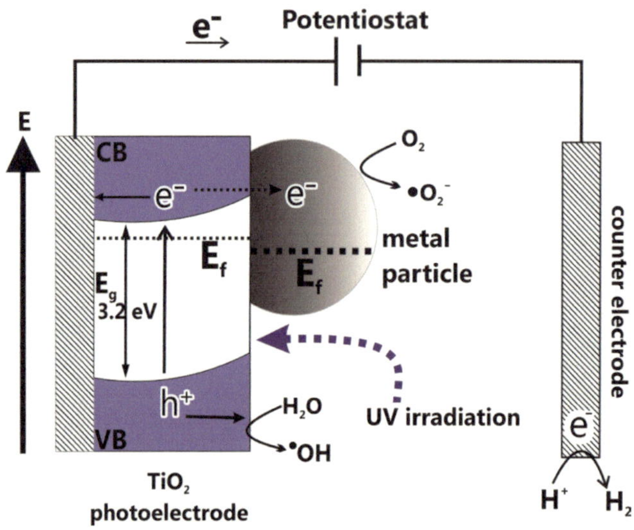

Figure 9. Metal coupling on TiO_2 surface and the mechanism of charge separation in a photoelectrochemical system

electrochemical deposition using 0.2 g L^{-1} $AgNO_3$ in 2.5 g L^{-1} EDTA solution applying −0.1 V for 1–20 s.

Xing et al. [157] produced Pt-nanoparticles-decorated TiO_2 NTs by cyclic voltammetry electrodeposition in 19.3 mM H_2PtCl_6 solution from −0.4 to 0.5 V at a scan rate of 10 mV s^{-1} (controlling the number of cycles). Yin and colleagues [158] also prepared Pt/TiO_2 NTs using an electrochemical approach, but using AC electrodeposition at 2–4 V for 5–30 min in solution containing 1 mmol L^{-1} of H_2PtCl_6.

In the paper of Qin and colleagues [159] Pd particles were deposited onto the TiO_2 NTs electrode by a pulse electrodeposition technique in $PdCl_2$ (2 g L^{-1}) electrolyte solution (pH 1.5). Cheng et al. [160] prepared Pd/TiO_2 NTs through an electrochemical deposition method at a constant potential of −0.8 V using $PdCl_2$ solution (1 mM) in 0.5 mol L^{-1} NaCl electrolyte. The Pd/TNTs sample displayed absorption between 540 nm and 700 nm and presented transient photocurrent density of about 0.094 mA cm^{-2}, higher than that of TNTs (0.067 mA m^{-2}) under xenon lamp irradiation, indicating that decoration with Pd improves the charge separation, according to the Schottky barrier formation mechanism.

All these materials have been demonstrated to massively improve photoelectrocatalytic oxidation processes. Works dealing with water contaminated by a wide range of compounds are discussed below and summarized in Table 2.

5. Application of nanostructured materials in photoelectrocatalysis

As the complexity of contaminants increases, the efficiency of photoelectrocatalytic treatment methods needs to be enhanced by the use of different strategies, as they pose a potential risk to the environment. Most reported work tackles the oxidation of organic pollutants, such as dyes of different classes and industry uses, hormones, pharmaceuticals, pesticides, etc. Oxidation of biological microorganisms such as bacteria and fungus has also been investigated. In all these studies, oxidation is promoted by •OH action generated at the interface photoanode/electrolyte. As discussed previously, these hydroxyl radicals are generated on n-type semiconductors when the holes (h^+) on the electrode surface react with water and/or hydroxyl ions.

The reduction of inorganic contaminants has been studied as well. The main contaminants described have been bromide, nitrate, nitrite and CO_2. In this case, the reduction takes place at a p-type semiconductor [12]. The reduction of toxic metals (Cr^{6+} to Cr^{3+}) has also been described [61] in a photoelectrocatalytic process where the cathode is Pt but is conjugated in a system where the organic molecules are oxidized simultaneously in a photocathode such as Ti/TiO_2 and the electrons are forwarded to the counter electrode, where the reduction of Cr (VI) takes place [167]. Solar conversion of CO_2 to hydrocarbon fuels seems promising to reduce global warming for improved sustainability. Solar fuels include hydrogen, carbon monoxide, methane and methanol [168].

More recently, the application of semiconductor materials has received a great deal of attention in a re-emerging field: the generation of hydrogen as a clean energy carrier. Studies have described the direct water splitting process and the degradation of organic pollutants in order to obtain hydrogen [16]. For this purpose, the use of n and p-type semiconductor materials using the photoelectrocatalysis method was investigated. The choice of the semiconductor material for hydrogen generation purposes depends on the valence and conduction-band energy levels, which are pH dependent (Figure 10).

The lower edge of the conduction band needs to be greater than the energy level for H_2 evolution (according to Equation 11). For water-splitting purposes (Figure 10), the upper edge of the valence band needs to have enough energy to promote the H_2O/O_2 reaction (Equation 12), while for simultaneous organic-pollutant removal the energy level must be more electropositive than the OH^-/•OH level for hydroxyl radical formation (Equations 11,12):

$$2H^+ + 2e^- \rightarrow H_2 \tag{11}$$

$$2H_2O \rightarrow O_2 + 4e^- + 4H^+ \tag{12}$$

The use of solar light for hydrogen generation purposes has been desirable for the same reasons as for PEC purposes. Hence, the development of photoanodes that absorb light in the visible region (λ>400 nm) is necessary, and could be achieved by lowering the photoanode band-gap energy.

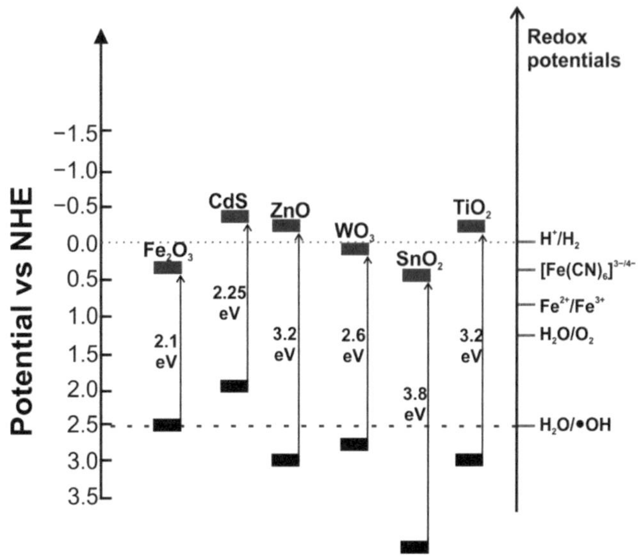

Figure 10. Schematic representation of E_g values (in eV) and position of CB and VB for the main semiconductors

5.1. TiO_2 nanomaterials applied to water treatment

The use of TiO_2-nanostructured materials in the removal of contaminants is undoubtedly a successful system in the treatment of wastewater. The use of nanotubes obtained from different routes has been described as an efficient alternative method to promote higher discoloration and partial mineralization of main organic pollutants, as they have a high and homogenous surface area and suitable photocurrent values.

The degradation of organochlorinated compounds [169, 170], pesticides [171, 172], aromatic amines [10], biological microorganisms [14, 15], hormones (endocrine disrupters) [173, 174], flameretardants [175] and mainly dyes [176] has been reported with high efficiency shown by nanotube materials acting as photoanodes in photoelectrocatalytic treatment.

TiO_2 NTs have proved to be more photoactive and to improve the efficiency of PEC degradation of pentachlorophenol under biased potential, with sodium sulphate as electrolyte (0.01 mol L^{-1}) and low pH of the original solution. The photoelectrocatalytic processes have been shown to be more efficient than electrocatalytic, photolytic and photocatalytic techniques [169]. Quan and colleagues [170] also observed the synergistic effect of photoelectrocatalysis compared to photocatalytic and electrochemical processes aiming at the degradation of pentachlorophenol in aqueous solution. They also reported that TiO_2 NTs under UV irradiation promoted higher mineralization than a conventional sol-gel film electrode.

The photoelectrocatalytic degradation of pesticides has been performed by TiO_2 thin films. Philippidis and colleagues achieved 82% of degradation of the pharmaceutical compound imidacloprid using Ti/TiO_2 electrodes prepared by the immobilization of P25 powder onto Ti substrate. The degradation efficiency increased with increased applied potential, following the first-order kinetics model after three hours of treatment. The method was proved to be more efficient than photocatalysis (63% removal) and photolysis (5% removal) operating under UV irradiation [171]. The pesticide Dipterex has been removed by using TiO_2 as a photoanode, prepared by a sol-gel method depositing over a nickel net. The method promoted a chemical oxygen demand (COD) removal and organophosphorous conversion of up to 82.6% and 83.5%, respectively, after 2 h of treatment under UV light [172].

The incomplete reduction of azo dyes and nitroaromatic compounds can usually promote aromatic amine formation, which can be released into the environment as potential carcinogens. This has been reported in drinking water treatment plants [177]. The use of TiO_2 NTs as photoanodes was proposed by Cardoso and colleagues. The method is efficient since it promotes the complete degradation and mineralization of 4,4-oxydianiline after 2 h of photoelectrocatalytic treatment under UV irradiation [10].

The PEC degradation of 4,4′-dibromobiphenyl used in flame retardants in the textile, and electronic industries, and in additives in plastics, has been performed using TiO_2 NTs as photoanodes. This class of compounds is described as toxic to human health and the environment. The photoelectrocatalytic process was more efficient than the photocatalytic and electrolytic process alone. Different anodes were compared: TiO_2, Zr/TiO_2 and Zr,N/TiO_2 NTs. The photoelectrocatalytic efficiency was significantly affected by the properties of the catalysts and the best performance was observed with TiO_2 doped with nitrogen and zirconium, as it had a higher photocurrent under UV irradiation by a 125 W mercury lamp [175].

Biological microorganisms can cause the contamination of water by spreading potential pathogens. TiO_2 nanotube arrays and Ag-loaded TiO_2 NTs have been employed in the disinfection of water containing *Mycobacterium smegmatis*. Under UV irradiation the photoelectrochemical treatment promoted 100% inactivation after 3 min. The effect of Ag on TiO_2 NTs has been observed in TOC removal, which reached 98% and 90% for Ag/TiO_2 and TiO_2, respectively, after 4 h of treatment [14]. The inactivation of *Mycobacterium kansasii* and *Mycobacterium avium* has also been conducted on TiO_2 and Ag/TiO_2 NTs electrodes by photoelectrocatalytic oxidation. The inactivation of both bacteriawas 100% after 3–5 minutes of treatment, faster than photocatalytic and photolytic treatment methods, indicating that the bias potential of the photoanode potentializes the treatment [15]. Egerton and colleagues described the PEC inactivation of wastewater containing *E. Coli* using TiO_2 irradiated by UV light. The method is also efficient for the removal of 4-nitrophenol and humic acid contaminants [178].

Endocrine disrupters have been reported as a class of compounds which can mimic or inhibit the natural actions of the endocrine system in animals and humans, such as synthesis, secretion, transport and binding. They can be either natural or synthetic compounds that come from different sources, such as pharmaceutical compounds, personal care products, disinfectionproducts and surfactants [173]. The literature [11] reports the removal of Bisphenol A from wastewater using TiO_2 NTs in a photoelectrocatalytic oxidation process under UV light and applied potential of +1.2 V. The removal was confirmed by HPLC/DAD analysis. The degradation of carbamazepine has been conducted with Ti/TiO_2 electrodes prepared by pulsed laser deposition. After 120 min of treatment, 73.5% pollutant removal was achieved, and 21.2% mineralization. Although complete degradation was not achieved the by-products were not toxic in the presence of *Vibrio Fisheri* [174]. The removal of these compounds is better than that achieved by other methods, such as photocatalysis [179], activated sludge [180] and biological treatment [181].

Different activities in the textile, paper, pharmaceutical, leather and food industries, among others, release a huge amount of dyes in effluents that can reach drinking water treatment plants if they are not appropriately treated. There are serious concerns over these compounds – many are potential carcinogens, or have xenobiotic or toxic properties that can harm the environment and living organisms [176].

The PEC oxidation of methyl orange [182], methylene blue [183] and rhodamine B [184] dyes has been reported. The photoelectrochemical method promoted 100% discoloration and high reduction of the toxicity of dispersed and indigoid organic dyes [185-187].

Recently, the main target of PEC studies has been the visible light activation of materials [188]. The relevance of reactors for photoelectrocatalytic treatment has also been described. It has been mentioned that the use of solar cells to supply the energy in PEC systems could reduce the cost of batch reactors by making it unnecessary to purchase electricity –electricity costshave been pointed out as the main disadvantage of this process [189].

For hydrogen production, a lot of photocatalysts have been studied in the literature,though mainly TiO_2 and modified TiO_2. Lianos described the use of TiO_2 supported on ITO and FTO and TiO_2 doped with N, C and S as well as the use of photocatalysts combined with noble metals such as Pt, Pd and Au and the coupled semiconductors TiO_2/SnO_2, TiO_2/WO_3, TiO_2/RuO_2, TiO_2/V_2O_5 in an attempt to use visible light irradiation [16]. Pure TiO_2 nanotube arrays have also been described in photoelectrochemical water splitting and simultaneous degradation of methylene blue [78]. The PEC experiments were conducted using an artificial sunlight simulator. The higher photoconversion efficiency for hydrogen generation and the degradation efficiency of MB were attributed to the better electron transfer process observed for two-step TiO_2 NTs over one-step TiO_2 NTs. CdS/TiO_2 nanotubes for photoelectrochemical hydrogen production have also been described: the doped material presented a better performance in the H_2 generation rate than the pure TiO_2 NTs under solar light illumination [190].

Zhao and colleagues carried out simultaneous photoelectrochemical destruction. They obtained contaminant and nickel recovery on the cathode. The deposition of TiO_2 film was performed by dip-coating [167]. Paschoal and colleagues promoted the photoelectrochemical reduction of bromate under Ti/TiO_2 coated as a photocathode. Photoelectrocatalytic reduction of BrO_3^- to Br^- can reach 70% at neutral pH under biased potential of −0.20 V after 75 minutes of treatment [191]. Table2 shows a summary of the selected studiesusingdopedand undoped TiO_2 photoanodes used in photoelectrocatalytic applications.

Photoanode	PEC application	Reference
TiO_2 NTs	Organics degradation	[10], [169], [170], [11], [192], [175], [193]
	Water splitting	[194, 195], [196], [197]
TiO_2 NTs coupled with other semiconductors	Organics degradation	[9], [198], [199], [200], [186], [201], [162]
	Water splitting	[202]
Anion-doped TiO_2 NTs	Organics degradation	[203], [143], [133], [148], [149]
	Water splitting	[204], [147], [93]
Cation-doped TiO_2 NTs	Organics degradation	[127], [154], [205]
	Water splitting	[128]
TiO_2 NTs coupled with noble metals	Organics degradation	[206], [207]
	Water splitting	[208], [158]
	Disinfection	[14], [15], [209]
TiO_2 thin film	Organics degradation	[171], [172], [189]
	Water splitting	[210], [211]
	Disinfection	[212], [62]
Doped TiO_2 thin film	Organics degradation	[213], [214], [215], [205], [216]

Table 2. Photoelectrocatalytic applications of doped and undoped TiO_2-based nanostructured semiconductors

5.2. Application of doped, decorated and composite of TiO_2 nanomaterials PEC

N-doped TiO_2 coatings prepared by radiofrequency magnetron sputtering has been employed on the degradation of the antibiotic chlortetracycline under 0.6 A of current intensity and solar simulator irradiation during 180 min, leading to 99% degradation. This is more efficient than pure Ti/TiO_2. This process has also shown to be efficient in the inactivation of faecal coliform, which is an indicator pathogen [217]. Wu and Zhang [204] prepared nitrogen-doped double-wall TiO_2 NTs, which under simulated solar light presented a high photoelectrochemical water splitting performance due to the high surface areas and absorbance in the visible light region. Sun et al. [203] prepared N-doped TiO_2 NTs, which presented better efficiency in Rhodamine B PEC degradation.

Boron-doped TiO_2 NTs have also been studied as photoanodes prepared by chemical vapour deposition. The electrode was applied in the degradation of methyl orange dye under visible light irradiation promoting 100% discoloration under applied potential of +2.0 V and UV irradiation [192]. In the studies by Lu and colleagues [133] and Li et al. [148] boron-doped TiO_2 NTs were prepared and applied in the PEC degradation of atrazine and phenol, respectively.

TiO_2 has been doped with nickel and used as a photocatalyst in the degradation of Acid Red 88 dye. The photoanode powder was prepared by the sol-gel method and 95% COD and TOC removal was obtained after 35 min of treatment under UV and solar irradiation. The colour removal was 72% for photocatalytic treatment and 97% for photoelectrocatalytic treatment under +1.6 V [189]. Gong and colleagues prepared W-doped TiO_2 NTs and applied these in simultaneous Rhodamine B degradation and production of hydrogen [127]; tungsten-doped TiO_2 films were also applied in dodecyl-benzenesulfonate removal by PEC [213].

Arrays of porous iron-doped TiO_2 as photoelectrocatalyst with controllable pore size have been synthesized by using polystyrene spheres as templates. It was found that photoelectrochemical hydrogen generation was favoured by a shift in the flat-band potential from −0.38V to −0.55 V vs. SCE and an increase of photocurrent by 80% [218].

Pt-deposited TiO_2 photoanodes have been prepared by a sol-gel method, where the amount of Pt was shown to interfere with the photoelectrochemical response for glucose oxidation. The increased Pt lowered the photocurrent but the overall oxidation efficiency of the PEC process was better than the PC process, for both TiO_2 and Pt-TiO_2 films [219]. Ye et al. [208] prepared TiO_2 NTs sensitized by palladium quantum dots (Pd QDs), which exhibit highly efficient photoelectrocatalytic hydrogen generation. Zhang and colleagues [206] prepared TiO_2 NTs loaded with Pd nanoparticles, and the PEC activity was investigated with degradation of methylene blue and Rhodamine B.

CdS-ZnS/TiO_2 composite material has been investigated in the production of electricity. The band-gap energy can be tuned between that of ZnS (3.5 eV) and that of CdS (2.3 eV) by varying Cd (or Zn) content. Photocatalytic and photoelectrocatalytic processes in basic electrolyte with ethanol as a sacrificial electron donor was also investigated. The performance of CdS-ZnS, Pt/(CdS-ZnS), Pt/(CdS-ZnS)/TiO_2 and Pt/TiO_2 photoanodes was compared and 75% CdS–25% ZnS over pure TiO_2 presented better electrocatalyst effect than 100% CdS over TiO_2 [220]. CdS nanocrystallites-decorated TiO_2 nanotube array photoelectrodes were prepared through anodization and electrodeposition strategies. Enhancement of photoelectrocatalytic degradation of Rhodamine B was achieved under Xenon light irradiation [198].

Georgieva and colleagues described the use of bicomponent anodes of TiO_2/WO_3 for the photoelectrocatalytic oxidation of organic species. WO_3 is a promising additive for TiO_2 since it modifies its photochemical properties in a favourable manner, both with respect to reduced recombination and visible light activity because of the lower band-gap energy. The coupling of semiconductor oxides leads to electron and hole transfer between the two materials in opposite directions, thus limiting recombination of the photogenerated species in the same material [29]. These materials have been employed in the degradation of 2,3-dichlorophenol under visible light irradiation [199], the removal of the hair dye Basic Red 51 under UV and visible light source [200] and the PEC oxidation of indigo carmine dye [186].

The use of heterojunctions was studied by Christensen and colleagues, who conducted the PEC degradation of *E. Coli* under UV irradiation using Si/TiO_2/Au as photoanode. The experiments were performed in water and air [221]. The silicon nanowire/TiO_2 heterojunction arrays were employed on the PEC degradation of phenol under simulated solar light irradiation. The kinetic constant and total organic carbon (TOC) removal were 1.7 times and two times as great as those of n-Si/TiO_2, respectively [222].

The PEC degradation of flame-retardants has been described under macroporous silicon/graphene (MPSi/Gr) heterostructure. The experiments were conducted under visible light irradiation and compared to photocatalytic degradation. The photoelectrocatalytic degradation five times faster than PC degradation [223].

CdTe nanotubes have been produced by using ZnO as a template on an ITO surface. These were then used with the photoelectrocatalytic degradation of the Acid Blue 80 dye. This study provided a good strategy for the design of visible light-responsive photocatalysts that can be recycled and possess high efficiency, extremely low mass and high chemical stability [224].

The PEC remediation of 2,4-dichlorophenol by visible-light-enhanced WO_3 has also been described. The degradation process achieved 74% pollutant removal after a period of 24 hours, monitored by both chemical analysis and a bacterial biosensor (*Escherichia coli*) toxicity assay [225].

For hydrogen production, photocatalysts reported in the literature apart from TiO_2 include ZnO, Fe_2O_3, and $SrTiO_3$, which has the energy levels necessary to create active radical species that could efficiently carry out photodegradation process [16]. Under visible light irradiation some n-type materials have been described: nanoporous WO_3, α-Fe_2O_3 or haematite and nanocrystalline $BiVO_4$ [18].

The Cu/Cu_2O system as photocathode has been described in relation to nitrate removal under UV irradiation and biased potential. The material was prepared by electrodeposition and long-term stability was achieved. 93% nitrate removal was achieved after 75 min under the best experimental conditions. Nitrate reduction on Cu/Cu_2O photoelectrodes occurs in the cathodic compartment cell via electrons generated under UV irradiation, as expected for a p-type electrode, leading to 42% of remaining nitrite and 52% gaseous nitrogen derived, respectively [12].

Zanoni and colleagues employed TiO_2 NTs in the photoelectrocatalytic oxidation of an organic synthetic dye (reactive black 5) and the simultaneous hydrogen generation. The photoanode was irradiated with UV light and biased at +1.0 V. Complete dye degradation and 72% mineralization was achieved after 2 h of treatment. The estimated overall hydrogen generation was around 44%, which corresponds to 0.6 mL cm^{-2} [226].

6. Final remarks

Photoelectrocatalysis is an emerging field with many applications, such as organics oxidation, inorganics reduction, biological materials and production of electricity and hydrogen.

The technique could be described as a multidisciplinary field, where the basic concept is the irradiation by light ($h\nu \geq E_g$) of the semiconductor surface. There is the generation of electron/hole pairs (e^-/h^+) by the promotion of an electron from the valence band (lower energy level) to the conduction band (higher energy level). The electrons are forwarded to the counter electrode under positive anodic bias (n-type) in order to minimize the recombination of these pairs due to the short life-time. When immersed in electrolyte the adsorbed water molecules and/or hydroxyl ions react with the holes on the valence band to generate hydroxyl radicals (•OH), which are a powerful oxidizing agent.

Titanium dioxide (TiO_2) is a classic example of an n-type semiconductor widely used as a catalyst for heterogeneous photocatalysis and photoelectrochemical applications. It has received a great deal of attention due to its good chemical and thermal stability, non-toxicity, low cost, high photoactivity and other advantageous properties. It is a typical n-type semiconductor mainly composed of anatase and rutile allotropic forms whose band-gap energy is 3.2 and 3.0 eV, respectively. The anatase phase is the desired form as it is more photoactive than the other forms.

The degradation of organic pollutants by photoelectrocatalysis has been described in the literature as one of the most effective treatments among advanced oxidative processes (AOPs) in the oxidation of recalcitrant compounds, as they are harmful to the environment and human health. The contamination of water is an increasing concern because pollutants can accumulate in the environment and are mutagenic and genotoxic.

The architecture of nanostructures used in the electrode construction has deeply influenced the results of PEC. Nanotube, nanowire, nanofibre, nanorod, and nanowall morphologies can be easily obtained by electrochemical methods. These kinds of nanostructures have improved efficiently organic contaminants degradation, especially due to their high surface area and ability to minimize charge recombination. The use of nanotube arrays has received a great deal of attention especially because it is the structure with the highest surface area/geometric area ratio; moreover, it is of a highly oriented and organized nature, leading to efficient charge transport as it has a unique and effective direct interfacial direction, decreasing the charge recombination effect. Among all TiO_2 NTs preparation routes, the electrochemical anodization method presents the greatest advantages, since they are cheaper, simpler and allow precise control of dimensions, presenting highly ordered nanotube arrays. The first generation of nanotube materials applied in PEC materials were obtained in aqueous solutions with the addition of HNO_3, H_2SO_4 and H_3PO_4 to HF acid as electrolyte. The second generation of nanotube arrays was obtained in buffered electrolytes. Aiming for better quality and performance, the third generation was obtained in organic medium as ethylene glycol, diethylene glycol, glycerol and NH_4F. Non-fluoride-based electrolytes are classified as the fourth generation, where HCl, H_2O_2 and a combination of both are used as electrolyte. Nanotube array photoanodes have presented good results on the water decontamination of organic contaminants and also water disinfection.

Recently, studies have addressed the challenge of obtaining PEC materials which can be activated by visible light, with the aim of using solar light to promote photoactivation, not only to reduce cost but also to establish an environmentally friendly method. For this purpose, different strategies are discussed in the literature to improve photoactivity and shift the PEC material absorption to the visible region, such as the use of photoanodes decorated with Ag and Pt, or combinations of semiconductors like ZnO/TiO_2, CdS/TiO_2, WO_3/TiO_2 in order to obtain composite and bicomponent materials; doping with metals (Fe, Mn, Cr), non-metals (B, C, Si) and co-doping (N-F, N-C) has also been thoroughly described.

http://dx.doi.org/10.5772/58333

Therefore, the use of TiO_2 and other materials is of huge relevance to photoelectrocatalysis applied to water treatment, and the success of photoanodes and photocathodes depends on the synthesis process and a better understanding of materials' properties.

7. Summary

The importance of photoelectrocatalysis has been discussed, with emphasis on recent advances in TiO_2-based materials and strategies of electrochemical synthesis and modification. Currently, TiO_2 nanotube arrays occupy a prominent position. These can be prepared by electrochemical anodization of titanium plates in fluoride-containing electrolytes. In the search for catalysts that can be photoactivated with visible radiation, doping or modification of these materials can be easily performed by electrochemical techniques. The use of these photocatalysts immobilized on conducting substrates employed in photoelectrochemical reactors is a viable strategy for increasing the efficiency of water splitting or to promote efficient degradation of organic compounds.

Author details

Guilherme Garcia Bessegato, Thaís Tasso Guaraldo and Maria Valnice Boldrin Zanoni*

*Address all correspondence to: boldrinv@iq.unesp.br

Department of Analytical Chemistry, Institute of Chemistry, Universidade Estadual Paulista (Unesp), Araraquara, Brazil

References

[1] Heller A. Conversion of sunlight into electrical power and photoassisted electrolysis of water in photoelectrochemical cells. Accounts of Chemical Research 1981; 14(5): 154-62.

[2] Tryk DA, Fujishima A, Honda K. Recent topics in photoelectrochemistry: achievements and future prospects. ElectrochimicaActa 2000; 45(15-16): 2363-76.

[3] Bard AJ. Photoelectrochemistry. Science 1980; 207(4427): 139-44.

[4] Daghrir R, Drogui P, Robert D. Photoelectrocatalytic technologies for environmental applications. Journal of Photochemistry and Photobiology A: Chemistry 2012; 238: 41-52.

[5] Linsebigler AL, Lu GQ, Yates JT. Photocatalysis on TiO_2 Surfaces-Principles, Mechanisms, and Selected Results. Chemical Reviews 1995; 95(3): 735-58.

[6] Kronik L, Shapira Y. Surface photovoltage phenomena: theory, experiment, and applications. Surface Science Reports 1999; 37(1-5): 1-206.

[7] Zhang H, Chen G, Bahnemann DW. Photoelectrocatalytic materials for environmental applications. Journal of Materials Chemistry 2009; 19(29): 5089-121.

[8] Fujishima A, Honda K. Electrochemical Photolysis of Water at a Semiconductor Electrode. 1972; 238(5358): 37-8.

[9] Bessegato GG, Cardoso JC, Silva BF, Zanoni MVB. Enhanced photoabsorption properties of composites of Ti/TiO_2 nanotubes decorated by Sb_2S_3 and improvement of degradation of hair dye. Journal of Photochemistry and Photobiology A: Chemistry 2013; 276: 96-103.

[10] Cardoso JC, Lizier TM, Boldrin Zanoni MV. Highly ordered TiO_2 nanotube arrays and photoelectrocatalytic oxidation of aromatic amine. Applied Catalysis B-Environmental 2010; 99(1-2): 96-102.

[11] Brugnera MF, Rajeshwar K, Cardoso JC, Boldrin Zanoni MV. Bisphenol A removal from wastewater using self-organized TiO_2nanotubular array electrodes. Chemosphere 2010; 78(5): 569-75.

[12] Monteiro Paschoal FM, Nunez L, de Vasconcelos Lanza MR, Boldrin Zanoni MV. Nitrate Removal on a Cu/Cu_2O Photocathode under UV Irradiation and Bias Potential. Journal of Advanced Oxidation Technologies 2013; 16(1): 63-70.

[13] LaTempa TJ, Rani S, Bao N, Grimes CA. Generation of fuel from CO_2 saturated liquids using a p-Si nanowire parallel to n-TiO_2 nanotube array photoelectrochemical cell. Nanoscale 2012; 4(7): 2245-50.

[14] Brugnera MF, Miyata M, Zocolo GJ, Fujimura Leite CQ, Boldrin Zanoni MV. Inactivation and disposal of by-products from Mycobacterium smegmatis by photoelectrocatalytic oxidation using Ti/TiO_2-Ag nanotube electrodes.Electrochimica Acta 2012; 85: 33-41.

[15] Brugnera MF, Miyata M, Zocolo GJ, Fujimura Leite CQ, Boldrin Zanoni MV. A photoelectrocatalytic process that disinfects water contaminated with Mycobacterium kansasii and Mycobacterium avium. Water Research 2013; 47(17): 6596-605.

[16] Lianos P. Production of electricity and hydrogen by photocatalytic degradation of organic wastes in a photoelectrochemical cell: The concept of the Photofuelcell: A review of a re-emerging research field. Journal of Hazardous Materials 2011; 185(2-3): 575-90.

[17] Paulauskas IE, Katz JE, Jellison GE, Jr., Lewis NS, Boatner LA. Photoelectrochemical studies of semiconducting photoanodes for hydrogen production via water dissociation. Thin Solid Films 2008; 516(22): 8175-8.

[18] Abe R. Recent progress on photocatalytic and photoelectrochemical water splitting under visible light irradiation. Journal of Photochemistry and Photobiology C: Photochemistry Reviews 2010; 11(4): 179-209.

[19] Paramasivam I, Jha H, Liu N, Schmuki P. A Review of Photocatalysis using Self-organized TiO_2 Nanotubes and Other Ordered Oxide Nanostructures. Small 2012; 8(20): 3073-103.

[20] Roy P, Berger S, Schmuki P. TiO_2 Nanotubes: Synthesis and Applications. AngewandteChemie-International Edition 2011; 50(13): 2904-39.

[21] Andreozzi R, Caprio V, Insola A, Marotta R. Advanced oxidation processes (AOP) for water purification and recovery. Catalysis Today 1999; 53(1): 51-9.

[22] Pichat P, Disdier J, Hoang-Van C, Mas D, Goutailler G, Gaysse C. Purification/deodorization of indoor air and gaseous effluents by TiO_2 photocatalysis. Catalysis Today 2000; 63(2-4): 363-9.

[23] Villa RD, Trovo AG, Pupo Nogueira RF. Soil remediation using a coupled process: soil washing with surfactant followed by photo-Fenton oxidation. Journal of Hazardous Materials 2010; 174(1-3): 770-5.

[24] Rajeshwar K. Fundamentals of Semiconductor Electrochemistry and Photoelectrochemistry. In Licht S. (ed.) Semiconductor Electrodes and Photoelectrochemistry, Encyclopedia of Electrochemistry,Weinheim: Wiley 2007; 6: 1-51. Available at http://www.wiley-vch.de/bard/eoe/pdf/v06_1.pdf (accessed 15 November 2013).

[25] Rajeshwar K, Osugi ME, Chanmanee W, et al. Heterogeneous photocatalytic treatment of organic dyes in air and aqueous media. Journal of Photochemistry and Photobiology C: Photochemistry Reviews 2008; 9(4): 171-92.

[26] Egerton TA. Does photoelectrocatalysis by TiO_2 work? Journal of Chemical Technology and Biotechnology 2011; 86(8): 1024-31.

[27] Pelaez M, Nolan NT, Pillai SC, et al. A review on the visible light active titanium dioxide photocatalysts for environmental applications. Applied Catalysis B-Environmental 2012; 125: 331-49.

[28] Vinodgopal K, Hotchandani S, Kamat PV. Electrochemically Assisted Photocatalysis-TiO_2 Particulate Film Electrodes for Photocatalytic Degradation of 4-Chlorophenol. Journal of Physical Chemistry 1993; 97(35): 9040-4.

[29] Georgieva J, Valova E, Armyanov S, Philippidis N, Poulios I, Sotiropoulos S. Bi-component semiconductor oxide photoanodes for the photoelectrocatalytic oxidation of

organic solutes and vapours: A short review with emphasis to TiO_2-WO_3 photoanodes. Journal of Hazardous Materials 2012; 211: 30-46.

[30] Rajeshwar K. Photoelectrochemistry and the environment. Journal of Applied Electrochemistry 1995; 25(12): 1067-82.

[31] Finklea HO. Semiconductor electrodes. New York: Elsevier, 1988.

[32] Chong MN, Jin B, Chow CWK, Saint C. Recent developments in photocatalytic water treatment technology: A review. Water Research 2010; 44(10): 2997-3027.

[33] Zhang Y, Xiong X, Han Y, et al. Photoelectrocatalytic degradation of recalcitrant organic pollutants using TiO_2 film electrodes: An overview. Chemosphere 2012; 88(2): 145-54.

[34] Rostami I, Juhasz AL. Assessment of Persistent Organic Pollutant (POP) Bioavailability and Bioaccessibility for Human Health Exposure Assessment: A Critical Review. Critical Reviews in Environmental Science and Technology 2011; 41(7): 623-56.

[35] Horáková M, Klementová Š, Kříž P, et al. The synergistic effect of Advanced Oxidation Processes to eliminate resistant chemical compounds.Surface and Coatings Technology.Available at: http://dx.doi.org/10.1016/j.surfcoat.2013.10.068 (accessed 1 February 2014).

[36] Martínez-Huitle CA, Brillas E. Decontamination of wastewaters containing synthetic organic dyes by electrochemical methods: A general review. Applied Catalysis B: Environmental 2009; 87(3-4): 105-45.

[37] Feng L, van Hullebusch ED, Rodrigo MA, Esposito G, Oturan MA. Removal of residual anti-inflammatory and analgesic pharmaceuticals from aqueous systems by electrochemical advanced oxidation processes. A review. Chemical Engineering Journal 2013; 228(0): 944-64.

[38] Shan AY, Ghazi TIM, Rashid SA. Immobilisation of titanium dioxide onto supporting materials in heterogeneous photocatalysis: A review. Applied Catalysis A: General 2010; 389(1-2): 1-8.

[39] Shankar K, Basham JI, Allam NK, et al. Recent Advances in the Use of TiO_2 Nanotube and Nanowire Arrays for Oxidative Photoelectrochemistry. Journal of Physical Chemistry C 2009; 113(16): 6327-59.

[40] Carp O, Huisman CL, Reller A. Photoinduced reactivity of titanium dioxide. Progress in Solid State Chemistry 2004; 32(1-2): 33-177.

[41] Shao C, Zhou G, Li Z, Wu Y, Xu D, Sun B. Fabrication of large-diameter tube-like mesoporous TiO_2 via homogeneous precipitation and photocatalytic decomposition of papermaking wastewater. Chemical Engineering Journal 2013; 230: 227-35.

[42] He F, Li J, Li T, Li G. Solvothermal synthesis of mesoporous TiO_2: The effect of morphology, size and calcination progress on photocatalytic activity in the degradation of gaseous benzene. Chemical Engineering Journal 2014; 237(0): 312-21.

[43] Guaraldo TT, Pulcinelli SH, Zanoni MVB. Influence of particle size on the photoactivity of Ti/TiO_2 thin film electrodes, and enhanced photoelectrocatalytic degradation of indigo carmine dye. Journal of Photochemistry and Photobiology A: Chemistry 2011; 217(1): 259-66.

[44] Zubieta CE, Soltero-Martinez JFA, Luengo CV, Schulz PC. Preparation, characterization and photoactivity of TiO_2 obtained by a reverse microemulsion route. Powder Technology 2011; 212(3): 410-7.

[45] Rajashekhar KE, Devi LG. Polymorphic phase transformation of Degussa P25 TiO_2 by the chelation of diaminopyridine on TiO_6^{2-} octahedron: Correlation of anatase to rutile phase ratio on the photocatalytic activity. Journal of Molecular Catalysis a-Chemical 2013; 374: 12-21.

[46] Andronic L, Duta A. TiO_2 thin films for dyes photodegradation. Thin Solid Films 2007; 515(16): 6294-7.

[47] Kim DH, Anderson MA. Photoelectrocatalytic degradation of formic-acid using a porous TiO_2 thin-film electrode. Environmental Science & Technology 1994; 28(3): 479-83.

[48] Qin X, Jing L, Tian G, Qu Y, Feng Y. Enhanced photocatalytic activity for degrading Rhodamine B solution of commercial Degussa P25 TiO_2 and its mechanisms. Journal of Hazardous Materials 2009; 172(2–3): 1168-74.

[49] Lee H, Song MY, Jurng J, Park Y-K. The synthesis and coating process of TiO_2 nanoparticles using CVD process. Powder Technology 2011; 214(1): 64-8.

[50] Rathouský J, Wessels K, Wark M, Oekermann T. Texture properties of nanoporous TiO_2 films prepared by anodic electrodeposition using a structure-directing agent. In: From Zeolites to Porous MOF Materials.. Studies in Surface Science and Catalysis, Elsevier2007; 170:1494-501.

[51] Firdaus CM, Rizam MSBS, Rusop M, Hidayah SR. Characterization of ZnO and ZnO: TiO_2 Thin Films Prepared by Sol-Gel Spray-Spin Coating Technique. Procedia Engineering 2012; 41(0): 1367-73.

[52] Zanoni MVB, Sene JJ, Anderson MA.Photoelectrocatalytic degradation of Remazol Brilliant Orange 3R on titanium dioxide thin-film electrodes. Journal of Photochemistry and Photobiology A: Chemistry 2003; 157(1): 55-63.

[53] Wu C-Y, Lee Y-L, Lo Y-S, Lin C-J, Wu C-H.Thickness-dependent photocatalytic performance of nanocrystalline TiO_2 thin films prepared by sol-gel spin coating. Applied Surface Science 2013; 280: 737-44.

[54] Wang X, Shi F, Gao X, Fan C, Huang W, Feng X. A sol-gel dip/spin coating method to prepare titanium oxide films. Thin Solid Films 2013; 548(0): 34-9.

[55] Mechiakh R, Ben Sedrine N, Chtourou R, Bensaha R. Correlation between microstructure and optical properties of nano-crystalline TiO_2 thin films prepared by sol-gel dip coating. Applied Surface Science 2010; 257(3): 670-6.

[56] Vargas-Florencia D, Edvinsson T, Hagfeldt A, Furo I. Pores in nanostructured TiO_2 films. Size distribution and pore permeability. Journal of Physical Chemistry C 2007; 111(21): 7605-11.

[57] Gaya UI, Abdullah AH. Heterogeneous photocatalytic degradation of organic contaminants over titanium dioxide: A review of fundamentals, progress and problems. Journal of Photochemistry and Photobiology C: Photochemistry Reviews 2008; 9(1): 1-12.

[58] Carneiro PA, Osugi ME, Sene JJ, Anderson MA, Zanoni MVB. Evaluation of color removal and degradation of a reactive textile azo dye on nanoporous TiO_2 thin-film electrodes.ElectrochimicaActa 2004; 49(22-23): 3807-20.

[59] Zhou M, Ma X. Efficient photoelectrocatalytic activity of TiO_2/Ti anode fabricated by metalorganic chemical vapor deposition (MOCVD). Electrochemistry Communications 2009; 11(4): 921-4.

[60] Liu Y, Gan X, Zhou B, et al. Photoelectrocatalytic degradation of tetracycline by highly effective TiO_2nanopore arrays electrode. Journal of Hazardous Materials 2009; 171(1-3): 678-83.

[61] MonteiroPaschoal FM, Anderson MA, Zanoni MVB.Simultaneous removal of chromium and leather dye from simulated tannery effluent by photoelectrochemistry. Journal of Hazardous Materials 2009; 166(1): 531-7.

[62] Fraga LE, Anderson MA, Beatriz MLPMA, Paschoal FMM, Romao LP, Zanoni MVB. Evaluation of the photoelectrocatalytic method for oxidizing chloride and simultaneous removal of microcystin toxins in surface waters.ElectrochimicaActa 2009; 54(7): 2069-76.

[63] Tang H, Prasad K, Sanjines R, Schmid PE, Levy F. Electrical and optical-properties of TiO_2 anatase thin-films. Journal of Applied Physics 1994; 75(4): 2042-7.

[64] Rouquerol J, Avnir D, Everett DH, et al. Guidelines for the characterization of porous solids. Studies in Surface Science and Catalysis 1994; 87: 1-9.

[65] Pan JH, Zhao XS, Lee WI. Block copolymer-templated synthesis of highly organized mesoporous TiO_2-based films and their photoelectrochemical applications. Chemical Engineering Journal 2011; 170(2-3): 363-80.

[66] Hepel M, Hazelton S. Photoelectrocatalytic degradation of diazo dyes on nanostructured WO_3 electrodes. ElectrochimicaActa 2005; 50(25-26): 5278-91.

[67] Thongsuriwong K, Amornpitoksuk P, Suwanboon S. Structure, morphology, photocatalytic and antibacterial activities of ZnO thin films prepared by sol-gel dip-coating method. Advanced Powder Technology 2013; 24(1): 275-80.

[68] Mahadik MA, Shinde SS, Rajpure KY, Bhosale CH. Photocatalytic oxidation of Rhodamine B with ferric oxide thin films under solar illumination. Materials Research Bulletin 2013; 48(10): 4058-65.

[69] Prakasam HE, Varghese OK, Paulose M, Mor GK, Grimes CA. Synthesis and photoelectrochemical properties of nanoporous iron (III) oxide by potentiostatic anodization. Nanotechnology 2006; 17(17): 4285-91.

[70] Liu X, Wang F, Wang Q. Nanostructure-based WO_3 photoanodes for photoelectrochemical water splitting. Physical Chemistry Chemical Physics 2012; 14(22): 7894-911.

[71] Rehman S, Ullah R, Butt AM, Gohar ND. Strategies of making TiO_2 and ZnO visible light active. Journal of Hazardous Materials 2009; 170(2-3): 560-9.

[72] Minggu LJ, Daud WRW, Kassim MB. An overview of photocells and photoreactors for photoelectrochemical water splitting. International Journal of Hydrogen Energy 2010; 35(11): 5233-44.

[73] Ahmed S, Rasul MG, Martens WN, Brown R, Hashib MA. Advances in Heterogeneous Photocatalytic Degradation of Phenols and Dyes in Wastewater: A Review. Water Air and Soil Pollution 2011; 215(1-4): 3-29.

[74] Song X-M, Wu J-M, Yan M. Photocatalytic and photoelectrocatalytic degradation of aqueous Rhodamine B by low-temperature deposited anatase thin films. Materials Chemistry and Physics 2008; 112(2): 510-5.

[75] Zainal Z, Lee CY, Hussein MZ, Kassim A, Yusof NA. Electrochemical-assisted photodegradation of dye on TiO_2 thin films: investigation on the effect of operational parameters. Journal of Hazardous Materials 2005; 118(1-3): 197-203.

[76] Hashimoto K, Irie H, Fujishima A. TiO_2 photocatalysis: A historical overview and future prospects. Japanese Journal of Applied Physics Part 1-Regular Papers Brief Communications & Review Papers 2005; 44(12): 8269-85.

[77] Ni M, Leung MKH, Leung DYC, Sumathy K. A review and recent developments in photocatalytic water-splitting using TiO_2 for hydrogen production. Renewable & Sustainable Energy Reviews 2007; 11(3): 401-25.

[78] Wu H, Zhang Z. Photoelectrochemical water splitting and simultaneous photoelectrocatalytic degradation of organic pollutant on highly smooth and ordered TiO_2 nanotube arrays. Journal of Solid State Chemistry 2011; 184(12): 3202-7.

[79] Baram N, Starosvetsky D, Starosvetsky J, Epshtein M, Armon R, Ein-Eli Y. Enhanced inactivation of E-coli bacteria using immobilized porous TiO_2 photoelectrocatalysis. ElectrochimicaActa 2009; 54(12): 3381-6.

[80] Paschoal FMM, Anderson MA, Zanoni MVB. The photoelectrocatalytic oxidative treatment of textile wastewater containing disperse dyes. Desalination 2009; 249(3): 1350-5.

[81] Mohamed AER, Rohani S. Modified TiO_2 nanotube arrays (TNTAs): progressive strategies towards visible light responsive photoanode, a review. Energy & Environmental Science 2011; 4(4): 1065-86.

[82] Iijima S. Helical microtubules of graphitic carbon. Nature 1991; 354(6348): 56-8.

[83] Zhang Z, Yuan Y, Shi G, et al. Photoelectrocatalytic activity of highly ordered TiO_2 nanotube arrays electrode for azo dye degradation. Environmental Science & Technology 2007; 41(17): 6259-63.

[84] Zhang Q, Jing Y, Shiue A, et al. Photocatalytic degradation of malathion by TiO_2 and Pt-TiO_2 nanotube photocatalyst and kinetic study. Journal of Environmental Science and Health Part B-Pesticides Food Contaminants and Agricultural Wastes 2013; 48(8): 686-92.

[85] Liang H-C, Li X-Z. Effects of structure of anodic TiO_2 nanotube arrays on photocatalytic activity for the degradation of 2,3-dichlorophenol in aqueous solution. Journal of Hazardous Materials 2009; 162(2-3): 1415-22.

[86] Smith YR, Kar A, Subramanian V. Investigation of Physicochemical Parameters that Influence Photocatalytic Degradation of Methyl Orange over TiO_2 Nanotubes. Industrial & Engineering Chemistry Research 2009; 48(23): 10,268-76.

[87] Grandcolas M, Cottineau T, Louvet A, Keller N, Keller V. Solar light-activated photocatalytic degradation of gas phase diethylsulfide on WO_3-modified TiO_2 nanotubes. Applied Catalysis B-Environmental 2013; 138: 128-40.

[88] Sennik E, Colak Z, Kilinc N, Ozturk ZZ. Synthesis of highly-ordered TiO_2 nanotubes for a hydrogen sensor. International Journal of Hydrogen Energy 2010; 35(9): 4420-7.

[89] Zhao R, Xu M, Wang J, Chen G. A pH sensor based on the TiO_2 nanotube array modified Ti electrode. Electrochimica Acta 2010; 55(20): 5647-51.

[90] Mun K-S, Alvarez SD, Choi W-Y, Sailor MJ. A Stable, Label-free Optical Interferometric Biosensor Based on TiO_2 Nanotube Arrays.Acs. Nano 2010; 4(4): 2070-6.

[91] Mor GK, Shankar K, Paulose M, Varghese OK, Grimes CA. Use of highly-ordered TiO_2 nanotube arrays in dye-sensitized solar cells. Nano Letters 2006; 6(2): 215-8.

[92] Patrick CE, Giustino F. Structural and Electronic Properties of Semiconductor-Sensitized Solar-Cell Interfaces. Advanced Functional Materials 2011; 21(24): 4663-7.

[93] Park JH, Kim S, Bard AJ. Novel carbon-doped TiO_2 nanotube arrays with high aspect ratios for efficient solar water splitting. Nano Letters 2006; 6(1): 24-8.

[94] Grimes CA, Mor GK. TiO_2 nanotube arrays: Synthesis, Properties, and Applications. New York: Springer, 2009.

[95] Zhang Q, Ackerman E, Li Y. Photocatalytic reduction of CO_2 to fuels by novel TiO_2 nanotubes. Abstracts of Papers of the American Chemical Society 2011; 241.

[96] Popat KC, Eltgroth M, La Tempa TJ, Grimes CA, Desai TA. Titania nanotubes: A novel platform for drug-eluting coatings for medical implants. Small 2007; 3(11): 1878-81.

[97] Popat KC, Leoni L, Grimes CA, Desai TA. Influence of engineered titaniananotubular surfaces on bone cells. Biomaterials 2007; 28(21): 3188-97.

[98] Miao Z, Xu DS, Ouyang JH, Guo GL, Zhao XS, Tang YQ. Electrochemically induced sol-gel preparation of single-crystalline TiO_2 nanowires. Nano Letters 2002; 2(7): 717-20.

[99] Yu J, Yu H, Cheng B, Zhao X, Zhang Q. Preparation and photocatalytic activity of mesoporous anatase TiO_2nanofibers by a hydrothermal method. Journal of Photochemistry and Photobiology A: Chemistry 2006; 182(2): 121-7.

[100] Liu B, Aydil ES.Growth of Oriented Single-Crystalline Rutile TiO_2Nanorods on Transparent Conducting Substrates for Dye-Sensitized Solar Cells. Journal of the American Chemical Society 2009; 131(11): 3985-90.

[101] Wu JJ, Yu CC. Aligned TiO_2nanorods and nanowalls. Journal of Physical Chemistry B 2004; 108(11): 3377-9.

[102] Zhu K, Neale NR, Miedaner A, Frank AJ.Enhanced charge-collection efficiencies and light scattering in dye-sensitized solar cells using oriented TiO_2 nanotubes arrays. Nano Letters 2007; 7(1): 69-74.

[103] Liu Z, Zhang X, Nishimoto S, et al. Highly ordered TiO_2 nanotube arrays with controllable length for photoelectrocatalytic degradation of phenol. Journal of Physical Chemistry C 2008; 112(1): 253-9.

[104] Grimes CA. Synthesis and application of highly ordered arrays of TiO_2 nanotubes. Journal of Materials Chemistry 2007; 17(15): 1451-7.

[105] Mor GK, Varghese OK, Paulose M, Shankar K, Grimes CA. A review on highly ordered, vertically oriented TiO_2 nanotube arrays: Fabrication, material properties, and solar energy applications. Solar Energy Materials and Solar Cells 2006; 90(14).

[106] Macak JM, Tsuchiya H, Ghicov A, et al. TiO_2 nanotubes: Self-organized electrochemical formation, properties and applications. Current Opinion in Solid State & Materials Science 2007; 11(1-2).

[107] Nah Y-C, Paramasivam I, Schmuki P. Doped TiO_2 and TiO_2 Nanotubes: Synthesis and Applications. Chemphyschem 2010; 11(13): 2698-713.

[108] Ou H-H, Lo S-L. Review of titania nanotubes synthesized via the hydrothermal treatment: Fabrication, modification, and application. Separation and Purification Technology 2007; 58(1): 179-91.

[109] Zhang M, Bando Y, Wada K. Sol-gel template preparation of TiO_2 nanotubes and nanorods. Journal of Materials Science Letters 2001; 20(2): 167-70.

[110] Bavykin DV, Friedrich JM, Walsh FC. Protonated titanates and TiO_2 nanostructured materials: Synthesis, properties, and applications. Advanced Materials 2006; 18(21): 2807-24.

[111] Zwilling V, Darque-Ceretti E, Boutry-Forveille A, David D, Perrin MY, Aucouturier M. Structure and physicochemistry of anodic oxide films on titanium and TA6V alloy. Surface and Interface Analysis 1999; 27(7): 629-37.

[112] Gong D, Grimes CA, Varghese OK, et al. Titanium oxide nanotube arrays prepared by anodic oxidation. Journal of Materials Research 2001; 16(12): 3331-4.

[113] Cai QY, Paulose M, Varghese OK, Grimes CA. The effect of electrolyte composition on the fabrication of self-organized titanium oxide nanotube arrays by anodic oxidation. Journal of Materials Research 2005; 20(1): 230-6.

[114] Ruan CM, Paulose M, Varghese OK, Mor GK, Grimes CA. Fabrication of highly ordered TiO_2 nanotube arrays using an organic electrolyte. Journal of Physical Chemistry B 2005; 109(33): 15,754-9.

[115] Paulose M, Shankar K, Yoriya S, et al. Anodic growth of highly ordered TiO_2 nanotube arrays to 134 μm in length. Journal of Physical Chemistry B 2006; 110(33): 16,179-84.

[116] Paulose M, Prakasam HE, Varghese OK, et al. TiO_2 nanotube arrays of 1000 μm length by anodization of titanium foil: Phenol red diffusion. Journal of Physical Chemistry C 2007; 111(41): 14,992-7.

[117] Shankar K, Mor GK, Fitzgerald A, Grimes CA. Cation effect on the electrochemical formation of very high aspect ratio TiO_2 nanotube arrays in formamide-Water mixtures. Journal of Physical Chemistry C 2007; 111(1): 21-6.

[118] Richter C, Wu Z, Panaitescu E, Willey RJ, Menon L. Ultrahigh-aspect-ratio titania nanotubes. Advanced Materials 2007; 19(7): 946-948.

[119] Allam NK, Grimes CA. Formation of vertically oriented TiO_2 nanotube arrays using a fluoride free HCl aqueous electrolyte. Journal of Physical Chemistry C 2007; 111(35): 13,028-32.

[120] Mor GK, Varghese OK, Paulose M, Mukherjee N, Grimes CA. Fabrication of tapered, conical-shaped titania nanotubes. Journal of Materials Research 2003; 18(11): 2588-93.

[121] Bauer S, Kleber S, Schmuki P. TiO_2 nanotubes: Tailoring the geometry in H_3PO_4/HF electrolytes. Electrochemistry Communications 2006; 8(8): 1321-5.

[122] LaTempa TJ, Feng X, Paulose M, Grimes CA. Temperature-Dependent Growth of Self-Assembled Hematite (alpha-Fe2O3) Nanotube Arrays: Rapid Electrochemical Synthesis and Photoelectrochemical Properties. Journal of Physical Chemistry C 2009; 113(36): 16,293-8.

[123] Lai CW, Abd Hamid SB, Sreekantan S. A Novel Solar Driven Photocatalyst: Well-Aligned Anodic WO_3 Nanotubes. International Journal of Photoenergy 2013; 2013. doi:10.1155/2013/745301 (accessed 25 November 2013).

[124] Park J, Kim K, Choi J. Formation of ZnO nanowires during short durations of potentiostatic and galvanostatic anodization. Current Applied Physics 2013; 13(7): 1370-5.

[125] Ghicov A, Schmidt B, Kunze J, Schmuki P. Photoresponse in the visible range from Cr doped TiO_2 nanotubes. Chemical Physics Letters 2007; 433(4-6): 323-6.

[126] Ferreira VC, Nunes MR, Silvestre AJ, Monteiro OC. Synthesis and properties of Co-doped titanate nanotubes and their optical sensitization with methylene blue. Materials Chemistry and Physics 2013; 142(1): 355-62.

[127] Gong J, Pu W, Yang C, Zhang J. Novel one-step preparation of tungsten loaded TiO_2 nanotube arrays with enhanced photoelectrocatalytic activity for pollutant degradation and hydrogen production. Catalysis Communications 2013; 36: 89-93.

[128] Li Y, Xiang Y, Peng S, Wang X, Zhou L. Modification of Zr-doped titania nanotube arrays by urea pyrolysis for enhanced visible-light photoelectrochemical H-2 generation. ElectrochimicaActa 2013; 87: 794-800.

[129] Xu Z, Yu J. Visible-light-induced photoelectrochemical behaviors of Fe-modified TiO_2 nanotube arrays.Nanoscale 2011; 3(8): 3138-44.

[130] Sun L, Cai J, Wu Q, Huang P, Su Y, Lin C. N-doped TiO_2 nanotube array photoelectrode for visible-light-induced photoelectrochemical and photoelectrocatalytic activities. Electrochimica Acta 2013; 108(0): 525-31.

[131] Yu Y, Wu H-H, Zhu B-L, et al. Preparation, characterization and photocatalytic activities of F-doped TiO_2 nanotubes. Catalysis Letters 2008; 121(1-2): 165-71.

[132] Yan G, Zhang M, Hou J, Yang J. Photoelectrochemical and photocatalytic properties of N plus S co-doped TiO_2 nanotube array films under visible light irradiation. Materials Chemistry and Physics 2011; 129(1-2): 553-7.

[133] Lu N, Zhao H, Li J, Quan X, Chen S. Characterization of boron-doped TiO_2 nanotube arrays prepared by electrochemical method and its visible light activity. Separation and Purification Technology 2008; 62(3): 668-73.

[134] Paramasivalm I, Macak JM, Schmuki P. Photocatalytic activity of TiO_2-nanotube layers loaded with Ag and Au nanoparticles. Electrochemistry Communications 2008; 10(1): 71-5.

[135] Li J, Lin C-J, Li J-T, Lin Z-Q. A photoelectrochemical study of CdS modified TiO_2 nanotube arrays as photoanodes for cathodic protection of stainless steel. Thin Solid Films 2011; 519(16): 5494-502.

[136] Gan J, Zhai T, Lu X, Xie S, Mao Y, Tong Y. Facile preparation and photoelectrochemical properties of CdSe/TiO_2 NTAs. Materials Research Bulletin 2012; 47(3): 580-5.

[137] Sakthivel S, Shankar MV, Palanichamy M, Arabindoo B, Bahnemann DW, Murugesan V. Enhancement of photocatalytic activity by metal deposition: characterisation and photonic efficiency of Pt, Au and Pd deposited on TiO_2 catalyst. Water Research 2004; 38(13): 3001-8.

[138] Henglein A. Photochemistry of colloidal cadmium-sulfide.2.Effects of adsorbed methyl viologen and of colloidal platinum. Journal of Physical Chemistry 1982; 86(13): 2291-3.

[139] Shankar K, Tep KC, Mor GK, Grimes CA. An electrochemical strategy to incorporate nitrogen in nanostructured TiO_2 thin films: modification of bandgap and photoelectrochemical properties. Journal of Physics D-Applied Physics 2006; 39(11): 2361-6.

[140] Antony RP, Mathews T, Ajikumar PK, Krishna DN, Dash S, Tyagi AK. Electrochemically synthesized visible light absorbing vertically aligned N-doped TiO_2 nanotube array films. Materials Research Bulletin 2012; 47(12): 4491-7.

[141] Zhou X-y, Shao J, Wan B.A One-Step Electrochemical Method for the Production of $TiO_{2-x}N_x$ Nanotubes. Journal of the Electrochemical Society 2013; 160(6): H335-H7.

[142] Kim D, Fujimoto S, Schmuki P, Tsuchiya H. Nitrogen doped anodic TiO_2 nanotubes grown from nitrogen-containing Ti alloys. Electrochemistry Communications 2008; 10(6): 910-3.

[143] Su Y, Zhang X, Zhou M, Han S, Lei L. Preparation of high efficient photoelectrode of N-F-codoped TiO_2 nanotubes. Journal of Photochemistry and Photobiology A: Chemistry 2008; 194(2-3): 152-60.

[144] Xua Z, Yanga W, Lia Q, Gaoa S, Shanga JK. Passivated n–p co-doping of niobium and nitrogen into self-organized TiO_2 nanotube arrays for enhanced visible light photocatalytic performance.2014; 144: 343-52.

[145] Liu H, Liu G, Shi X. N/Zr-codoped TiO_2 nanotube arrays: Fabrication, characterization, and enhanced photocatalytic activity. Colloids and Surfaces a-Physicochemical and Engineering Aspects 2010; 363(1-3): 35-40.

[146] Milad AMH, Minggu LJ, Kassim MB, Daud WRW. Carbon doped TiO_2 nanotubes photoanodes prepared by in-situ anodic oxidation of Ti-foil in acidic and organic medium with photocurrent enhancement. Ceramics International 2013; 39(4): 3731-9.

[147] Krengvirat W, Sreekantan S, Noor A-FM, et al. Carbon-incorporated TiO_2photoelectr-odes prepared via rapid-anodic oxidation for efficient visible-light hydrogen generation. International Journal of Hydrogen Energy 2012; 37(13): 10,046-56.

[148] Li J, Lu N, Quan X, Chen S, Zhao H. Facile method for fabricating boron-doped TiO_2 nanotube array with enhanced photoelectrocatalytic properties. Industrial & Engineering Chemistry Research 2008; 47(11): 3804-8.

[149] Zhou X, Peng F, Wang H, Yu H, Yang J. Preparation of B,N-codoped nanotube arrays and their enhanced visible light photoelectrochemical performances. Electrochemistry Communications 2011; 13(2): 121-4.

[150] Ruan CM, Paulose M, Varghese OK, Grimes CA. Enhanced photo electrochemical-response in highly ordered TiO_2 nanotube-arrays anodized in boric acid containing electrolyte. Solar Energy Materials and Solar Cells 2006; 90(9): 1283-95.

[151] Das C, Paramasivam I, Liu N, Schmuki P. Photoelectrochemical and photocatalytic activity of tungsten doped TiO_2 nanotube layers in the near visible region. ElectrochimicaActa 2011; 56(28): 10,557-61.

[152] Sun M, Cui X. Anodically grown Si-W codoped TiO_2 nanotubes and its enhanced visible light photoelectrochemical response. Electrochemistry Communications 2012; 20: 133-6.

[153] Liu H, Liu G, Zhou Q. Preparation and characterization of Zr doped TiO_2 nanotube arrays on the titanium sheet and their enhanced photocatalytic activity. Journal of Solid State Chemistry 2009; 182(12): 3238-42.

[154] Nie J, Mo Y, Zheng B, Yuan H, Xiao D. Electrochemical fabrication of lanthanum-doped TiO_2 nanotube array electrode and investigation of its photoelectrochemical capability. ElectrochimicaActa 2013; 90: 589-96.

[155] Xie K, Sun L, Wang C, et al. Photoelectrocatalytic properties of Ag nanoparticles loaded TiO_2 nanotube arrays prepared by pulse current deposition. ElectrochimicaActa 2010; 55(24): 7211-8.

[156] Zhang S, Peng F, Wang H, et al. Electrodeposition preparation of Ag loaded N-doped TiO_2nanotube arrays with enhanced visible light photocatalytic performance. Catalysis Communications 2011; 12(8): 689-93.

[157] Xing L, Jia J, Wang Y, Zhang B, Dong S. Pt modified TiO_2 nanotubes electrode: Preparation and electrocatalytic application for methanol oxidation. International Journal of Hydrogen Energy 2010; 35(22): 12,169-73.

[158] Yin Y, University T, Tan X, et al. Efficient synthesis of titania nanotubes and enhanced photoresponse of Pt decorated TiO_2 for water splitting. Frontiers of Chemical Engineering in China 2013; 3(3): 298-304.

[159] Qin Y-H, Yang H-H, Lv R-L, Wang W-G, Wang C-W.TiO_2 nanotube arrays supported Pd nanoparticles for ethanol electrooxidation in alkaline media. ElectrochimicaActa 2013; 106: 372-7.

[160] Cheng X, Liu H, Chen Q, Li J, Wang P. Preparation and characterization of palladium nano-crystallite decorated TiO_2nano-tubes photoelectrode and its enhanced photocatalytic efficiency for degradation of diclofenac. Journal of Hazardous Materials 2013; 254: 141-8.

[161] Zhang X, Lin S, Liao J, et al. Uniform deposition of water-soluble CdS quantum dots on TiO_2 nanotube arrays by cyclic voltammetric electrodeposition: Effectively prevent aggregation and enhance visible-light photocatalytic activity. 2013; 108: 296-303.

[162] Feng H, Tran TT, Chen L, Yuan L, Cai Q. Visible light-induced efficiently oxidative decomposition of p-Nitrophenol by CdTe/TiO_2 nanotube arrays. Chemical Engineering Journal 2013; 215: 591-9.

[163] Tsui L-k, Zangari G. Modification of TiO_2 nanotubes by Cu_2O for photoelectrochemical, photocatalytic, and photovoltaic devices. 2013. http://dx.doi.org/10.1016/j.electacta.2013.09.150 (accessed 1 January 2014).

[164] Asahi R, Morikawa T, Ohwaki T, Aoki K, Taga Y. Visible-light photocatalysis in nitrogen-doped titanium oxides. Science 2001; 293(5528): 269-71.

[165] Li S, Lin S, Liao J, Pan N, Li D, Li J. Nitrogen-Doped TiO_2 Nanotube Arrays with Enhanced Photoelectrochemical Property. International Journal of Photoenergy 2012; 2012. http://dx.doi.org/10.1155/2012/794207 (accessed 01 January 2014).

[166] Serpone N. Is the band gap of pristine TiO_2 narrowed by anion-and cation-doping of titanium dioxide in second-generation photocatalysts? Journal of Physical Chemistry B 2006; 110(48): 24,287-93.

[167] Zhao X, Guo L, Hu C, Liu H, Qu J. Simultaneous destruction of Nickel (II)-EDTA with TiO_2/Ti film anode and electrodeposition of nickel ions on the cathode. Applied Catalysis B: Environmental 2014; 144(0): 478-85.

[168] Tahir M, Amin NS.Advances in visible light responsive titanium oxide-based photocatalysts for CO_2 conversion to hydrocarbon fuels. Energy Conversion and Management 2013; 76(0): 194-214.

[169] Quan X, Ruan X, Zhao H, Chen S, Zhao Y. Photoelectrocatalytic degradation of pentachlorophenol in aqueous solution using a TiO_2 nanotube film electrode. Environmental Pollution 2007; 147(2): 409-14.

[170] Quan X, Yang SG, Ruan XL, Zhao HM. Preparation of titania nanotubes and their environmental applications as electrode. Environmental Science & Technology 2005; 39(10): 3770-5.

[171] Philippidis N, Sotiropoulos S, Efstathiou A, Poulios I. Photoelectrocatalytic degradation of the insecticide imidacloprid using TiO_2/Ti electrodes. Journal of Photochemistry and Photobiology A: Chemistry 2009; 204(2-3): 129-36.

[172] Fang T, Yang C, Liao L. Photoelectrocatalytic degradation of high COD dipterex pesticide by using TiO_2/Ni photo electrode. Journal of Environmental Sciences-China 2012; 24(6): 1149-56.

[173] Chang H-S, Choo K-H, Lee B, Choi S-J.The methods of identification, analysis, and removal of endocrine disrupting compounds (EDCs) in water. Journal of Hazardous Materials 2009; 172(1): 1-12.

[174] Daghrir R, Drogui P, Dimboukou-Mpira A, El Khakani MA.Photoelectrocatalytic degradation of carbamazepine using Ti/TiO_2 nanostructured electrodes deposited by means of a pulsed laser deposition process. Chemosphere 2013; 93(11): 2756-66.

[175] Liu H, Liu G, Fan J, et al. Photoelectrocatalytic degradation of 4,4′-dibromobiphenyl in aqueous solution on TiO_2 and doped TiO_2 nanotube arrays. Chemosphere 2011; 82(1): 43-7.

[176] Martinez-Huitle CA, Brillas E. Decontamination of wastewaters containing synthetic organic dyes by electrochemical methods: A general review. Applied Catalysis B: Environmental 2009; 87(3-4): 105-45.

[177] Jurado-Sanchez B, Ballesteros E, Gallego M. Occurrence of aromatic amines and N-nitrosamines in the different steps of a drinking water treatment plant. Water Research 2012; 46(14): 4543-55.

[178] Egerton TA, Christensen PA, Kosa SAM, Onoka B, Harper JC, Tinlin JR. Photoelectrocatalysis by titanium dioxide for water treatment. International Journal of Environment and Pollution 2006; 27(1-3): 2-19.

[179] Pardeshi SK, Patil AB. Solar photocatalytic degradation of resorcinol a model endocrine disrupter in water using zinc oxide. Journal of Hazardous Materials 2009; 163(1): 403-9.

[180] Stasinakis AS, Kordoutis CI, Tsiouma VC, Gatidou G, Thomaidis NS.Removal of selected endocrine disrupters in activated sludge systems: Effect of sludge retention time on their sorption and biodegradation. Bioresource Technology 2010; 101(7): 2090-5.

[181] Balest L, Lopez A, Mascolo G, Di Iaconi C. Removal of endocrine disrupter compounds from municipal wastewater using an aerobic granular biomass reactor. Biochemical Engineering Journal 2008; 41(3): 288-94.

[182] Zheng Q, Lee C. Visible light photoelectrocatalytic degradation of methyl orange using anodized nanoporous WO_3. ElectrochimicaActa 2014; 115(0): 140-5.

[183] Li J, Zheng L, Li L, Xian Y, Jin L. Fabrication of TiO_2/Ti electrode by laser-assisted anodic oxidation and its application on photoelectrocatalytic degradation of methylene blue. Journal of Hazardous Materials 2007; 139(1): 72-8.

[184] Song H, Shang J, Zhu T, Ye J, Li Q, Teng F.The improved photoelectrocatalytic degradation of rhodamine B driven by the half-rectified square wave.Electrochimica Acta 2013; 102: 375-80.

[185] Osugi ME, Rajeshwar K, Ferraz ERA, de Oliveira DP, Araújo ÂR, Zanoni MVB. Comparison of oxidation efficiency of disperse dyes by chemical and photoelectrocatalytic chlorination and removal of mutagenic activity. Electrochimica Acta 2009; 54(7): 2086-93.

[186] Guaraldo TT, Zanoni TB, de Torresi SIC, et al. On the application of nanostructured electrodes prepared by Ti/TiO_2/WO_3"template": A case study of removing toxicity of indigo using visible irradiation. Chemosphere 2013; 91(5): 586-93.

[187] Carneiro PA, Oliveira DP, Umbuzeiro GA, Boldrin Zanoni MV. Mutagenic activity removal of selected disperse dye by photoeletrocatalytic treatment. Journal of Applied Electrochemistry 2010; 40(3): 485-92.

[188] Pelaez M, Nolan NT, Pillai SC, et al.A review on the visible light active titanium dioxide photocatalysts for environmental applications. Applied Catalysis B: Environmental 2012; 125(0): 331-49.

[189] Olya ME, Pirkarami A. Cost-effective photoelectrocatalytic treatment of dyes in a batch reactor equipped with solar cells. Separation and Purification Technology 2013; 118(0): 557-66.

[190] Liu Y, Zhou H, Zhou B, et al. Highly stable CdS-modified short TiO_2 nanotube array electrode for efficient visible-light hydrogen generation. International Journal of Hydrogen Energy 2011; 36(1): 167-74.

[191] MonteiroPaschoal FM, Pepping G, Boldrin Zanoni MV, Anderson MA. Photoelectrocatalytic Removal of Bromate Using Ti/TiO_2 Coated as a Photocathode. Environmental Science & Technology 2009; 43(19): 7496-502.

[192] Su Y, Han S, Zhang X, Chen X, Lei L. Preparation and visible-light-driven photoelectrocatalytic properties of boron-doped TiO_2 nanotubes. Materials Chemistry and Physics 2008; 110(2-3): 239-46.

[193] Xin Y, Liu H, Han L, Zhou Y. Comparative study of photocatalytic and photoelectrocatalytic properties of alachlor using different morphology TiO_2/Ti photoelectrodes. Journal of Hazardous Materials 2011; 192(3): 1812-8.

[194] Palmas S, Da Pozzo A, Mascia M, et al. Effect of the preparation conditions on the performance of TiO_2 nanotube arrays obtained by electrochemical oxidation. International Journal of Hydrogen Energy 2011; 36(15): 8894-901.

[195] Li Y, Yu H, Song W, Li G, Yi B, Shao Z. A novel photoelectrochemical cell with self-organized TiO_2 nanotubes as photoanodes for hydrogen generation. International Journal of Hydrogen Energy 2011; 36(22): 14,374-80.

[196] Sun K-C, Chen Y-C, Kuo M-Y, et al. Synthesis and characterization of highly ordered TiO_2 nanotube arrays for hydrogen generation via water splitting. Materials Chemistry and Physics 2011; 129(1-2): 35-9.

[197] Zhang Z, Hossain MF, Takahashi T. Photoelectrochemical water splitting on highly smooth and ordered TiO_2 nanotube arrays for hydrogen generation. International Journal of Hydrogen Energy 2010; 35(16): 8528-35.

[198] Cheng X, Pan G, Yu X, Zheng T. Preparation of CdS NCs decorated TiO_2nano-tubes arrays photoelectrode and its enhanced photoelectrocatalytic performance and mechanism. Electrochimica Acta 2013; 105: 535-41.

[199] Yang E-l, Shi J-j, LiangH-c, Cheuk W-K. Coaxial WO_3/TiO_2 nanotubes/nanorods with high visible light activity for the photodegradation of 2,3-dichlorophenol. Chemical Engineering Journal 2011; 174(2-3): 539-45.

[200] Fraga LE, Franco JH, Orlandi MO, Zanoni MVB.Photoelectrocatalytic oxidation of hair dye basic red 51 at W/WO_3/TiO_2bicomposite photoanode activated by ultraviolet and visible radiation. Journal of Environmental Chemical Engineering 2013; 1(3): 194-9.

[201] Cong Y, Li Z, Zhang Y, Wang Q, Xu Q. Synthesis of alpha-Fe_2O_3/TiO_2 nanotube arrays for photoelectro-Fenton degradation of phenol. Chemical Engineering Journal 2012; 191: 356-63.

[202] Lin C-J, Lu Y-T, Hsieh C-H, Chien S-H. Surface modification of highly ordered TiO_2 nanotube arrays for efficient photoelectrocatalytic water splitting. Applied Physics Letters 2009; 94(11).

[203] Sun L, Cai J, Wu Q, Huang P, Su Y, Lin C. N-doped TiO_2 nanotube array photoelectrode for visible-light-induced photoelectrochemical and photoelectrocatalytic activities. ElectrochimicaActa 2013; 108: 525-31.

[204] Wu H, Zhang Z. High photoelectrochemical water splitting performance on nitrogen doped double-wall TiO_2 nanotube array electrodes. International Journal of Hydrogen Energy 2011; 36(21): 13481-7.

[205] Rauf MA, Meetani MA, Hisaindee S. An overview on the photocatalytic degradation of azo dyes in the presence of TiO_2 doped with selective transition metals. Desalination 2011; 276(1-3): 13-27.

[206] Zhang Z, Yu Y, Wang P. Hierarchical Top-Porous/Bottom-Tubular TiO_2 Nanostructures Decorated with Pd Nanoparticles for Efficient Photoelectrocatalytic Decomposition of Synergistic Pollutants. Acs. Applied Materials & Interfaces 2012; 4(2): 990-6.

[207] Li Z, Cui X, Lin Y. Electrochemically Synthesized Ordered TiO_2 and Platinum Nanocomposite Electrode: Preparation, Characterization, and Application to Photoelectrocatalytic Methanol Oxidation. Journal of Nanoscience and Nanotechnology 2009; 9(4): 2297-302.

[208] Ye M, Gong J, Lai Y, Lin C, Lin Z. High-Efficiency Photoelectrocatalytic Hydrogen Generation Enabled by Palladium Quantum Dots-Sensitized TiO_2 Nanotube Arrays. Journal of the American Chemical Society 2012; 134(38): 15,720-3.

[209] Kang Q, Lu QZ, Liu SH, et al.A ternary hybrid CdS/Pt-TiO_2 nanotube structure for photoelectrocatalytic bactericidal effects on Escherichia Coli. Biomaterials 2010; 31(12): 3317-26.

[210] Seferlis AK, Neophytides SG.On the kinetics of photoelectrocatalytic water splitting on nanocrystalline TiO_2 films. Applied Catalysis B-Environmental 2013; 132: 543-52.

[211] Palmas S, Mascia M, Vacca A, Tredici I. Photoelectrocatalytic Performances of Nanostructured/Decorated TiO_2 Electrodes: Effect of Wavelength and Cell Configuration. International Journal of Photoenergy 2013; 2013. http://dx.doi.org/10.1155/2013/173760 (accessed 1 February 2014).

[212] Selcuk H. Disinfection and formation of disinfection by-products in a photoelectrocatalytic system. Water Research 2010; 44(13): 3966-72.

[213] Gong J, Yang C, Pu W, Zhang J. Liquid phase deposition of tungsten doped TiO_2 films for visible light photoelectrocatalytic degradation of dodecyl-benzenesulfonate. Chemical Engineering Journal 2011; 167(1): 190-7.

[214] Gong J, Pu W, Yang C, Zhang J. Tungsten and nitrogen co-doped TiO_2 electrode sensitized with Fe-chlorophyllin for visible light photoelectrocatalysis. Chemical Engineering Journal 2012; 209:94-101.

[215] Daghrir R, Drogui P, Delegan N, El Khakani MA.Electrochemical degradation of chlortetracycline using N-doped Ti/TiO_2 photoanode under sunlight irradiations. Water Research 2013; 47(17): 6801-10.

[216] Xie Y-B, Li X-Z. Degradation of bisphenolA in aqueous solution by $H2O_2$-assisted photoelectrocatalytic oxidation. Journal of Hazardous Materials 2006; 138(3): 526-33.

[217] Daghrir R, Drogui P, Delegan N, El Khakani MA.Removal of chlortetracycline from spiked municipal wastewater using a photoelectrocatalytic process operated under sunlight irradiations. Science of The Total Environment 2014; 466-467(0): 300-5.

[218] Lei J, Li X, Li W, Sun F, Lu D, Yi J. Arrayed porous iron-doped TiO_2 as photoelectrocatalyst with controllable pore size. International Journal of Hydrogen Energy 2011; 36(14): 8167-72.

[219] Gan WY, Friedmann D, Amal R, Zhang S, Chiang K, Zhao H. A comparative study between photocatalytic and photoelectrocatalytic properties of Pt deposited TiO_2 thin films for glucose degradation. Chemical Engineering Journal 2010; 158(3): 482-8.

[220] Antoniadou M, Daskalaki VM, Balis N, Kondarides DI, Kordulis C, Lianos P. Photocatalysis and photoelectrocatalysis using (CdS-ZnS)/TiO_2 combined photocatalysts. Applied Catalysis B-Environmental 2011; 107(1-2): 188-96.

[221] Christensen PA, Egerton TA, Lin WF, Meynet P, Shao ZG, Wright NG.A novel electrochemical device for the disinfection of fluids by OH radicals. Chemical Communications 2006; (38): 4022-3.

[222] Yu H, Chen S, Quan X, Zhao H, Zhang Y. Silicon nanowire/TiO_2 heterojunction arrays for effective photoelectrocatalysis under simulated solar light irradiation. Applied Catalysis B-Environmental 2009; 90(1-2): 242-8.

[223] Su J, Yu H, Chen S, Quan X, Zhao Q. Visible-light-driven photocatalytic and photoelectrocatalytic debromination of BDE-47 on a macroporous silicon/grapheneheterostructure. Separation and Purification Technology 2012; 96: 154-60.

[224] Wang X, Li G, Zhu H, Yu JC, Xiao X, Li Q. Vertically aligned CdTe nanotube arrays on indium tin oxide for visible-light-driven photoelectrocatalysis. Applied Catalysis B: Environmental 2014; 147(0): 17-21.

[225] Scott-Emuakpor EO, Kruth A, Todd MJ, Raab A, Paton GI, Macphee DE. Remediation of 2,4-dichlorophenol contaminated water by visible light-enhanced WO_3 photoelectrocatalysis. Applied Catalysis B: Environmental 2012; 123: 433-9.

[226] Zanoni MVB, Guaraldo T. Photoelectrochemical Hydrogen Generation and Concomitant Organic Dye Oxidation under TiO_2 Nanotube. ECS transactions 2013; 50(36): 63-70.

Chapter 11

The Use of Electrochemical Noise in the Study of Nanometric Coatings

C. Gaona Tiburcio, I. López Cázares,
J.A. Cabral Miramontes, P. Zambrano Robledo and
J. Uruchurtu Chavarín

Additional information is available at the end of the chapter

http://dx.doi.org/10.5772/57205

1. Introduction

Nanostructured materials are a new class of materials that are distinguished from conventional polycrystalline materials by their extremely fine crystallite size from 100 nm down to several nm. Interest to nanostructured materials has been growing rapidly due to reports showing that they posses remarkable mechanical, magnetic and optical properties.

R. Smalley, R. Curl and H. Kroto discovered a new form of carbon (carbon-60). Thereafter, in 1993, came the discovery of "carbon nanotubes", by S. Iijima and D. Bethune. Today nanotubes have been joined by nanocrystals, nanowires, nanoroads and other nanoforms.

Highly sophisticated surface related properties, such as optical, magnetic, electronic, catalytic, mechanical, tribological, chemical as well as magnetic, electronic and optical can be obtained by advanced nanostructured coatings, making them attractive for many modern industrial applications. There are many types of design models for nanostructured coatings, such as nano-composite coatings, nano-modulated multilayer coating, nano-graded coatings. Design of nanostructured coatings needs consideration of many factors, e.g. the interface volume, crystallite size, single layer thickness, surface and interfacial energy, texture, epitaxial stress and strain, etc., all of which depend significantly on materials selection, deposition methods, processing parameters and what is specially important service condition of coated parts [1].

Reduction in the weight of engines is a key factor in improving the fuel efficiency. The use of lightweight materials has become more prevalent as car manufacturers strive to reduce vehicle weight to improve performance, lower fuel and oil consumption, and to reduce emissions [2]. Employment of nanotechnology in current and future automotive, aero and other engines will

INTECH
open science | open minds

© 2014 The Author(s). Licensee InTech. This chapter is distributed under the terms of the Creative Commons Attribution License (http://creativecommons.org/licenses/by/3.0), which permits unrestricted use, distribution, and reproduction in any medium, provided the original work is properly cited.

go long way in solving energy crisis [3]. Most manufacturers have replaced cast iron engine blocks with lightweight and low-cost aluminum-silicon crankcases. Several Al based alloys and metal-matrix composites are in use.

The function of a protective coating is to prevent industrial vapors, liquids, solids or gases from coming into contact with metallic structures. Paints have been used which offer only a physical barrier. Engine components are subjected to severe environmental stimuli for corrosion. The higher temperature, fuel and combustion products mixed with oxidizing atmosphere and thermal shock add to the corrosive media. Diesel engine and internal combustion (IC) engine even produces sulfuric and formic acid as a product of combustion under certain conditions such as cold weather [4].

Energy producing fossil steam turbine and boiler components operate in a severe environment at elevated temperatures. To protect these components from degradation by the operating environment, MCr (M: Fe, Ni or Co) coatings are used in the industry [5]. In particular Fe-Cr or Ni-Cr are predominantly used. These coatings provide corrosion / oxidation protection by formation of a stable and slowly growing protective oxide scale such as Cr_2O_3 on the external surface, which separates the component substrate in contact with the corrosive environment.

Coatings, particularly nanocoatings can help to improve performance and life of engines. Higher efficiency is obtained from various aspects of coatings, like reduced: dimension, weight, surface finishing, friction, etc., as presented in figure 1.

Thin films are fabricated by the deposition of individual atoms on a substrate. A thin film is defined as a low dimensional material created by condensing atomic – molecular – ionic species of matter. Thin films have been used for more than a half-century in making many devices and coatings, however, thin film technology is still being developed on a daily basis since it is a key in the development of new materials such as nanometer materials [6]. Thin film processing also saves on energy consumption in production and is considered an environmentally being material technology for the next century.

Thin films are deposited on a substrate by thermal evaporation, chemical decomposition or the evaporation of source materials by the irradiation of energetic species. The thin films origin starts with a random nucleation process followed by nucleation and growth stages. Nucleation and growth are dependent upon various deposition conditions, such as growth temperature, rate and substrate chemistry. Film microstructure, defects and film stress depend on the deposition conditions at the nucleation stage.

The deposition process using the irradiation of energetic species is known as sputtering, that was first observed by Bunsen and Grove 150 years ago. However, this process has a low level efficiency and an optimum sputtering design is necessary for each material.

The sputtered structure modifies the material properties, they may show features that are different from the bulk materials in terms of mechanical strength, carrier transportation, superconductivity, ferroelectricity, magnetic and optical properties. These films may be characterized by a strong internal stress and a number of lattice defects. The defect density [4] in nanoscale materials is very high, but not high enough as in amorphous. As depicted in fig.

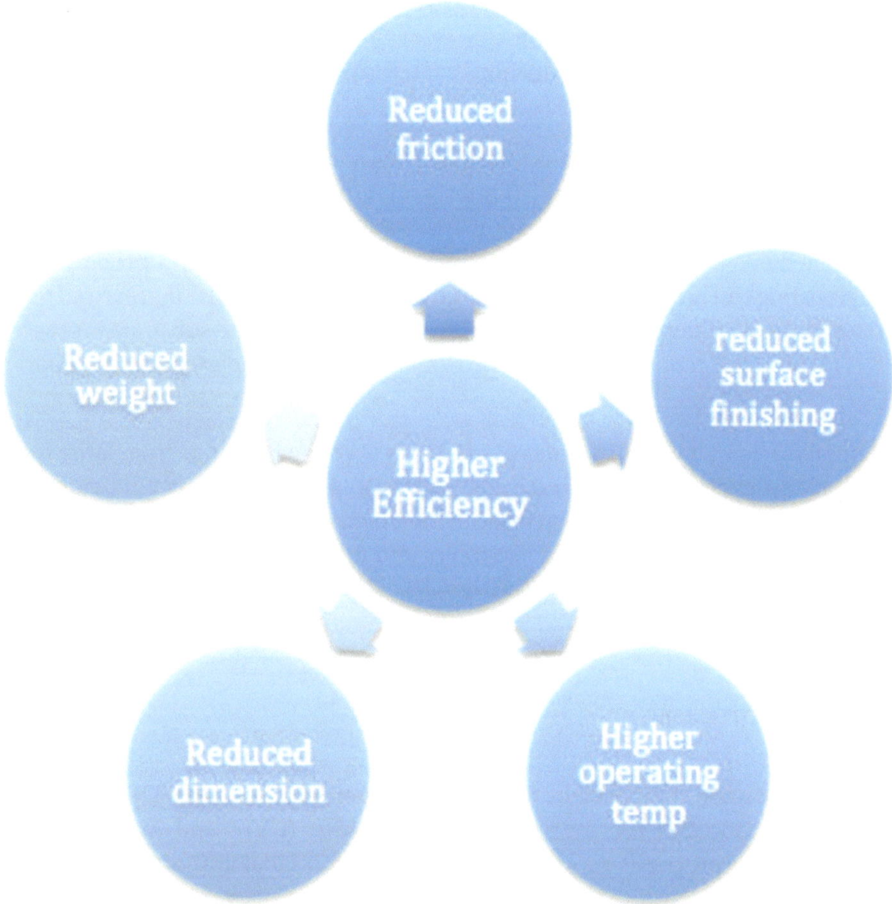

Figure 1. Aspects that provides higher efficiency to coatings.

(1-j), hardness as a function of grain size, Hall-Petch relation (hardness for a polycrystal with average grain diameter d, H_d=hardness of single crystal, $H_0+kd^{-1/2}$) predicts increase of hardness and flow stress as the grain size decreases. However, as the grain size is very small (in the range of 100 nm), the deformation mechanism changes from dislocation controlled slip to grain boundary sliding increasing plasticity at the same time. When the grain size further reduces almost to become amorphous, the material behaves in visco-elastic manner. This provides a global maximum in properties such as hardness, flow stress, toughness, ductility and thermal insulation (beause the conductivity of nanoscale material is much less in certain metallic system such as aluminum due to phonon scattering by high defect density) when the grain size is in nanoscale.

Thin film process is also essential for making nanometer materials. Nanomaterials are defined as materials thereof in other materials, having one or more dimensions of nanometer size (1nm = 10^{-7} cm = 10 A). Nanomaterials are classified in three types: Zero-dimensional (three dimensions of nanometer size), One-dimensional (two dimensions of nanometer size) and two-dimensional (one dimension of nanometer size). Nanostructured thin film and coatings possess properties different to homogenous materials due to the deliberate engineering of nanoscale features into the structure [7, 8].

Much of research activity is focused on the synthesis and processing of nanostructured bulk materials [9, 10, 11]. In spite of growing interest and research activities to the nano-structured multifunctional coatings in the last years there is still a lack in understanding the effect of processing methods and processing parameters on microsctructure and properties of nano-structured coatings including those that are used or are intended for use in wear and oxidation / corrosion service conditions. Attack of the corrosive medium on the substrate is severe due to the defects within the coating (such as micro-cracks, pores, pinholes, grain boundaries, etc.) [12, 13, 14].

The potential market for coatings and coated objects is very wide range from large scale Steel, Ni and Ti alloys sheet products to cutting and processing tools; wear resistant parts for automotive, aircraft / space and chemical industries; biocompatible and wear resistant surgical implants down to miniature parts for electronics and microelectronics. Another alternative is the use of multilayer metallic coatings, but very few studies are reported [15,16, 17, 18, 19, 20].

2. Deposition process

The physical process is composed of the physical vapor deposition (PVD) and the chemical processes are composed of the chemical vapor deposition (CVD) and the chemical solvent deposition. The PVD process is divided in two categories: thermal evaporation and sputtering. The thermal evaporation comprises evaporating source materials in a vacuum chamber below $1X10^{-6}$ torr ($1.3 X 10^{-4}$ Pa) and condensing the evaporated particles on a substrate. The thermal evaporation is conventionally called vacuum deposition [7, 21, 22, 23].

In the sputtering when a solid surface is bombarded with energetic particles such as accelerated ions, surface atoms of the solid are scattered backward due to collisions between the surface atoms and the energetic particles (figure 2). This phenomenon is called sputtering. When a thin foil is bombarded with energetic particles, some of the scattered atoms transmit through the foil. Cathode sputtering is a cathode disintegration and the can say impact evaporation. Several sputtering systems exists: diode, magnetron and ion-beam (McClanahan and Laegreid, historical review of sputtering deposition) [24].

In magnetron sputtering, a magnetic field is superposed on the cathode and glow discharge, which is parallel to the cathode surface [25, 26, 27]. The electrons in the glow discharge show cycloidal motion, and the center of the orbit drifts in the direction of ExB with the drift velocity of E/B, where E and B denote the electric field in the discharge and the superposed transverse magnetic field, respectively. The magnetic field is oriented such that these drift paths for electrons form a closed loop. This electron-trapping effect increases the collision rate between the electrons and the sputtering gas molecules. This enables one to lower the sputtering gas

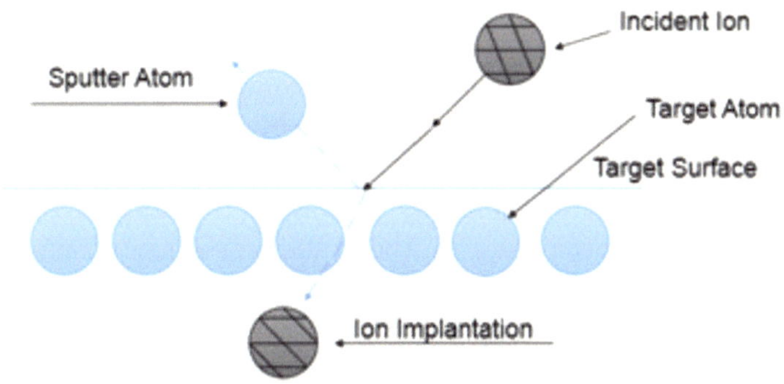

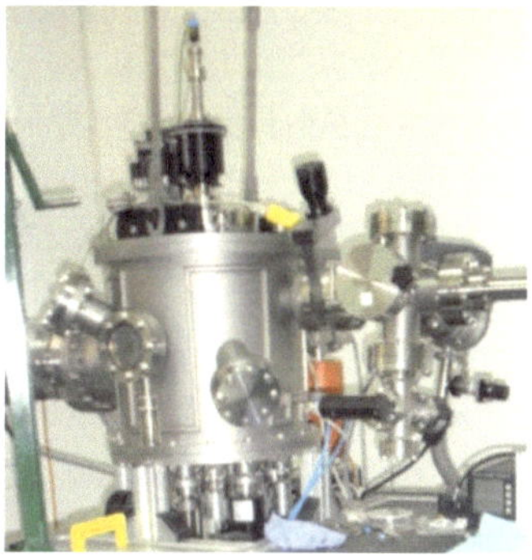

Figure 2. Sputtering process and sputtering chamber.

pressure as low as 10^{-5} torr. In the magnetron system, the magnetic field increases the plasma density, which leads to increases in the current density at the cathode target, effectively increasing the sputtering rate at the target. Due to the gas's low working pressure, the sputtered particles traverse the discharge space without collisions, which results in a high deposition rate.

Sputtering is accomplished by an energetic particle bombarding a target surface with sufficient energy (50 eV – 1000 eV) to result in the ejection of one or more atoms from the target. Typically, the target, which is called the cathode and is the material to be deposited, is connected to a negative voltage supply [28]. The anode is the substrate. In the sputtering process, the positive ions from ionizing gas strike the target surface and remove neutral target. The neutral atoms leave the target, condense on the substrate surface and form into thin films. In magnetron sputtering a magnetic field confines the plasma around the surface of the target.

Gheriani R. [29] analyzed the effect of substrate surface ion bombardment etching on reaction between chromium thin films (magnetron sputtering deposited) and Steel substrates (1 wt.% C). The study was carried out by X-ray diffraction, Auger electron spectroscopy, SEM (scanning electron microscopy) and Vickers micro-hardness measurements. They found that the reaction between carbon and chromium atoms leads to the formation and growth of Cr_7C_3 and $Cr_{23}C_6$ carbides in the considered temperature range. A significant increase in micro-hardness and adhesion of the coating films is associated with ion pre-cleaning and with increasing annealing temperature.

The recent evolution of coating technology allows depostition of multilayer coatings, with enhanced mechanical and corrosion behaviors. In 2010, Cekada M., et.al. [30], did an analysis of the diffusion processes in Al/Cr, Al/Fe and Cr/Fe bimetallic multilayers. They deposited 6 layers by sputtering deposited, and applied different annealings 240 – 650 ºC, for measure depth distributions of elements in the films. Auger depth profiling was conducted, and finds that in Al/Fe and Cr/Fe systems the dominant contribution to the depth was roughness. The activation energies are rather low, which can be attributed to the porous microstructure of films.

Grips W. [31] studied the electrochemical behavior of single layer CrN, TiN, TiAlN coatings and TiAlN/CrN multilayer coatings prepared by magnetron sputtering. They analyzed the corrosion behavior in 3.5% NaCl solution by potentiodynamic polarization and electrochemical impedance spectroscopy (EIS). The potentiodynamic polarization measurements showed that for all the coatings the corrosion potential shifted to higher values as compared to the uncoated substrate. Similarly, the corrosion current density decreased for coated samples, indicating better corrosion resistance of the coated samples. The multilayer coatings exhibited superior corrosion behavior as compared to the single layer coatings. The values of the charge transfer resistance and the pore resistance showed gradual increase from TiN – CrN – TiAlN/CrN multilayers. For all the coated samples, the coating capacitance and the double layer capacitance did not show significant variations, indicating that the surface of the coated substrate was less affected.

The thermal spraying process has commonly been accepted as the most effective and economic method for the fabrication process of the nanostructured coatings, and thus has contributed in a wide variety of high performance applications requiring improved wear, erosion, corrosion, and thermal insulation resistance [32, 33, 34]. In thermal spraying, feedstock materials in the form of powder or wire are fed into a heat source of spray equipment, where they are fully or partially melted and accelerated in a gas stream toward a substrate to be coated. The high-temperature exposure and the following rapid quenching, which intrinsically involved thermal spraying, can either improve or deteriorate the nano-scale microstructures of the coatings. Therefore, several challenges still remain in terms of feedstock preparation and processing itself, and an intense research. [35] Studied the properties of nanostructured thermal spray coatings WC-Co. They present Cr_2O_3 based solid-lubricant coating as one of new applications of the nanostructured coatings and its mechanical properties are compared to those of conventional counterpart. With the results, they finds that the advanced nanostructured WC-Co feedstock powders having desirable morphology was successfully fabricated and the resultant coatings showed significant improvement of wear-resistance in comparison with the conventionals. Also concludes that fine mixture of multi-components inside each powder particle of the nanostructured feedstock results In not only homogenous microstructure but also high performance of the coatings.

Electroplating is the application of a metal coating to a metallic or other conducting surface by an electrochemical process [36, 37, 38]. The article to be plated is made the cathode of an electrolysis cell through which a direct electric current is passed. The article is immersed in an aqueous solution containing the required metal in an oxidized form, either as an aquated cation or as a complex ion. The anode is usually a bar of the metal being plated. During electrolysis metal is deposited on to the work and metal from the bar dissolves.

Cr electroplating is regularly employed to coat the piston rings in engine. Different types of ferrous-based powders, containing C, Si, Sn, Ni, Cr, Mo, Cu, Ti, V and B, etc., are also employed to coat Al alloys for diesel engine applications [4]. Atmospheric plasma spray "APS" and Laser Surface Engineering "LSE" are extensively used for thermal barrier coating of nickel based super alloys component of aero and gas turbine engine components [39,40].

3. Corrosion

The high corrosion resistance of hard coatings used for decorative and / or wear protection applications is well known. However, when deposited on less noble materials like Steel, brass, Al or Mg alloys, the coated parts suffer from severe corrosive attack due to inherent coating defects or in homogeneities (cracks, pores, transient grain boundaries) [41, 42]. They opens possible paths for the corrosive media to reach the substrate. In the case of a less noble substrate material galvanic corrosion at the substrate will occur. This kind of corrosion is localized to the defect area and is characterized by the anodic dissolution of the substrate material with a high anodic current density at the defect site. It is also called galvanic corrosion and in the case of pores and pinholes as defects it is called pitting corrosion. The intensity of the corrosive attack depends strongly on the potential difference of the coating and the substrate material in the respective electrolyte. If this potential difference is large enough, galvanic corrosion will occur. A lot of research has been undertaken to get rid of this problem. There are several ways to improve the corrosion resistance of coated parts: thicker coatings [43], dense coatings (fine grained or amorphous), elimination of droplets (by modification of the depositation process: pulsed magnetron sputtering, filtered cathodic arc, etc), multilayers [44], duplex coatings [45], intermediate etching [46], alloying of the hard coating with less noble elements [47].

Fenker M. [48], studied the corrosion behavior of Ti and Cr multilayer coatings in sodium chloride media by circuit potential measurements, potentiodynamic corrosion tests and salt spray tests. The CrN, Nb_2N, Nb-C:H, TiN, TiMgN multilayer depositions were carried out. The substrate material was high speed steel. The corrosion behavior was electrochemically studied using open-circuit-potential measurements and potentiodynamic corrosion (polarization) tests. In their results they show that all measures lead to an improvement of the corrosion behavior in potentiodynamic polarization test in NaCl environment. Only one measure was able to avoid red rust sites after 24 h of salt spray testing. This was the alloying of TiN with magnesium with Mg contents up 17 at.%. The electrochemical effect of this alloying is a reduction of the electrochemical potential with increasing Mg content in a NaCl solution. As a consequence, the potential difference between the coating and the steel substrate is reduced and, therefore, the driving force for galvanic corrosion can be diminished.

Al coatings are candidates for the cathodic protection of construction steel due to their good corrosion resistance in aggressive media. However the mechanical properties and pitting corrosion sensitivity of some Al in chloride media have limited its application. Different alternatives have been investigated to improve mechanical resistance and sacrificial behavior of Al based coatings on steels. The addition of transition metals like Cr or Ti have been studied, and the best result there are been obtained for Al-Cr alloys deposited by magnetron sputtering, and is has been reported that the addition of Mo or Mg could limit the sensitivity to pitting [49]. Creus, et.al. [50] developed a study to improve the mechanical properties of Al based coatings, conserving a sacrificial protection characteristics. Al/Cr multilayer architectures, deposited by dc magnetron sputtering, were tested. Three architectures composed of a bilayer, a multilayer and a nanolayer were compared. The mechanical properties and corrosion resistance in saline solution were discussed and compared to single Al and Cr coatings. The open circuit potential and polarization resistance (Rp) of the coated substrate versus time were recorded during long immersion tests. A cyclic polarization technique was used to estimate the pitting resistance. They show that configuration parameters like composition of the layers and the thickness influence the mechanical and corrosion properties. The Al/Cr bilayer coating does not lead to significant improvement of mechanical and corrosion properties compared to single Al coating. The mechanical properties are increased for the multilayer configurations leading to the best behavior for the nanolayer coating, and multilayer architectures seem to be a way to improve the mechanical properties of Al based coating.

The long term oxidation behavior of nanocrystalline 304 SS and 304-4Al coatings has been investigated by Cheruvu and Wei [51]. The coatings were deposited on 304 SS samples using a plasma enhanced magnetron sputtering process. Cyclic oxidation tests were conducted on the coated and uncoated samples at a peak temperature of 750ºC for up to 1000 one-hour thermal cycles between the peak and room temperature. The addition of 4% Al stabilized the αFe structure and prevented the formation of the σ phase. The external oxide scale on the coated samples exhibited good spallation resistance compared to the scale on the uncoated samples. The ultrafine grain structure of 304SS coating had promoted selective oxidation of Cr and the addition of Al to the 304SS coating promoted selective oxidation Al over Cr.

Liu, et. at. [52] have investigated the effect of grain size on the protective external oxide scale formation on Ni-20Cr-3Al nanocrystalline coatings. Their results showed that the external oxide formed on the fine grained coatings varied from Cr_2O_3 to Al_2O_3 as the grain size decreased from 500 to 50 nm. The ultra fine grain structure promotes grain boundary diffusion, which facilitates selective oxidation of Al over Cr. The critical Al or Cr required for the formation of a continuous Al_2O_3 or Cr_2O_3 is significantly lower for the ultra-fine grains coatings than that for conventional materials.

Corrosion generally refers to the electrochemical attack process that takes place on most of metals. Metals are susceptible to suffer this attack because they have free electrons and are capable of establishing electrochemical cells in their structures. But the thermodynamic energy or chemical energy stored in they, varies from one metal to other. The energy is higher for numerous metals (Mg, Fe, and Al), and relatively low for other metals like Ag, Au, Cu, etc. Many metals experiments corrosion when in contact of an electrolyte and any contaminant like Cl. From an engineering point of view, corrosion is a progressive and destructive process that compromises structural integrity and represents an enormous economic loss [53].

During corrosion, at least two electrochemical reactions, an oxidation and a reduction reaction, occur at a metal-electrolyte interface. Because corrosion is due to an electrochemical mechanism, it is clear that electrochemical techniques can be used to study corrosion reactions and mechanisms. Then many techniques have been used to monitor the rate of corrosion and to study both, the corrosion and its inhibition mechanisms.

3.1. Electrochemical Noise (EN)

In a typical experimental system, an instrument known as a potentiostat has control of the voltage across the working electrode – counter electrode pair, and it adjusts this voltage to maintain the potential difference between the working and reference electrodes (which it senses through a high-impedance feedback loop) in accord with the program defined by a function generator, see figure 3.

The potential and the current can be measured, stored, and analyzed with this method. The evaluation of spontaneous fluctuations of the current and potential of corroding electrodes, wich is commonly refered to as electrochemical noise "EN" analysis, has been studied extensively [54, 55, 56]. EN is a generic term for low-level random fluctuations, usually less than 1 milivolt and nano to micro amps, in a corroding material.

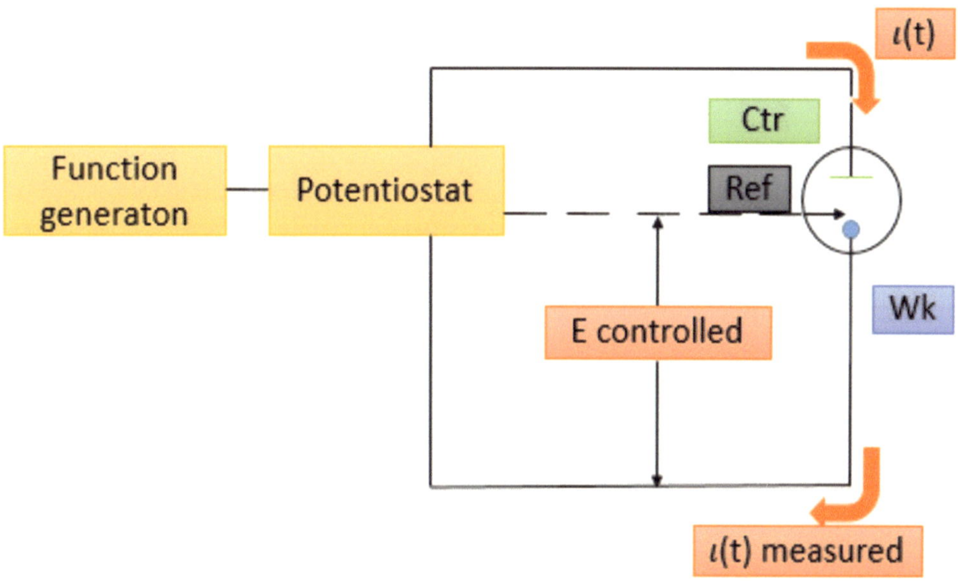

Figure 3. Experimental arrangement for controlled potential experiments.

Eden in 1986 [57], introduced a new parameter that depends on voltage and current standard deviations, correlate the EN resistance (Rn) with polarization resistance (Rp):

$$Rn = \frac{(\beta a)(\beta c)}{2.3(icorr)(\beta a + \beta c)} \tag{1}$$

where:

Rp = polarization resistance

Ba = anodic Tafel constant

Bc = cathodic Tafel constant

Icorr = corrosion current (A cm^{-2})

The corrosion rate is related to the polarization resistance by Stern-Geary:

$$i_{corr} = \frac{B}{Rp} \quad (2)$$

where B is the Stearn-Geary coefficient polarization resistance is the slope of the steady-state potential current curve at the corrosion potential. In 1993, Mansfeld and Xiao [58] have reported some data and correlations between Rp and Rn.

To measure be use three identical electrodes placed in an electrochemical cell. One electrode is connected like working electrode WE, another like reference RE, and the remaining like ground, GE. Then the potential noise is recorded with the RE and current is recorded with the WE. The GE enables the passing of the current. The current and potential transients are recorded simultaneously. However, EN requires monitoring of very small signal fluctuation, which may be susceptible to outsider sources of noise if improperly used.

The EN technique has some advantages related other corrosion monitoring techniques (e.g. linear polarization resistance or potentiodinamyc curves), mainly the absence of an applied voltage or current perturbation, and that the measurements can be taken at very short intervals. The difficulty in the interpretation of time series, graphs and data is the major disadvantage.

This analysis technique is so versatile that it can be applied to various materials, applications and processes. Recently Estupiñan L., et.al [59], performed a statistical analysis of electrochemical noise transients to correlate with the morphologies observed by SEM in different steels, in order to identify the type of corrosion developed; and Almeraya C., used EN to study Ni-Mo alloys fabricated by mechanical alloying and densified by spark plasma sintering [60]. García H., et.al [53] performer in 2009 a comparative study on carbon steel, of three corrosion resistant hybrid coatings, by using a loop reactor to measure the corrosion by means of the EN technique. They obtain the time series responses of current and potential, and by means of these signals, the EN resistance (Rn) was calculated according to Rn equation. Find that the hybrid-bi-component coating produced the highest resistance and consequently exhibited the lowest corrosion rate.

3.1.1. Study of nanometric coatings

In this chapter we present the results of a study in order to know the characteristics and corrosion resistance of nanostructured bilayer coatings. As discussed before, Al alloys are candidates in aerospace industry due to their corrosion resistance. However, their pitting corrosion sensitivity limits its application in many marine environments. Have been investigated that addition of Cr improve the mechanical resistance and sacrificial behavior. In this work, electrochemical noise has been carried out to investigate the corrosion behavior of Al and Cr coatings on Aluminum alloys AA6061-T6 and AA2024-T3, with different architecture organization (Al/Cr y Cr/Al).

3.1.2. *Experimental methodology*

Coatings consisted of Al and Cr deposited by magnetron sputtering on aluminum alloys AA6061-T6 and AA2024-T3 (2 cm x 2.5 cm). On each substrate was deposited an array of bilayers in a different order (1 micron thick), one with an initial layer of Al and an outer layer of Cr (Al / Cr), and another with an initial layer of Cr and a coating over it of Al (Cr / Al). We also used non-coated substrates (blank). The nomenclature used was: substrate / inner layer / outer layer.

3.1.3. *Microscopy analysis (MEB)*

Can be seen in figure 4 the micrographs of the systems on the substrate Al2024. In 2024AlCr few pits can be observed on the substrate, center-starry shaped, as is common in this material. Not corroded area presents a surface with large grains and dripped aspect. According to the mappings, appear Cr and oxygen, which can be an oxide, it is a common topography in coatings with outer layer of transition metal. In contrast to the previous system, in 2024CrAl pits are not observed. In this case the topography is more "smooth" because the nodules are larger, but fewer, as reported for outer layer coatings of Al. Regarding to Al corrosion products, appear to have been dissolved, since there is nothing on the surface.

Can be seen in figure 5 the micrographs of 6061AlCr and 6061CrAl. Observed in the first few and very small pits. The coating appears to have poor adhesion, associated to the Al substrate or even the stresses produced by the corrosion products of Al, leading to partial delamination of the coating.

In the 6061CrAl the surface is smooth and there aren't visible pits, but gives an appearance that the coating becomes brittle, since this coating is susceptible to internal stresses that exceed the cohesive forces of the aluminum substrate. In the case of both coatings don't observe corrosion product layer, so it could have been dissolved

3.1.4. *Electrochemical tests*

Was used a conventional pitting cell, as shown in figure 6. Electrochemical tests were conducted by electrochemical noise, according to equation 3 and ASTM G-199-09 standard [61], with a data acquisition rate of one data per second (1024 points). The results were analyzed from time series and statistical parameters.

$$Rn = \frac{\sigma E}{\sigma I} \tag{3}$$

where:

Rn = noise resistance

σE = standard deviation of potential

σi= standard deviation of current

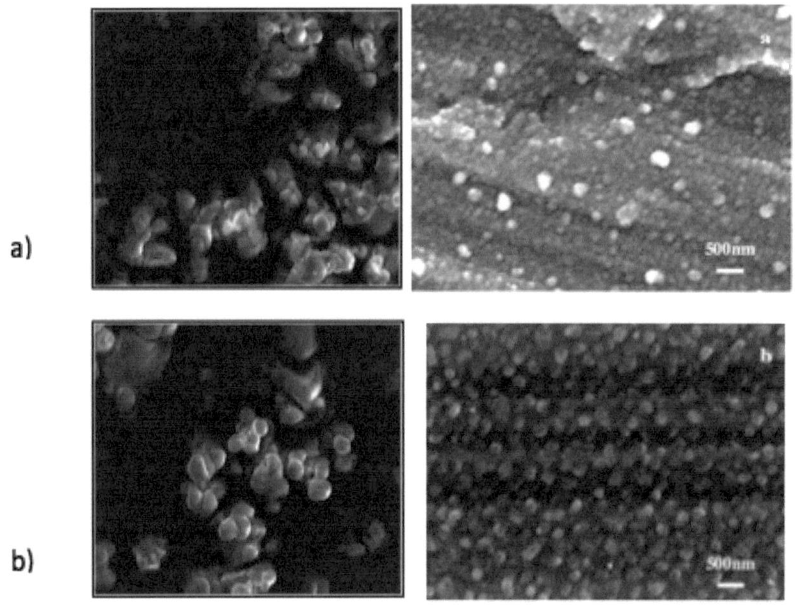

Figure 4. Micrographs of 2024AlCr a) y 2024CrAl b).

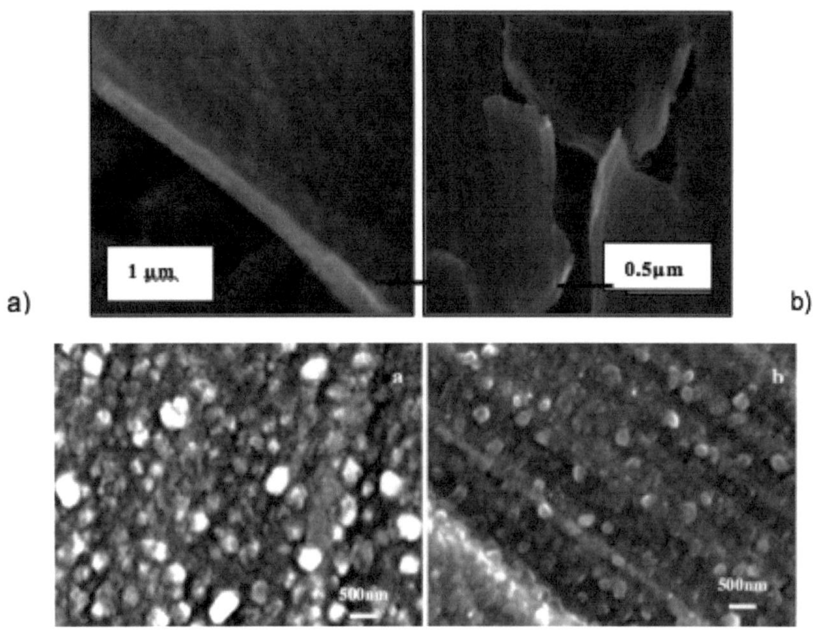

Figure 5. Micrographs of 6061AlCr a) y 6061CrAl b).

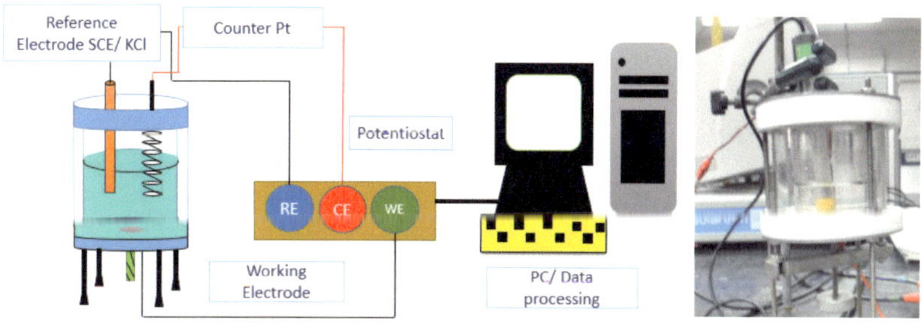

Figure 6. Experimental arrangement for electrochemical tests.

3.1.5. Analysis of non coated substrates

In the time series, figure 7, the transients have great amplitude and long term, with a fast rise and exponential fall, characteristic of metastable pits. In statistical analysis resulted an Rn of 1.07 $x10^3$ Ohm/cm^2 and 1.054 $x10^3$ Ohm/cm^2, to blanks 2024 and 6061 respectively, due to blocking of pits by corrosion products of the substrate, corresponding to that reported by other researchers in previous works [62,63].

3.1.6. Analysis of coated substrates

There were few events, all of low amplitude and high frequency (Figures 8 and 9), characteristic of a uniform corrosion process [64]. Potential noise is very stable in both, while in the current noise there are very fast fluctuations.

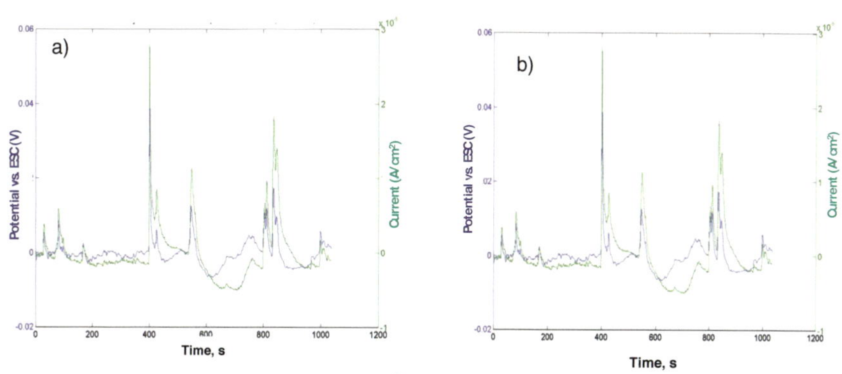

Figure 7. Potential and current time series, for uncoated substrates: a) blank 2024) blank 6061.

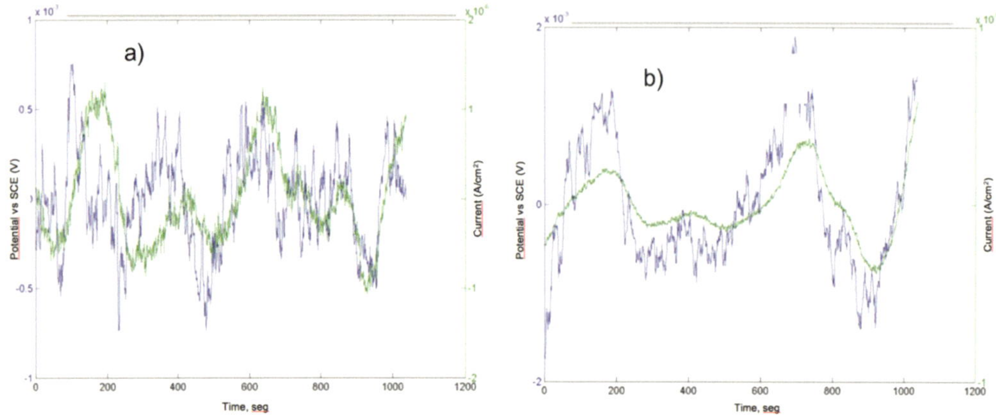

Figure 8. Potential and current time series for systems: a) 2024Al/Cr and b) 2024Cr/Al.

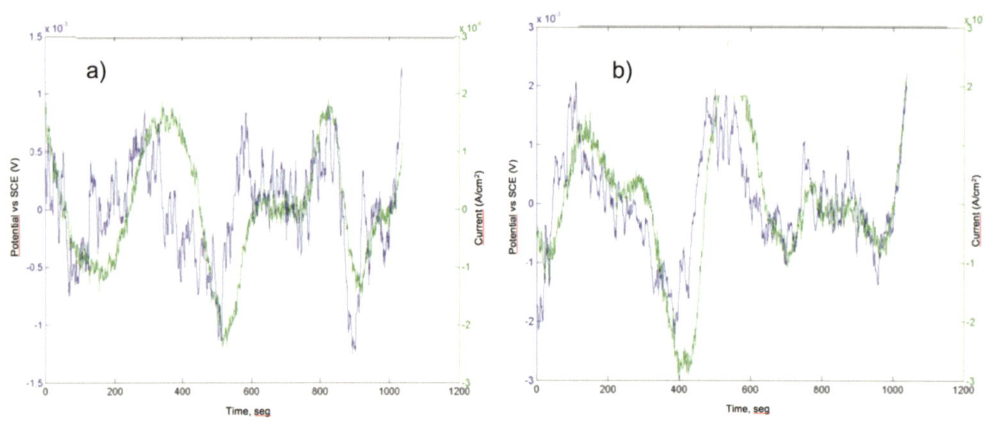

Figure 9. Potential and current time series for systems: a) 6061Al/Cr and b) 6061Cr/Al.

The statistical parameters (Tables 1 and 2). In the analysis, the SDC is higher in systems 6061AlCr and 2024AlCr that in 2024CrAl and 6061CrAl, because corrosion processes occur more rapidly in systems AlCr. SDV is higher in 6061AlCr than in 6061CrAl and 2024CrAl. In the first case due to corrosion is even more localized and passive processes are lower. In the second case is because the Rn in interface is greater. Regarding SDAV, for all systems is approximately 10.5 V / SCE indicating uniform corrosion process, except for 6061CrAl, which has a value of 4.10 V / SCE, which is characteristic of a slight corrosion or passivation. This interpretation agrees with that made by direct inspection of the time series. Regarding the LI (equation 4), indicates that the corrosion type for the system AlCr is mixed in both substrates, and is uniform for CrAl system. Which means that you may see some localized pitting, and uniform corrosion as the dominant form. Again, this parameter leads to the same information obtained by direct analysis of SDAV. Icorr is higher in the arrangement AlCr for both substrates, as a result of charge transfer processes, facilitated to the system with the Cr surface layer. Resume, the statistical analysis indicated that corrosion processes occur faster in arrangements Al / Cr, consistent with other studies reporting [65,66, 67], and the dominant mechanisms were uniform corrosion and light passivation, similar to that found by Hladky [68] y Castillo [69].

$$LI = \frac{\sigma I}{I_{rms}} \quad (4)$$

where, I_{smr} = root mean square.

Rn is the noise resistance (ohm/cm^2), Icorr is the corrosion current (A/cm^2), SDV is the standard deviation of the potential (V), SDC is the standard deviation of current (A), SDAV is the standard deviation of the average potential (V) and LI is the location index.

system	Rn	Icorr	SDV	SDC	SDAV	LI
2024Al/Cr	424.52	6.12 E^{-5}	1.43E^{-4}	8E^{-7}	5.11E^{-5}	0.0141
2024Cr/Al	768.04	3.4E^{-5}	4.2E^{-4}	5.44E^{-7}	8.75E^{-5}	0.0086

Table 1. Statistical parameters obtained for systems 2024 Al/Cr and 2024Cr/Al.

system	Rn	Icorr	SDV	SDC	SDAV	LI
6061Al/Cr	353.7	7.34 E^{-5}	5.33E^{-4}	1.5E^{-7}	8.2E^{-5}	0.0229
6061Cr/Al	684.8	3.8E^{-5}	4.2E^{-4}	6.2^{-7}	1.43E^{-5}	0.0085

Table 2. Statistical parameters obtained for systems 6061 Al/Cr and 6061cr/Al.

According to the results, it is possible to establish a mechanism for both systems Al / Cr, Cr / Al, regardless of the substrate.

In systems Al / Cr, once breaking the passive film of Cr, Cl^- ions can enter through these defects to reach the interface with the Al, such that due to the area ratio, the Al area smaller behaves as an anode and forms a galvanic couple, causing it to begin to dissolve the Al, and the corrosion products cause internal stresses at the interface, causing a delamination of the coating. Thus also showed higher Icorr values and mixed type corrosion.

In coatings Cr / Al, once breaking the oxide film of Al, the Cl^- can penetrate, with the difference that here the ratio of areas is different. Al acts as the anode and has a much larger area, then dissolves and corrosion products blocking Cr defects of the inner layer, so that increase the Rn and decrease the Icorr. Both systems however decreased density of pits, compared with those presented in the uncoated substrates. The types of corrosion that occurred in the systems were mixed, uniform and also a slight passivation.

4. Conclusions

In study of ECN for nanocoatings, all systems are considered cathodic, because they were dependent on the defects present in the coating. CrAl systems are the most stable, regardless of the substrate, due to the outer layer of Al which provides cathodic protection, and a second barrier constituted by the corrosion products. AlCr systems are less resistant to corrosion due to the corrosion product layer formed in the interface, which causes internal stresses leading to coating delamination. All studied systems decreased propagation of pitting, leading to a state of passivation or uniform corrosion, each by different mechanisms. Coatings possess superior corrosion resistance over the individually substrate.

5. Summary of previous works

Year	Ref	Author	Work
1991	44	J. Munemasa	Research the effect of the surface roughness of substrates on the corrosion properties of films coated by physical vapour deposition. Ti and TiN were deposited on substrates by magnetron sputtering method.
1995	45	A. Matthews	The characteristics of Vacuum plasma based coating and treatment methods are discussed, especially with regard to their beneficial influence on wear or corrosion.
1998	52	Liu	Have investigated the effect of grain size on the protective external oxide scale formation on Ni-20Cr-3Al nanocrystalline coatings.
2000	46	H.S. Park	A thick CrN film was deposited by magnetron sputtering on high-speed steel substrates, but the coating process was interrupted several times. The corrosion resistance was characterized by electrochemical potentiodynamic and potentiostatic measurements.
2004	29	R. Gheriani	Analyzed the effect of substrate surface ion bombardment etching on reaction between chromium thin films (magnetron sputtering deposited) and Steel substrates. The study was carried out by X-ray diffraction, Auger electron spectroscopy, SEM and Vickers micro-hardness measurements.
	49	Griguceviciene	The addition of transition metals like Cr or Ti have been studied, and the best result there are been obtained for Al-Cr alloys deposited by magnetron sputtering.
2006	31	Grips	Studied the electrochemical behavior of single layer CrN, TiN, TiAlN coatings and TiAlN/CrN multilayer coatings prepared by magnetron sputtering. They analyzed the corrosion behavior by potentiodynamic polarization and electrochemical impedance spectroscopy (EIS).
	48	Fenker	Studied the corrosion behavior of Ti and Cr multilayer coatings in NaCl media by circuit potential measurements, potentiodynamic corrosion and salt spray tests. The CrN, Nb_2N, Nb-C:H, TiN, TiMgN multilayer depositions were carried out.
2008	50	Creus	Al/Cr multilayer architectures, deposited by dc magnetron sputtering, were tested. The open circuit potential and polarization resistance of the

Year	Ref	Author	Work
			coated substrate were recorded during long immersion tests. A cyclic polarization technique was used to estimate the pitting resistance.
	51	Cheruvu	The long-term oxidation behavior of nanocrystalline 304 SS and 304-4Al coatings has been investigated. The coatings were deposited on 304 SS samples using a plasma enhanced magnetron sputtering process. Cyclic oxidation tests were conducted on the coated and uncoated samples.
2009	53	García H.	Comparative study on carbon steel, of three corrosion resistant hybrid coatings, by using a loop reactor to measure the corrosion by means of the EN technique.
2010	30	Cekada M.	Analysis of the diffusion processes in Al/Cr, Al/Fe and Cr/Fe bimetallic multilayers.
2011	59	Estupiñan L.	Performed a statistical analysis of electrochemical noise transients to correlate with the morphologies observed by SEM in different steels, in order to identify the type of corrosion developed.
2012	60	Almeraya C.	Used electrochemical noise to study Ni-Mo alloys fabricated by mechanical alloying and densified by spark plasma sintering.
	69	Castillo M.	Electrochemical Impedance Spectroscopy Behavior of Nanometric Al-Cr and Cr-Al Coatings by Magnetron Sputtering. Find that bilayers had good homogeneity and adherence with substrates. The Cr/Al coatings were more stables.

Acknowledgements

We thank to PROMEP program (PROMEP/103.5/12/3585 PTC-586). We also thank to the Academic Body UANL-CA-316 "Deterioration and integrity of composite materials".

Author details

C. Gaona Tiburcio[1*], I. López Cázares[1], J.A. Cabral Miramontes[1], P. Zambrano Robledo[1] and J. Uruchurtu Chavarín[2]

*Address all correspondence to: citlalli.gaona@gmail.com

1 Universidad Autónoma de Nuevo León, UANL, Facultad de Ingeniería Mecánica y Eléctrica, FIME, Centro de Investigación e Innovación en Ingeniería Aeronáutica, CIIIA, Aeropuerto Internacional del Norte, Apocada, Nuevo León, México

2 Universidad Autónoma del Estado de Morelos, UAEM, FCQI-CIICAP. Av. Universidad S/N, Cuernavaca, Mor., México

References

[1] Nanostructured Coatings. Nanostructure Science and Technology. Ed. Albano Cavaleiro, J.T.M. de Hosson. Springer, 2007.

[2] H.L. MacLean, L.B. Lave. Pror. Energy Combust. Sci. 29 (1) (2003) 1.

[3] Beyond the Nanohype. Technology Quarterly. The economist. London. 13th March 2003. www.economist.com/science.

[4] Narendra B. Dahotre, S. Nayak. Nanocoatings for engine application. Surf. Coat. Technol. 194 (2005) 58-67.

[5] M.A. Uusitalo, P.M.J. Vuoristo, T.A. Mantyla. Surf. Coat. Technol. 161 (2002) 275.

[6] Kiyotaka Wasa, Makoto Kitabatake, Hideaki Adachi. Thin film materials technology: sputtering of control compound materials. Thin film materials and devices. Chap. 1. Ed. Springer, 2004.

[7] P. Xiao and R. Dorey. Nanostructured Thin Films and Coatings. Jour. of Nanomaterials. Vol. 2008, ID 931380.

[8] Materials Chemistry. SpringerLink: Springer e-Books. Bradley D. Fahlman. Springer, 2007. 485 pag.

[9] Synthesis and Processing of Nanostructured Materials: Ceramic Engineering and Science Proceedings, Volume 27, Issue 8. Andrew Wereszczak (Editor), Edgar Lara-Curzio (Editor), William M. Mullins (Editor). 138 pag. November 2006.

[10] Nanostructured Materials - Processing, Properties and Potential Applications. Ed. Koch, Carl C. William Andrew Publishing/Noyes (2002).

[11] E R.W. Siegel. Chap 4. Synthesis and Processing of Nanostructured Materials. In: Mechanical Properties and Deformation Behavior of Materials Having Ultra-Fine Microstructures. Springer Netherlands (1993).

[12] H. A. Jehn. Surf. Coat. Technol. 125 (2000) 212.

[13] Wlasny et al. "Role of graphene defects in corrosion of graphene-coated Cu(111) surface" Appl. Phys. Lett. 102, 111601 (2013).

[14] P. Gselman, T. Bonina, F. Zupani et.al. Characterization of Defects in PVD TiAlN Hard Coatings. Mat. and tech. 46 (2012) 4, 351–354

[15] W. D. Münz, L. A. Donohue, P. E. Hovsepian. Surf. Coat. Technol. 125 (2000) 269.

[16] C. Charrier, P. Jacquot, E. Denisse, J.P. Millet, H. Mazille. Surf. Coat. Tecnnol. 90 (1997) 29.

[17] G.V. Strukov, G.K. Strukova, E.D. Shoo, et.al. Instruments and Experimental Techniques, 2009, Vol. 52, No. 5, 727–730.

[18] I.A. Kartsonakis, E.P. Koumoulos, A.C. Balaskas, et.al. Hybrid organic – inorganic multilayer coatings including nanocontainers for corrosion protection of metal alloys. Corr. Sci. Vol 57 (2012).

[19] Patent Multilayer metal-coated steel sheet. US 5206093 A (1993).

[20] Yun Tu et al. Fabrication and thermal characteristics of multilayer metal-coated regenerated grating sensors for high-temperature sensing. Smart Mater. Struct. 22 075026 (2013).

[21] H.G Prengel, P.C Jindal, K.H Wendt, A.T Santhanam, P.L Hegde, R.M Penich, A new class of high performance PVD coatings for carbide cutting tools, Surf. Coat. Technol., Vol 139, Issue 1, 1 May 2001, 25-34

[22] B. Navinšek, P. Panjan, I. Milošev, PVD coatings as an environmentally clean alternative to electroplating and electroless processes, Surf. Coat. Technol., Vol 116–119, September 1999, 476-487

[23] B. Schuhmacher, C. Schwerdt, U. Seyfert, O. Zimmer. Innovative steel strip coatings by means of PVD in a continuous pilot line: process technology and coating development, Surf. Coat. Technol., Vol 163–164, 30 January 2003, 703-709

[24] McClanahan and Laegreid. Production of thin films by controlled deposition of sputtered material. Sputtering by particle bombardment III. Topics in Applied Physics, Vol.64. Springer 1991, 339-377.

[25] P.J Kelly, R.D Arnell, Magnetron sputtering: a review of recent developments and applications, Vacuum, Vol 56, Issue 3, March 2000, 159-172

[26] J. Böhlmark. Dissertation: Fundamentals of High Power Impulse Magnetron Sputtering. Inst. Tech., Linkoping Univ. Sweden (2005).

[27] C.H. Shon, J.K. Lee. Modeling of magnetron sputtering plasmas. App. Surf. Sci. 192 (2002) 258-269.

[28] Z.L. Wang, Y.Liu, Z. Kluwer. Handbook of Nanophase and nanostructured materials: Synthesis / Characterization / Materials Systems and Applications II. Vol. 1. Springer 2002.

[29] R. Gheriani, R. Halimi and R. Bensaha. Effect of substrate surface ion bombardment etching on reaction between chromium thin films and steel substrates. Surf. Coat. Technol. Vol 180-181 (2004) 49-52.

[30] M. Cekada, M. Panjan, D. Cimpric, et.al. Analysis of the diffusion processes in Al/Cr, Al/Fe and Cr/Fe multilayers using the MRI model. Vacuum 84 (2010) 147-151.

[31] V.K.W. Grips, H.C. Barshilia, V.E. Selvi, Kalavati, K.S. Rajam. Electrochemical behavior of single layer CrN, TiN, TiAlN coatings and nanolayered TiAlN/CrN multilayer coatings prepared by reactive direct current magnetron sputtering. Thin Solid Films 514 (2006) 204-211.

[32] Y.F. Qiao, T.E. Fischer. A. Dent. The effects of Fuel Chemistry and Feedstock Powder Structure on the Mechanical and Tribological Properties of HVOF Thermal Sprayed WC-Co Coatings with Very Fine Structures. Surf. Coat. Technol. Vol 172 (2003).

[33] J. Stokes. Theory and Application of the High Velocity Oxy-Fuel (HVOF) Thermal Spray Process. Dublin City Univ. (2008).

[34] Handbook of thermal spray technology. ASM International (2004).

[35] J. Kim, H. Yang, K. Baik, et.al. Development and properties of nanostructured thermal spray coatings. Curr. App. Phys. 6 (2006) 1002-1006.

[36] Electro-Plating. Paul N. Hasluck Ed. General Books (2012).

[37] Modern Electroplating. Vol 55. 5 ed. John Wiley & Sons (2011).

[38] N. Kanani. Electroplating: Basic Principles, Processes and Practice. Elsevier (2004).

[39] V.D.N. Rao, D.M. Kabat, D. Yeager, B. Zizitter. Engine Studies of Solid Film Lubricant Coated Pistons. SAE Paper 970009 (1997).

[40] J.R. Nicholls. Advances in Coating Design for High-performance Gas Turbines. MRS Bull. 28 (9) (2003) 659.

[41] J. Creus, H. Mazille, H. Idrissi, Porosity evaluation of protective coatings onto steel, through electrochemical techniques, Surf. and Coat. Technol. 130 (2000) 224-232

[42] H.A. Jehn, I. Pfeiffer-Schaller. Galvanotechnik 84 (10) (1993) 3283.

[43] H.A. Jehn. Improvement of the corrosion resistance of PVD hard coating–substrate systems. Surf. Coat. Technol. 125 (2000) 212.

[44] J. Munemasa, T. Kumakiri. Effect of the surface roughness of substrates on the corrosion properties of films coated by physical vapour deposition. Surf. Coat. Technol. 49 (1991) 496

[45] A. Matthews, A. Leyland, et.al. Plasma-based surface engineering processes for wear and corrosion protection. Vac. Sci. Tecnol. A 13 (3) (1995) 1202.

[46] H.S. Park, H. Kappl, et.al. Structure modification of magnetron-sputtered CrN coatings by intermediate plasma etching steps. Surf. Coat. Technol. 133-134 (2000) 176.

[47] J. Creus, A. Billard, F. Sanchette. Corrosion behaviour of amorphous Al–Cr and Al–Cr–(N) coatings deposited by dc magnetron sputtering on mild steel substrate. Thin Solid Films 466 (2004) 1.

[48] M. Fenker, M. Balzer and H. Kappl. Corrosion behavior of decorative and wear resistant coatings on steel deposited by reactive magnetron sputtering – tests and improvements. Thin Solid Films. 515 (2006) 27-32.

[49] A. Griguceviciene, K. Leinartas, R. Juskenas, E. Juzeliunas. J. Electroanal. Chem. 565 (2004) 203.

[50] J. Creus, E.H. Top, C. Savall, Ph. Refait, et. al. Mechanical and corrosion properties of dc magnetron sputtered Al/Cr multilayers. Surf. Coat. Technol. 202 (2008) 4047-4055.

[51] N.S. Cheruvu, R. Wei, M.R. Govindaraju and D.W. Gandy. Microstructure and oxidation resistance of nanocrystalline 304 SS-Al coatings. Sur. Coat. Technol. 204 (2009) 751-755.

[52] Z. Liu, W. Gao, K.L. Dahm, F. Wang. Oxidation behaviour of sputter-deposited Ni–Cr–Al micro-crystalline coatings. Acta Mater. 46 (1998) 1691.

[53] García-Hernández, Hernández-Sandoval, Regalado-González, et.al. Hybrid corrosion-resistant coatings analyzed using electrochemical noise. Anti-corr. Meth. and Mat. 56/6 (2009) 310-315.

[54] K. Hladky and J.L. Dawson. The measurement of corrosion using electrochemical 1/f noise. Corr. Sci. Vol. 22, pp 231. (1982)

[55] J. Uruchurtu, JL Dawson. Noise analysis of pure aluminum under different pitting conditions. Corrosion, Vol 43, No 1 (1987) 19-26.

[56] J. Uruchurtu-Chavarín and J.M. Malo. Electrochemical noise as a powerful electrochemical technique for corrosion studies. Trends in corrosion research. Vol. 2 pp 49-58 (1997).

[57] D.A. Eden, K. Hladky, D.G. John, J.L. Dawson. Electrochemical noise simultaneous monitoring of potential and current noise signals from corroding electrodes. Corrosion 86, Paper 274. NACE, 1986.

[58] F. Mansfeld and H. Xiao. Electrochemical noise analysis of iron exposed to NaCl solution of different corrosivity. J. Electrochem. Soc. Vol. 140 No 8 2005.

[59] F.H. Estupiñán-López, F. Almeraya-Calderón, et. al., Int. J. Electrochem. Sci., Vol 6, (2011) 1785-1796.

[60] F. Almeraya-Calderon, R.G. Bautista-Margulis. et al. , Int. J. Electrochem. Sci., Vol. 7 (2012) 12999 – 13007.

[61] Standard Guide for Electrochemical Noise Measurements. G199-09; ASTM, 2009

[62] R. A. Cottis y Turgoose S. Electrochemical Impedance an Noise Testing Made Easy. NACE In Houston EEUU. 1999.

[63] R.A. Cottis, M.A. Al-Ansari, G. Bayley, A. Pettiti "Electrochemical Noise Measurements for corrosión studies" Mat. Sci. Forum. 2 (1998).

[64] A. Legat, V. Dolecek "Corrosion Monitoring System Based on Measurement and Analysis of Electrochemical Noise. Corrosion. 51, 4 (1995).

[65] A.J. Bard, L.R. Faulkner. Electrochemical Methods. Fundamentals and Applications. J. Wiley & Sons. 2nd ed. 2001.

[66] P.C. Pistouris. Electrochemical Noise Measurements for corrosion Applications. Ed. 85. ASTM West Conshohocken EU. 343-358.

[67] J.R. Kearns, D.A. Eden. M.R. Yafee, J.V. Fahey, D.L. Reichert, y D.C. Silverman. Electrochemical Noise Measurements for corrosion Applications. Ed ASTM West Conshohocken EEUU. (1996) 446-470.

[68] K. Hladky, J.L. Dawson. The measurement of localized corrosion using electrochemical noise. Corr. Sci. 21(4), (1981) 317-330.

[69] M.P. Castillo † (et al) "Electrochemical Impedance Spectroscopy Behavior of Nanometric Al-Cr and Cr-Al Coatings by Magnetron Sputtering. Int. J. Electrochem. Sci., 7(2012) 1125-1133.